Case Studies in Superconducting Magnets

Design and Operational Issues

SELECTED TOPICS IN SUPERCONDUCTIVITY

Series Editor: Stuart Wolf
Naval Research Laboratory
Washington, D.C.

CASE STUDIES IN SUPERCONDUCTING MAGNETS
Design and Operational Issues
Yukikazu Iwasa

INTRODUCTION TO HIGH-TEMPERATURE SUPERCONDUCTIVITY
Thomas P. Sheahen

Case Studies in Superconducting Magnets

Design and Operational Issues

Yukikazu Iwasa

**Francis Bitter National Magnet Laboratory
and Department of Mechanical Engineering
Massachusetts Institute of Technology
Cambridge, Massachusetts**

Plenum Press • New York and London

Library of Congress Cataloging in Publication Data

Iwasa, Yukikazu
  Case studies in superconducting magnets: design and operational
  issues / Yukikazu Iwasa.
      p.   cm. — (Selected topics in superconductivity)
  Includes bibliographical references and index.
  ISBN 0-306-44881-5
  1. Superconducting magnets.   I. Title.
QC761.3.I9   1994
  621.34—dc20   94-36837
                                                                        CIP

ISBN 0-306-44881-5

©1994 Plenum Press, New York
A Division of Plenum Publishing Corporation
233 Spring Street, New York, N.Y. 10013

Printed in the United States of America

To the Memory of My Father,
Seizaburo Iwasa

PREFACE

This book is based on *Superconducting Magnets*, a graduate course I started teaching in the Department of Mechanical Engineering at the Massachusetts Institute of Technology, in 1989, shortly after the discovery of high-temperature superconductors. The book, intended for graduate students and professional engineers, covers the basic concepts of superconducting magnet technology, focusing on design and operational issues.

My course consists of ten 3-hour lectures and eight homework sets, each set containing three to four "tutorial" problems to review lecture materials, to discuss topics in more depth than covered in the lecture, or to teach subjects not presented in the lecture at all. My colleague Emanuel Bobrov has helped me with the course, offering lectures on field computation and stress analysis. He has also created a few problems related to his lecture topics.

Because the use of tutorial problems accompanied, a week later, by solutions has been successful in the course, I have decided to use the same format for this book. Most problems require many steps in their solution and through these steps it is hoped that the reader will gain deeper insight. About 75% of the problems are based on those specifically created for the course's homework or quiz problems; the remainder are based on lecture materials. Because the principal magnet projects at the Francis Bitter National Magnet Laboratory (FBNML) have been high-field solenoidal magnets, problems directly related to other applications are not represented. However, important topics covered in this book, particularly on field distribution, magnets, force, thermal stability, dissipation, and protection, are sufficiently basic and generic in concept that solenoidal magnets are suitable examples.

In creating problems I have relied heavily on the magnet projects at FBNML and I am indebted to my colleagues in the Magnet Technology Division, specifically John Williams, Mat Leupold, Bob Weggel, and Emanuel Bobrov, with whom I have had the good fortune of working on these projects over a long period. Materials contributed by the other members of the Division, Alex Zhukovsky, Vlad Stejskal, Andy Szczepanowski, Dave Johnson, and Mel Vestal are also included and their contributions are acknowledged. I have also benefitted much through participation in the Technology Division of the Plasma Fusion Center (PFC) of MIT; I would particularly like to thank Bruce Montgomery from whom I have learned a great deal since my graduate student days. I would also like to thank Joe Minervini and Makoto Takayasu of the PFC, Dr. Larry Dresner of Oak Ridge National Laboratory, and Dr. Luca Bottura of Max-Planck Institut für Plasmaphysik, Garching, Germany, for advice on the creation of problems related to cable-in-conduit (CIC) conductors, and Dr. Ted Collings of the Battelle Memorial Institute, Ohio, for discussion on enabling technology *vs* replacing technology. In addition, I would like to thank many visiting scientists to FBNML, mostly from Japan—really too many to cite individually here—with whom I have collaborated with fruitful results, particularly in the areas of mechanical dissipation, magnet monitoring, and protection.

I would like to thank Don Stevenson, the retired Assistant Director of the FBNML, who read several versions of the manuscript and offered helpful suggestions, and Albe Dawson of PFC for suggestions on early chapters. Many of my former and present students helped me on this project and I express my deepest gratitude to them. Philip Michael combed through the three last editions and offered many insightful suggestions. Rick Nelson painstakingly read early drafts, checked and corrected solutions, and offered many suggestions on the phrasing of the questions, writing, and equation style; he also assisted me in the preparation of early editions of the Glossary. Mamoon Yunus created beautiful field plots and graphs; Hunwook Lim produced most of the figures and prepared the Index; Jun Beom Kim rechecked several derivations, collected much of the data presented in the Appendices, and produced most of the graphs; Abraham Udobot also prepared many figures and laid out all the figures.

I am also indebted to Dr. Hiroyasu Ogiwara of Toshiba Corporation for first suggesting a book based on my course materials, particularly on the problem sets. He has also arranged for me to offer lectures based on the course to magnet engineers at the Kanagawa Academy of Science and Technology, Kawasaki, Japan.

Thanks are also due to the National Science Foundation (FBNML's sponsor); the Department of Mechanical Engineering; the Department of Energy Office of Fusion Energy; the Department of Energy Office of Renewable Energy; the Department of Energy Office of Basic Sciences; and Daikin Industries, Ltd. for their support of this book project. I have used D.E. Knuth's indispensable TEX in typesetting the entire text, equations, tables, and even some figures.

Finally, I would express a word of appreciation to Kimiko who has made it possible for me to continue working on this project in the relaxed atmosphere of our home and thus to carry it forward to completion.

Yukikazu Iwasa

Weston, Massachusetts
August, 1994

"You know nothing till you prove it! FLY!" —*Jonathan Livingston Seagull*

Contents

ix

Chapter 1
Superconducting Magnet Technology

1.1 Introductory Remarks

Superconducting magnet technology comprises engineering aspects associated with the design, manufacture, and operation of superconducting magnets. In its bare essence, a superconducting magnet is a highly stressed device: it requires the best that engineering has to offer to ensure that it operates successfully, is reliable, and at the same time is economically viable. A typical 10-tesla magnet is subjected to an equivalent magnetic pressure of 40 MPa (400 atm), whether it is superconducting and operating at 4.2 K (liquid helium cooled) or 77 K (liquid nitrogen cooled), or resistive and operating at room temperature (water cooled). Superconducting magnet technology is interdisciplinary in that it requires knowledge and training in many fields of engineering, including mechanical, electrical, cryogenic, and materials.

Table 1.1 lists "first" events relevant to superconducting magnet technology. Particularly noteworthy events since the discovery of superconductivity in 1911 by Kamerlingh Onnes, who was also first to liquefy helium in 1908, are:

1. Development of water-cooled 10-T magnets by Francis Bitter in the 1930s;

2. Marketing of helium liquefiers, developed by Collins, in 1946;

3. Development in 1961 by Kunzler and others of magnet-grade superconductors;

4. Formulation, chiefly by Stekly, of design principles for cryostable magnets in the mid 1960s; and

5. Discovery of high-temperature superconductivity (HTS) in perovskite oxides by Müller and Bednorz in 1986.

We may safely state that Bitter initiated modern magnet technology. Although Bitter magnets are water cooled and resistive, resistive and superconducting magnets share many engineering requirements.

Soon after the availability of Collins liquefiers, liquid helium—until then a highly prized research commodity available only in a few research centers—became widely available and helped to propel the rapidly growing field of low temperature physics. Many important superconductors were discovered in the 1950s, leading to the development of magnet-grade superconductors.

The formulation of design principles for cryostable magnets by Stekly and others by the mid 1960s demonstrated the feasibility of building large superconducting magnets that operated reliably.

The discovery of HTS lifted superconducting magnet technology from the depth of a liquid helium well and ushered it into a new era with expanded options. It is estimated that the number of people involved in superconductivity jumped by an order of magnitude overnight after the discovery of HTS.

1

Table 1.1: "First" Events Relevant To Superconducting Magnet Technology

Decade	Event*
1930s	Meissner effect.
	Type II superconductors identified.
	Phenomenological theories of superconductivity.
	Bitter magnets generating fields up to 10 tesla.
1940s	Marketing of Collins helium liquefier.
1950s	Many more Type-II superconductors identified.
	GLAG and BCS theories of superconductivity.
	Small superconducting magnets (SCM).
1960s	Magnet-grade superconductors developed.
	International conference on high magnetic fields.
	National laboratory for magnetism and magnet technology.
	Bitter magnets generating fields up to 25 T.
	Flux jumps in SCM.
	Composite superconductors.
	Formulation of cryostability criteria.
	Large cryostable SCM (MHD and bubble chambers).
	Superconducting generators.
	Magnets wound with internally-cooled conductors.
	Multifilamentary Nb-Ti superconductors.
1970s	Multifilamentary Nb_3Sn superconductors.
	Maglev test vehicles.
	Superconducting dipoles for accelerators.
	Cable-in-conduit (CIC) conductors.
	Hybrid magnets generating 30 T.
	Commercial NMR systems using SCM.
1980s	Commercial MRI systems using SCM.
	Multinational experiments for fusion magnets.
	Submicron superconductors for 60-Hz applications.
	Superconducting accelerators.
	Discovery of HTS.

* Entries in each decade did not necessarily take place sequentially as listed. Acronyms are described in the Glossary (Appendix VI).

1.2 Superconductivity

The complete absence of electrical resistivity for the passage of direct current below a certain "critical" temperature (usually designated with the symbol T_c) is the basic premise of superconductivity. In addition to T_c, the critical field H_c and critical current density J_c are two other parameters that define a critical surface below which the superconducting phase can exist. T_c and H_c are thermodynamic properties that for a given superconducting material are invariant to metallurgical processing; J_c is not. Indeed the key contribution of Kunzler and others in 1961 was to demonstrate that for certain superconductors it is possible to enhance J_c dramatically by means of metallurgy alone. No formal theories of superconductivity, phenomenological or microscopic, will be presented in this book to explain relationships among T_c, H_c, or J_c; however, the magnetic behavior of superconductivity, which plays a key role in superconducting magnets, will be briefly reviewed by means of simple theoretical pictures.

Figure 1.1 shows the critical surface for a typical magnet-grade superconductor. On this critical surface, the following three important functions are used by the magnet engineer: $f_1(H, T, J = 0)$; $f_2(J, T, H_o = \text{constant})$; $f_3(J, H, T_o = \text{constant})$. f_1 is the H_c vs T_c plot; for "ideal" superconductors it is quite straightforward to derive a parabolic function of H_c on T_c from thermodynamics [1.1]:

$$H_c = H_o \left[1 - \left(\frac{T}{T_c} \right)^2 \right] \tag{1.1}$$

H_o is given by:

$$H_o = T_c \sqrt{\frac{\gamma_e}{2\mu_o}} \tag{1.2}$$

where γ_e is the electronic heat capacity constant in the normal state.

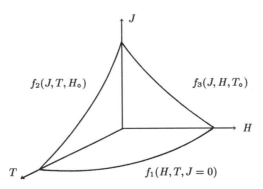

Fig. 1.1   Critical surface of a typical magnet-grade superconductor.

f_2 gives the J_c *vs* T plot, and for *all* superconductors of interest, J_c is a decreasing function of temperature. As we shall study in more detail in Chapter 5, this is the source of inherent instability in superconductors. We shall defer discussion of f_3 until after Sec. 1.2.3, where Type I and Type II superconductors are discussed.

1.2.1 Meissner Effect

Discovered by Meissner and Ochsenfeld in 1934, the Meissner effect describes the absence of magnetic field within the bulk of a superconductor. This complete diamagnetism of a superconductor is in fact more fundamental than the complete absence of electrical resistivity to the extent that a material's perfect diamagnetism *automatically* requires it to be a perfect electrical conductor. Unlike the complete absence of electrical resistivity, however, we do not benefit from perfect diamagnetism. The Meissner effect was in fact responsible for the single most important source of magnet failures in the early 1960s: flux jumping. Even today when flux jumping is no longer an issue due to an important innovation introduced in conductor design in the late 1960s, the Meissner effect is the basis for another important source of losses in the magnets—AC losses—that restricts the use of superconducting magnets primarily to DC applications.

1.2.2 London's Theory of Superconductivity

Although a microscopic theory of superconductivity by Bardeen, Cooper, and Schrieffer—known as the BCS theory—was not completed until 1957, development of phenomenological theories of superconductivity began in the 1930s. Among these is the electromagnetic theory of London (1935), in which the concept of penetration depth was introduced to account for the Meissner effect. Simply stated, a bulk superconductor is shielded completely from an external magnetic field by a supercurrent that flows within the penetration depth (λ) at the surface. According to London's theory, λ is given by:

$$\lambda = \sqrt{\frac{m}{\mu_o e^2 n_e}} \qquad (1.3)$$

where m, e, and n_e are, respectively, the electron's mass, charge, and concentration. μ_o is the permeability of free space. n_e in turn is given by:

$$n_e = \frac{2\varrho N_A}{W_A} \qquad (1.4)$$

where ϱ is the conductor's mass density, N_A is Avogadro's number, and W_A is its atomic weight. The factor 2 in Eq. 1.4, not in the original London theory, was inserted later because there are two "superelectrons" (a Cooper pair) for each atom. Values of λ and the superconductor's J_c, given by $e n_e v$ where v is the speed of sound, have been confirmed by experiment.

1.2.3 Type I and Type II Superconductors

Kamerlingh Onnes discovered superconductivity in pure mercury; subsequently other metals such as lead and indium were found to be superconductors. These materials, now called Type I (also known as "soft") superconductors, are unsuitable as magnet conductor materials because of their low H_c values: less than 10^5 A/m (corresponding to ~0.1 T). Magnet-grade superconductors trace their origin to the first Type II (also known as "hard") superconductor discovered by de Haas and Voogd in 1930 in an alloy of lead and bismuth [1.2].

A Type II superconductor may be modeled as a finely divided mixture of a Type I superconductor and normal conducting material. Indeed, in the early 1960s there were two physical models for this mixture: lamina and island (vortex). In the lamina model, proposed by Goodman, the hard superconductor consists of superconducting laminae separated by normal laminae. In the vortex model, proposed by Abrikosov at about the same time, and later experimentally verified by Essmann and Träuble [1.3], the superconductor consists of many hexagonally-arranged normal-state islands in a superconducting sea. For the hard superconductor to retain its bulk superconductivity well beyond 0.1 T, the width of each superconducting lamina or the radius of each normal island must be smaller than λ. The lamina's half width or the island's radius is the coherence length (ξ), an important spatial parameter, introduced by Pippard in 1953. ξ defines a distance over which the superconducting-normal transition takes place. According to the GLAG theory of superconductivity (after Ginsburg, Landau, Abrikosov, and Gorkov), formulated about the same time as Pippard's to account for the magnetic behavior of Type II superconductors, a superconductor is Type II if $\xi < \sqrt{2}\lambda$; it is Type I if $\xi > \sqrt{2}\lambda$. ξ decreases with alloying, which shortens the mean free path of the normal electrons; ξ is thus inversely proportional to the material's normal-state electrical resistivity. It is noted that the two magnet-grade superconductors—alloys of niobium titanium (Nb-Ti) and an intermetallic compound of niobium and tin (Nb$_3$Sn)—both have normal-state resistivities that are at least one order of magnitude greater than that of copper at room temperature. Incidentally, it has been noted that the HTS also have ξ much much shorter than λ.

1.2.4 Critical Current Density of Type II Superconductors

As mentioned earlier, J_c may be enhanced dramatically by means of metallurgical processing. The function $f_3(J, H, T_\circ)$ gives J_c vs H plots at a given temperature $T_\circ$ for conductors having enhanced J_c performance. This enhanced J_c performance is generally attributed to a "pinning" force that counteracts the $\vec{J_c} \times \vec{H}$ Lorentz force acting on the vortices. The pinning force is provided by "pinning" centers that are created in crystal structures by material impurities, metallurgical processes such as cold working in the form of dislocation cells, or heat treatment in the form of precipitations and grain boundaries. Kim and others, through their investigation of the magnetic behavior of Type II superconductors, obtained the basic J_c vs H equation by equating the $\vec{J_c} \times \vec{H}$ Lorentz force to the pinning force [1.4]:

$$J_c = \frac{\alpha_c}{H + H_\circ} \tag{1.5}$$

where α_c and $H_\circ$ are constants. Note that α_c essentially represents an asymptotic force density that balances the Lorentz force density for $H \gg H_\circ$. That is, Eq. 1.5 is really a simple force balance equation.

1.3 Magnet-Grade Superconductors

Although completely specifying a conductor for a given superconducting magnet
is an important task in the design phase, issues directly related to magnet-grade
superconductors are not specifically treated in this book. Magnet-grade superconductors are those conductors that meet rigorous specifications required for use in a
magnet, and are readily available commercially. What follows is a brief comment
to point out important differences between superconducting *materials* and *magnet-
grade superconductors*, and that it is a laborious task to develop a magnet-grade
superconductor from a material discovered in the laboratory.

1.3.1 Materials *vs* Magnet-Grade Superconductors

Table 1.2 lists the number of materials meeting certain criteria on superconductivity and illustrates that as the criteria move towards those required of a magnet-
grade superconductor, the number of materials meeting the criteria decreases *log-
arithmically*. H_{c2} is the "upper" critical field, relevant only to Type II superconductors. Indeed, of nearly 10,000 superconducting materials discovered to date,
at present (1994) there are basically only two magnet-grade superconductors, Nb-
Ti alloys and an intermetallic compound, Nb_3Sn. A drop of nearly four orders
of magnitude attests to the excruciatingly difficult task material scientists and
metallurgists face in transforming a material into a magnet-grade superconductor.

1.3.2 A Long Journey

It is a long journey to transform a superconducting material, discovered in the laboratory, into a magnet-grade superconductor. The journey consists of six stages,
given in Table 1.3: 1) the discovery of a superconducting material; 2) improvement
in J_c performance; 3) co-processing with matrix metal; 4) development of a multifilamentary conductor having I_c of at least $\sim$100 A; 5) production of a conductor
in length from $\sim$10 mm, typical in the material stage, to $\sim$1 km; and 6) meeting
other specifications of a magnet. The table also lists an approximate period for
the beginning of each stage with Nb_3Sn used as an example.

Despite more than a decade of intense research and development activity beginning
immediately after the development of Nb_3Sn conductors in 1961, Nb_3Sn must *still*
be custom-designed for each magnet application. Because of its extreme brittleness
and intolerance to a minute strain ($\sim$0.3%), the material is inherently difficult to
process and must be handled with great care.

Table 1.2: Superconducting Materials *vs* Conductors

Criterion	Number
1. Superconducting?	$\sim$10,000
2. $T_c > 10$ K ($\mu_o H_{c2} > 10$ T)?	$\sim$100
3. $J_c > 1$ GA/m² (@ $B > 5$ T)?	$\sim$10
4. A magnet-grade superconductor?	$\sim$1

Table 1.3: Material-to-Conductor Development Stages for Nb$_3$Sn

Stage	Event	Period
1	Discovery of superconducting material.	Early 1950s.
2	Improvement in J_c.	Early 1960s.
3	Co-processing with matrix metal.	Mid 1960s.
4	Multifilamentary conductor with $I_c > 100\,\text{A}$.	Early 1970s.
5	Long length, typically $\sim$1 km.	Mid 1970s.
6	Other specifications for magnets.	Late 1970s.

1.4 Magnet Design

In this section, important magnet design issues appropriate to the subject matter of the remainder of this book are briefly discussed.

1.4.1 Requirements and Key Issues

A magnet, whether it is experimental or a system component, must satisfy basic requirements on magnetic field, $\vec{H}(x, y, z, t)$, which include spatial distribution and temporal variation. Important parameters often given in the field specifications are: 1) H_o, the field at the magnet origin $(x = 0, y = 0, z = 0)$; 2) V_o, the volume within which $\vec{H}(x, y, z)$ is specified; and 3) $H(t)$, the field time variation. Chapters 2 and 3 discuss H_o and $\vec{H}(x, y, z)$ in some detail.

In addition to satisfying these basic field requirements, the magnet design must address the following key issues:

- Mechanical integrity. The magnet must be structurally strong to withstand large magnetic stresses, both under operating and fault conditions. Chapter 3 deals with magnetic forces and stresses.

- Operational reliability. The magnet must be stable in order to reach and stay at its operating point reliably. Chapters 5, 6, and 7 deal with this issue.

- Protection. In the event the magnet is driven into the normal state, it must remain undamaged and be capable of being energized to its operating point repeatedly. Chapter 8 is devoted to this subject.

- Conductor specification. For "small" magnets and those produced in a large quantity, the overall cost of a superconducting magnet system can be influenced to a large extent by the cost of the superconductor. For these magnets, it is important to improve "field efficiency" so that the field requirements can be met with the minimum amount of conductor.

- Cryogenics. Because it requires power to create and maintain the cryogenic environment for the operation of superconducting magnets, cryogenics also becomes an important issue for "small" and mass-produced magnets. When a superconducting magnet system is considered alone, cryogenics clearly plays a dominant role. It is for this reason that Chapter 4 is devoted to this subject; it is also for this reason that cryogenics is sometimes overemphasized relative to its importance to the *overall system*. It is worthwhile noting that in many important applications where superconducting magnets are to play critical roles, the magnets are but one component among many in the overall system and consequently, the cryogenics is a subcomponent. Indeed the power requirement for the cryogenic system is generally a fraction of the total power associated with the overall system.

Since the beginning of the HTS era, efforts to improve T_c have continued unabated. Indeed, it has recently been suggested that it may be possible to improve the critical temperature of a certain HTS from a nominal value of $\sim$100 K to the extraordinary value of $\sim$250 K by means of an intricate manufacturing process. Aside from its obvious importance in terms of understanding the physics of superconductivity, it is extremely doubtful that such a material would be used as magnet conductor. The *overall* cost of building a magnet made of such a 250-K T_c conductor and operating it at 200 K would most likely be substantially more than that of the same magnet made of a nominal 100-K T_c conductor operating at 80 K. To draw a parallel: diamond is the best electrical insulator, but it is rarely used for that purpose. Above a certain temperature that is still below room temperature, the savings in cryogenics becomes absolutely insignificant. The most important requirements for cryogenics are ease and reliability of operation rather than reduced capital and operating cost; kitchen refrigerators have already achieved an acceptable level of ease and reliability of operation.

1.4.2 Effect of Operating Temperature

Operation at temperatures substantially higher than 4.2 K—the base line temperature for superconducting magnets to date—has one of the following impacts on each of the key magnet issues: 1) virtually none; 2) makes it more difficult or costly; or 3) makes it less difficult or less costly. Figure 1.2 shows *qualitative* plots of "difficulty or cost" *vs* operating temperature for the five key issues—mechanical integrity, stability, protection, conductor, and cryogenics—over the temperature range likely in the near future.

Figure 1.2 indicates that difficulty in meeting mechanical integrity requirements is essentially independent of operating temperature. This statement is true for operating temperatures up to $\sim$100 K over which differential thermal expansions among most magnet materials are negligible. For a magnet of given field requirements, the necessary ampere-turns are independent of operating temperature. Because critical current density decreases with temperature universally among known superconductors, the conductor cost always increases with operating temperature; the expected benefit of a decrease in the cost of cryogenics with temperature must be compared with this expected increase in conductor cost. Operating temperature has profound impacts on stability and protection as will be discussed in Chapters 6 and 8. These chapters, through problems, illuminate the positive and negative impacts of increasing operating temperature over a wide span, feasible only with the use of high-T_c superconductors.

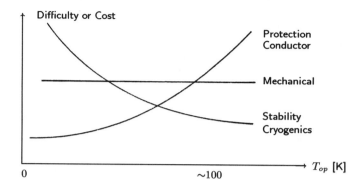

Fig. 1.2   Effects of temperature on five key magnet issues.

1.5 Class 1 and Class 2 Superconducting Magnets

Broadly speaking, superconducting magnets may be divided into two classes—Class 1 and Class 2. Generally, Class 1 magnets are physically large. Class 1 magnets most actively pursued presently are those required for magnetic confinement of plasma in fusion reactors. Superconducting magnet energy storage (SMES), presently not as actively pursued as fusion programs, also uses Class 1 magnets. Because their sheer sizes result in huge forces, the most pressing design issue in Class 1 magnets is mechanical integrity, which impacts other design issues, *e.g.* conductor. Based on the projections of Fig. 1.2, we may conclude that the impact of increased operating temperature is minimal in Class 1 magnets.

Class 2 magnets are small or densely constructed. Important Class 2 magnets include those for nuclear magnetic resonance (NMR) spectroscopy, magnetic resonance imaging (MRI) systems, superconducting generators, superconducting motors, dipole and quadrupole magnets for particle accelerators, maglev high-speed trains, and research requiring high magnetic fields. The first generation HTS magnets will most likely be of Class 2. The common feature of Class 2 is operation at high current densities, making the stability, protection, conductor, and cryogenics critical design issues. The impact of increased operating temperature on these issues, from Fig. 1.2, is mixed.

1.6 The Format of the Book

As stated in the Preface, the format adopted for this book is the use of tutorial problems accompanied by solutions. Each problem requires many steps in its solution, and through these steps it is hoped that the reader will gain deeper insight. Because most of the problems are solvable analytically in closed form, they deal chiefly with ideal cases; nevertheless, solutions to these problems are quite useful for real world problems with all the usual complexities. One needs the engineer's essential talent: the ability to transform a complex problem into an ideal case without losing vital points. Each problem is intended to develop the reader's ability to quickly grasp the zeroth or first order solution to a real world situation of a similar nature, thus enabling him to keep his bearing even after he enters into a numerical-analysis maze. In discussing design and operational issues, particularly field distribution, magnets, forces, thermal stability, dissipation, and protection, we focus primarily on solenoidal magnets. These issues are sufficiently basic and generic in concept that solenoidal magnets are suitable examples. Included also in selected problems are short presentations on topics relevant to the specific issues under discussion; these headings are typeset with a slanted bold font.

Chapters 2, 4, and 9 and the Appendices are briefly described here.

Chapter 2 is on static and quasi-static electromagnetic fields; the chapter starts with a presentation of Maxwell's equations. The problems cover useful field solutions, including some for magnetic shielding and induction heating.

Chapter 4 presents cryogenics. Its coverage is strictly for magnet designers and not for cryogenic engineers. The problems selected are thus intended to be useful for magnet applications.

Chapter 9 gives concluding remarks and looks at the prospect of superconducting magnets in the 21st century.

Many useful data are presented in Appendices. These include thermodynamic properties of helium and other cryogens. The properties are not extensive for cryogenic engineers but sufficient for magnet engineers to make quick, quantitative estimates. Other properties included in the Appendices are thermal and electrical properties of structural materials and conductive metals; critical properties of Nb-Ti and Nb$_3$Sn and BiPbSrCaCuO (2223); and selected properties of a few HTS materials. A glossary describes acronyms and terms that are of general interest in superconducting magnet technology and its areas of application but that are discussed only briefly or not at all in the main text of this book.

References

[1.1] A.B. Pippard, *The Elements of Classical Thermodynamics* (Cambridge University Press, Cambridge, 1966).

[1.2] J.K. Hulm and B.T. Matthias, "Overview of superconducting materials development," in *Superconductor Materials Science—Metallurgy, Fabrication, and Applications*, Eds., S. Foner and B.B. Schwartz (Plenum Press, New York, 1981), 1.

[1.3] U. Essmann and H. Träuble, "The direct observation of individual flux lines in Type II superconductors," *Phys. Lett.* **24A**, 526 (1967).

[1.4] Y.B. Kim, C.F. Hempstead, and A.R. Strnad, "Magnetization and critical supercurrents," *Phys. Rev.* **129**, 528 (1963).

"Why sir, there is a very good chance that you will soon be able to tax it."
— Michael Faraday's reputed reply to William Gladstone, the Prime Minister, who after being shown by Faraday a demonstration of the first dynamo, asked *"But, after all, what use is it?"*

CHAPTER 2
ELECTROMAGNETIC FIELDS

2.1 Introduction

In this chapter we review electromagnetic theory by presenting Maxwell's equations. This review is necessary to bring the reader's understanding of electromagnetic theory to a basic level to allow the main subject matter of this book—superconducting magnets—to be approached in as quantitative a manner as possible. After a presentation of Maxwell's equations and simple solutions that are not only tractable analytically but also useful in most magnet applications, several specific cases familiar to most magnet engineers will be presented and studied.

2.2 Maxwell's Equations

There are four basic Maxwell's equations: 1) Gauss's law; 2) Ampere's law; 3) Faraday's law; and 4) the law of magnetic induction continuity. In addition to these, we will make frequent use of the equations of charge conservation and other constituent relations. Each equation is briefly discussed below.

In this book SI units are used almost exclusively. Electromagnetic quantities used in this book are summarized in Table 2.1. There is a widespread practice in the magnet community of interchanging magnetic field H and magnetic induction (or magnetic flux density) B, expressing, for example, a magnetic field in the unit of tesla [T]. Although the practice is usually harmless and causes no confusion, care must be exercised, for example, when computing energy from an M vs H plot.

In SI units, the magnetic permeability of free space (μ_o) is by definition $4\pi \times 10^{-7}$ H/m; the electric permittivity of free space, ϵ_o, is approximately 8.85×10^{-12} C/m. Appendix I presents other important physical constants and selected conversion factors from "common" non-SI units to SI units.

Current density is by far the dominant source of the $\vec{H}$ fields associated with superconducting magnets. Thus, the relatively small time-varying $\vec{E}$ field contribution to the $\vec{H}$ field is not included in our presentation of Maxwell's equations.

Table 2.1: Electromagnetic Quantities

Symbol	Name	SI Units	
E	Electric field	volt/meter	[V/m]
H	Magnetic field	ampere/meter	[A/m]
B	Magnetic induction (or magnetic flux density)	tesla	[T]
J_f	Current density	ampere/(meter)^2	[A/m^2]
ρ_c	Charge density	coulomb/(meter)^3	[C/m^3]
ρ_e	Electrical resistivity	ohm meter	[Ω m]

2.2.1 Gauss's Law

In integral form, Gauss's law in free space is given by:

$$\oint_S \epsilon_o \vec{E} \cdot d\vec{A} = \int_V \rho_c \, dV \tag{2.1}$$

The surface integral of $\epsilon_o \vec{E}$ field is equal to the total electric charge within the volume enclosed by surface S. In differential form, Eq. 2.1 is given by:

$$\nabla \cdot \epsilon_o \vec{E} = \rho_c \tag{2.2}$$

Boundary Condition: At a surface with charge density σ_c [C/m²], the discontinuity in the normal component of electric field, from region 1 ($\vec{E}_1$) to region 2 ($\vec{E}_2$), is given by σ_c/ϵ_o:

$$\vec{n} \cdot (\vec{E}_2 - \vec{E}_1) = \frac{\sigma_c}{\epsilon_o} \tag{2.3}$$

The unit vector $\vec{n}$ is normal to the surface and points from region 1 to region 2.

2.2.2 Ampere's Law

In integral form, Ampere's law is given by:

$$\oint_C \vec{H} \cdot d\vec{s} = \int_S \vec{J}_f \cdot d\vec{A} \tag{2.4}$$

The equation states that the line integral of $\vec{H}$ field is equal to the total "free" electric current, *i.e.* not including magnetization currents, within the surface S enclosed by contour C. In differential form, Eq. 2.4 is given by:

$$\nabla \times \vec{H} = \vec{J}_f \tag{2.5}$$

Note that Eqs. 2.4 and 2.5 do not include $\partial \vec{E}/\partial t$ as a source of $\vec{H}$.

Boundary Condition: In the presence of a surface with surface current density $\vec{K}_f$ [A/m], there will be a discontinuity in the tangential component of magnetic field in passing through the surface from region 1 ($\vec{H}_1$) to region 2 ($\vec{H}_2$), given by:

$$\vec{n} \times (\vec{H}_2 - \vec{H}_1) = \vec{K}_f \tag{2.6}$$

2.2.3 Farady's Law

In integral form, Faraday's law is given by:

$$\oint_C \vec{E} \cdot d\vec{s} = -\frac{\partial}{\partial t} \int_S \vec{B} \cdot d\vec{A} \tag{2.7}$$

The equation states that the line integral of $\vec{E}$ field is equal to the time rate of change of the total magnetic flux over surface S enclosed by contour C. In differential form, Eq. 2.7 is given by:

$$\nabla \times \vec{E} = -\frac{\partial \vec{B}}{\partial t} \tag{2.8}$$

Boundary Condition: The tangential component of $\vec{E}$ field is always continuous in passing through a surface from region 1 ($\vec{E}_1$) to region 2 ($\vec{E}_2$). Namely:

$$\vec{n} \times (\vec{E}_2 - \vec{E}_1) = 0 \tag{2.9}$$

2.2.4 Magnetic Induction Continuity

In integral form, magnetic induction continuity is given by:

$$\oint_S \vec{B} \cdot d\vec{A} = 0 \tag{2.10}$$

The equation states that the surface integral of $\vec{B}$ field enclosing a volume is zero, or that there are no point sources of magnetic induction. In differential form, Eq. 2.10 is given by:

$$\nabla \cdot \vec{B} = 0 \tag{2.11}$$

Boundary Condition: The normal component of $\vec{B}$ field is always continuous in passing through a surface from region 1 ($\vec{B}_1$) to region 2 ($\vec{B}_2$). Namely:

$$\vec{n} \cdot (\vec{B}_2 - \vec{B}_1) = 0 \tag{2.12}$$

2.2.5 Charge Conservation

The current density $\vec{J}_f$ is related to the time rate of change of electric charge density. In integral form, the relation is given by:

$$\oint_S \vec{J}_f \cdot d\vec{A} = -\frac{\partial}{\partial t} \int_V \rho_c \, dV \tag{2.13}$$

In differential form, Eq. 2.13 is given by:

$$\nabla \cdot \vec{J}_f = -\frac{\partial \rho_c}{\partial t} \tag{2.14}$$

2.2.6 Magnetization and Constituent Relation

In homogeneous, isotropic, time invariant media, magnetic induction $\vec{B}$, magnetic field $\vec{H}$, and magnetization $\vec{M}$ are related by:

$$\vec{B} = \mu_\circ(\vec{H} + \vec{M}) \tag{2.15}$$

Note that $\vec{M}$ is also given in units of A/m. Note also that in homogeneous, isotropic, linear, time-invariant, "unsaturated" media, which we will always assume in this book unless otherwise indicated, $\vec{B} = \mu \vec{H}$.

In conductive materials such as metals, the presence of $\vec{E}$ field induces a current density $\vec{J}$ in the metal. The constituent relation between $\vec{J}$ and $\vec{E}$ is:

$$\vec{J} = \frac{\vec{E}}{\rho_e} \tag{2.16}$$

where ρ_e is the metal's electrical resistivity. Equation 2.16 may be expressed as $\vec{E} = \rho_e \vec{J}$, which is one form of Ohm's law.

2.3 Quasi-Static Case

The electric field $\vec{E}$ and magnetic induction $\vec{B}$ are coupled through Faraday's law (Eq. 2.7 or 2.8). In metals in which $\vec{J}$ is induced by $\vec{E}$ according to Eq. 2.16, a time-varying magnetic induction imposed on a conducting object can induce a current in the object, which in turn can generate a magnetic field. In general, the following complete set of field equations must be solved:

$$\nabla \cdot \epsilon_o \vec{E} = \rho_c \tag{2.17a}$$

$$\nabla \times \vec{H} = \vec{J}_f + \frac{\vec{E}}{\rho_e} \tag{2.17b}$$

$$\nabla \times \vec{E} = -\frac{\partial \vec{B}}{\partial t} \tag{2.17c}$$

The problem of solving the above equations for $\vec{E}$ and $\vec{H}$ can be greatly simplified if $\vec{E}$ and $\vec{H}$, coupled through Eqs. 2.17b and 2.17c, can be decoupled. Such "quasi-static" cases exist in many important practical applications in which coupled Maxwell's equations can be solved in terms of sets of static equations. In physical terms, if an induced magnetic field is negligible compared with the original magnetic field, then the quasi-static approximation may be used. Thus, in the zeroth-order approximation, we have:

$$\nabla \cdot \epsilon_o \vec{E}_0 = \rho_{c0} \tag{2.18a}$$

$$\nabla \times \vec{H}_0 = \vec{J}_f + \frac{\vec{E}_0}{\rho_e} \tag{2.18b}$$

$$\nabla \times \vec{E}_0 = 0 \tag{2.18c}$$

Note that here the zeroth-order E-field, $\vec{E}_0$, can be solved independently of H-field and once $\vec{E}_0$ is determined, $\vec{H}_0$ may be determined. (In the nonconducting case, $\vec{H}_0$ can also be solved independent of E-field.) In the 1st-order approximation, we have:

$$\nabla \cdot \epsilon_o \vec{E}_1 = \rho_{c1} \tag{2.19a}$$

$$\nabla \times \vec{H}_1 = \vec{J}_1 + \frac{\vec{E}_1}{\rho_e} \tag{2.19b}$$

$$\nabla \times \vec{E}_1 = -\frac{\partial \vec{B}_0}{\partial t} \tag{2.19c}$$

$$\nabla \cdot \vec{J}_1 = -\frac{\partial \rho_{c0}}{\partial t} \tag{2.19d}$$

The 1st-order E-field can still be determined independent of 1st-order H-field. In the absence of the 1st-order charge density, the zeroth-order B-field, $\vec{B}_0$, which is already known, becomes the sole source of $\vec{E}_1$.

The approximation process can continue indefinitely, but for the "low-frequency" cases of interest discussed in the Problem Section of this Chapter, we need to solve for only the zeroth and 1st order fields.

2.4 Poynting Vector

Poynting's theorem may be expressed as:

$$-\nabla \cdot \vec{S} = p + \frac{\partial w}{\partial t} \qquad (2.20)$$

where $\vec{S}$ is the Poynting vector given by: $\vec{S} = \vec{E} \times \vec{H}$. p is the power dissipation density and w is the energy density stored magnetically and electrically.

Equation 2.20 states that the negative of the divergence of the S-vector is equal to p plus the rate of change of energy storage. In practice p is always positive. If $\vec{S}$ is zero, then the stored electromagnetic energy within the volume decreases to make up for dissipation taking place within it; if w is zero, then $\vec{S}$ must flow inward, i.e. into the volume, to sustain the power dissipation.

2.4.1 Sinusoidal Case

When dealing with a sinusoidally time-varying electric field of complex amplitude $\vec{E}$ and a corresponding current density of complex amplitude $\vec{J} = \vec{E}/\rho_e$, the time-average dissipation power density $<p>$ is expressed by:

$$<p> = \tfrac{1}{2}\vec{E} \cdot \vec{J}^* = \frac{1}{2\rho_e}|E|^2 = \frac{\rho_e}{2}|J|^2 \qquad (2.21)$$

where $\vec{J}^*$ is the complex conjugate of $\vec{J}$.

In the sinusoidal case, the S-vector is given by:

$$\vec{S} = \tfrac{1}{2}\left(\vec{E} \times \vec{H}^*\right) \qquad (2.22a)$$

$$-\oint_S \vec{S} \cdot d\vec{A} = <P> + j2\omega(<W_m> - <W_e>) \qquad (2.22b)$$

where $<P>$, $<W_m>$, and $<W_e>$ are, respectively, the total time-averaged power dissipated, total time-averaged magnetic energy, and total time-averaged electric energy, each computed over the system volume.

$$<P> = \frac{1}{2\rho_e}\int_\mathcal{V}|E|^2\,d\mathcal{V} \qquad (2.23a)$$

$$<W_m> = \frac{\mu_o}{4}\int_\mathcal{V}|H|^2\,d\mathcal{V} \qquad (2.23b)$$

$$<W_e> = \frac{\epsilon_o}{4}\int_\mathcal{V}|E|^2\,d\mathcal{V} \qquad (2.23c)$$

The complex Poynting vector, $\vec{S}$, expanded up to the 1st-order fields, is given by:

$$\vec{S} = \tfrac{1}{2}\left(\vec{E}_0 \times \vec{H}_0^* + \vec{E}_0 \times \vec{H}_1^* + \vec{E}_1 \times \vec{H}_0^*\right) \qquad (2.24)$$

2.5 Field Solutions from the Scalar Potentials

The static electric field, because its curl is zero ($\nabla \times \vec{E} = 0$), is a conservative field and thus can always be given as the gradient of a scalar potential ϕ:

$$\vec{E} = -\text{grad}\,\phi = -\nabla\,\phi \tag{2.25}$$

Thus $\nabla \cdot \vec{E}$ may be given by:

$$\nabla \cdot \vec{E} = -\nabla \cdot \nabla\phi = -\nabla^2\phi \tag{2.26}$$

In the absence of charge density ($\rho_c = 0$), Eq. 2.2 reduces to:

$$\nabla \cdot \vec{E} = 0 \tag{2.27}$$

Combining Eqs. 2.26 and 2.27, we obtain:

$$\nabla^2\phi = 0 \tag{2.28}$$

Equation 2.28, known as Laplace's equation, expresses scalar potentials from which physically realizable $\vec{E}$ fields can be derived. Similarly, the magnetic field $\vec{H}$, in the absence of free current ($\nabla \times \vec{H} = 0$) in linear media in which $\vec{B} = \mu\vec{H}$, is derivable from the scalar potentials satisfying the Laplace's equation. Selected solutions of Laplace's equation in two-dimensional cylindrical coordinates and three-dimensional spherical coordinates are presented below.

2.5.1 Two-Dimensional Cylindrical Coordinates

For a two-dimensional potential in cylindrical coordinates, $\nabla^2\phi$ is given by:

$$\nabla^2\phi = \frac{1}{r}\frac{\partial}{\partial r}\left(r\frac{\partial\phi}{\partial r}\right) + \frac{1}{r^2}\frac{\partial^2\phi}{\partial\theta^2} = 0 \tag{2.29}$$

The standard technique to solve Eq. 2.29 is to express ϕ as the product of two functions, each a function of only one of the two coordinates:

$$\phi = R(r)\Phi(\theta) \tag{2.30}$$

The solutions to Eq. 2.30 have the following general forms:

$$\text{for } n = 0 \qquad \phi_0 = -A\ln r\,(C_1\theta + C_2) \tag{2.31a}$$

$$\text{for } n \neq 0 \qquad \phi_n = (A_1 r^n + A_2 r^{-n})(C_1\sin n\theta + C_2\cos n\theta) \tag{2.31b}$$

Special Cases

$n = 0$: The simplest form of field derivable from ϕ under this condition is one whose spatial dependence is $1/r$. Examples are the electric field due to a line charge ($\lambda = 2\pi\epsilon_o$) and the magnetic field associated with a current filament ($I = 2\pi$). Thus with a potential $[\phi_0]_E = 2\pi\epsilon_o \ln r$, we have: $\vec{E} = (1/r)\vec{i}_r$; with a potential $[\phi_0]_H = 2\pi\theta$, we have: $\vec{H} = (1/r)\vec{i}_\theta$.

$n = 1$: Both $\phi_1 = \sin\theta/r$ and $\phi_1 = \cos\theta/r$ are potentials associated with two-dimensional electric or magnetic dipoles. Note that either form has a singularity at the origin ($r = 0$); they are usually associated with dipole fields that do not include the origin. The choice of $\sin\theta$ or $\cos\theta$ depends on field orientation in the coordinate system. Also, the potentials $\phi_1' = r\sin\phi$ and $\phi_1' = r\cos\phi$ are associated with uniform vector fields.

$n = 2$: The potentials $\phi_2 = \cos 2\theta/r^2$ and $\phi_2' = r^2 \cos 2\theta$ are associated with two-dimensional quadrupole fields. The potential $\phi_2 = \cos 2\theta/r^2$, because of the singularity at the origin, is valid for space outside the origin; the potential $\phi_2' = r^2 \cos 2\theta$ is valid for space that includes the origin.

2.5.2 Spherical Coordinates

The solutions in spherical coordinates can also be expressed as the product of three functions each involving only one of the three coordinates:

$$\phi = R(r)\Theta(\theta)\Phi(\varphi) \tag{2.32}$$

Functions $R(r)$, $\Theta(\theta)$, and $\Phi(\varphi)$ have the following solutions:

$$R(r) = A_1 r^n + A_2 r^{-(n+1)} \tag{2.33a}$$

$$\Theta(\theta) = C P_n^m(\cos\theta) \qquad (m \leq n) \tag{2.33b}$$

$$\Phi(\varphi) = D_1 \sin m\varphi + D_2 \cos m\varphi \tag{2.33c}$$

$P_n^m(\cos\theta)$, known as the Legendre and Associated Legendre functions, are useful for analyzing fields in the central zone of uniform-field solenoidal magnets. They are tabulated in Tables 2.2 and 2.3.

Special Cases

$n = m = 0$: This case gives rise to the the simplest solution $\phi_0 = 1/r$, which results in, for example, the $\vec{E}$ field of a point charge of magnitude $1/4\pi\epsilon_o$.

$n = 1, m = 0$: There are two solutions, $\phi_1 = \cos\theta/r^2$ and $\phi_1' = r\cos\theta$. ϕ_1 results in a dipole field away from the origin, while ϕ_1' results in a corresponding dipole field near the origin.

In the Problem Section of this chapter we deal almost exclusively with two-dimensional cylindrical dipole and quadrupole fields and spherical dipole fields.

2.5.3 Del Operators

∇ operators in Cartesian, cylindrical, and spherical coordinates are given below:

$$\text{(Cartesian)} \qquad \nabla = \vec{\imath}_x \frac{\partial}{\partial x} + \vec{\imath}_y \frac{\partial}{\partial y} + \vec{\imath}_z \frac{\partial}{\partial z} \qquad (2.34)$$

$$\text{(Cylindrical)} \qquad \nabla = \vec{\imath}_r \frac{\partial}{\partial r} + \vec{\imath}_\theta \frac{1}{r} \frac{\partial}{\partial \theta} + \vec{\imath}_z \frac{\partial}{\partial z} \qquad (2.35)$$

$$\text{(Spherical)} \qquad \nabla = \vec{\imath}_r \frac{\partial}{\partial r} + \vec{\imath}_\theta \frac{1}{r} \frac{\partial}{\partial \theta} + \vec{\imath}_\varphi \frac{1}{r \sin \theta} \frac{\partial}{\partial \varphi} \qquad (2.36)$$

Table 2.2: Legendre Functions

$P_0^0(u) = 1$	
$P_1^0(u) = u$	$= \cos\theta$
$P_2^0(u) = \frac{1}{2}(3u^2 - 1)$	$= \frac{1}{4}(3\cos 2\theta + 1)$
$P_3^0(u) = \frac{1}{2}(5u^3 - 3)$	$= \frac{1}{8}(5\cos 3\theta + 3\cos\theta)$
$P_4^0(u) = \frac{1}{8}(35u^4 - 30u^2 + 3)$	$= \frac{1}{64}(35\cos 4\theta + 20\cos 2\theta + 9)$

Table 2.3: Associated Legendre Functions

$P_1^1(u) = (1-u^2)^{1/2}$	$= \sin\theta$
$P_2^1(u) = 3u(1-u^2)^{1/2}$	$= \frac{3}{2}\sin 2\theta$
$P_2^2(u) = 3(1-u^2)$	$= \frac{3}{2}(1 - \cos 2\theta)$
$P_3^1(u) = \frac{3}{2}(1-u^2)^{1/2}(5u^2 - 1)$	$= \frac{3}{8}(\sin\theta + 5\sin 3\theta)$
$P_3^2(u) = 15(1-u^2)u$	$= \frac{15}{4}(\cos\theta - 3\cos 3\theta)$
$P_3^3(u) = 15(1-u^2)^{3/2}$	$= \frac{15}{4}(3\sin\theta - \sin 3\theta)$
$P_4^1(u) = \frac{5}{2}(1-u^2)^{1/2}(7u^3 - 3u)$	$= \frac{5}{16}(2\sin 2\theta + 7\sin 4\theta)$
$P_4^2(u) = \frac{15}{2}(1-u^2)(7u^2 - 1)$	$= \frac{15}{16}(3 + 4\cos 2\theta - 7\cos 4\theta)$
$P_4^3(u) = 105(1-u^2)^{1/2}u$	$= \frac{105}{8}(2\sin 2\theta - \sin 4\theta)$
$P_4^4(u) = 105(1-u^2)^2$	$= \frac{105}{8}(3 - 4\cos 2\theta + \cos 4\theta)$

"It was absolutely marvelous working for Pauli. You could ask him anything. There was no worry that he would think a particular question was stupid, since he thought all questions were stupid." —Victor F. Weisskopf

Problem 2.1: Magnetized sphere in a uniform field

This problem deals with a magnetic sphere ($\mu > \mu_o$) exposed to a uniform external magnetic field. Although there is no net force acting on the sphere (because the background field is uniform) a field expression inside the sphere for the case $\mu \gg \mu_o$ is still useful in estimating the force on a ferromagnetic object placed in the fringing field generated by a magnet nearby. This force on an iron object due to the fringing field of a magnet will be discussed in more detail in Problem 3.13, Chapter 3.

Figure 2.1 shows a magnetic sphere of radius R and permeability μ in a uniform external magnetic field given by:

$$\vec{H}_\infty = H_0(-\cos\theta\,\vec{\imath}_r + \sin\theta\,\vec{\imath}_\theta) \tag{2.37}$$

a) Show that expressions for the magnetic inductions outside ($\vec{B}_1$) and inside ($\vec{B}_2$) the sphere are given by:

$$\vec{B}_1 = \mu_o H_0(-\cos\theta\,\vec{\imath}_r + \sin\theta\,\vec{\imath}_\theta)$$
$$+ \mu_o\left(\frac{\mu_o - \mu}{2\mu_o + \mu}\right)H_0\left(\frac{R}{r}\right)^3(2\cos\theta\,\vec{\imath}_r + \sin\theta\,\vec{\imath}_\theta) \tag{2.38a}$$

$$\vec{B}_2 = \frac{3\mu_o\mu H_0}{2\mu_o + \mu}(-\cos\theta\,\vec{\imath}_r + \sin\theta\,\vec{\imath}_\theta) \tag{2.38b}$$

Consider the following three limiting cases: $\mu/\mu_o = 0$, $\mu/\mu_o = 1$, $\mu/\mu_o = \infty$; make sure that the resulting expressions for the field in the sphere agree with the ones you would expect on physical grounds.

b) Make a *rough* sketch of $\vec{B}$ fields for each case, $\mu/\mu_o \ll 1$ and $\mu/\mu_o \gg 1$.

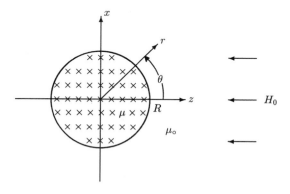

Fig. 2.1   Magnetized sphere in a uniform magnetic field.

Solution to Problem 2.1

a) This problem is most easily solved using the concept of scalar potentials discussed in the introductory section of this chapter. Namely, the magnetic potential ϕ is a scalar field such that:

$$\vec{H} = -\nabla\phi \qquad (S1.1)$$

In linear media, magnetic field and magnetic induction are related by:

$$\vec{B} = \mu\vec{H} \qquad (S1.2)$$

The problem is divided into two regions, region 1 $(r \geq R)$ and region 2 $(r \leq R)$. The appropriate potential for each region is given below:

$$\phi_1 = H_0 r \cos\theta + \frac{A}{r^2}\cos\theta \qquad (r \geq R) \qquad (S1.3a)$$

$$\phi_2 = Cr\cos\theta \qquad (r \leq R) \qquad (S1.3b)$$

Note that $\phi_1 \to H_0 r \cos\theta$ for $r \to \infty$, as required from Eq. 2.37, and ϕ_2 remains finite at $r = 0$.

Using the ∇ operator for spherical coordinates given by Eq. 2.36, we can derive $\vec{H}$ in each region.

$$\vec{H}_1 = -\frac{\partial}{\partial r}\left(H_0 r \cos\theta + \frac{A}{r^2}\cos\theta\right)\vec{i}_r - \frac{1}{r}\frac{\partial}{\partial\theta}\left(H_0 r \cos\theta + \frac{A}{r^2}\cos\theta\right)\vec{i}_\theta$$

$$= H_0(-\cos\theta\,\vec{i}_r + \sin\theta\,\vec{i}_\theta) + \frac{A}{r^3}(2\cos\theta\,\vec{i}_r + \sin\theta\,\vec{i}_\theta) \qquad (S1.4a)$$

$$\vec{H}_2 = -\frac{\partial}{\partial r}(Cr\cos\theta)\vec{i}_r - \frac{1}{r}\frac{\partial}{\partial\theta}(Cr\cos\theta)\vec{i}_\theta$$

$$= C(-\cos\theta\,\vec{i}_r + \sin\theta\,\vec{i}_\theta) \qquad (S1.4b)$$

Boundary Conditions

1) At $r = R$, the tangential component $(\vec{i}_\theta)$ of $\vec{H}$ is continuous since there is no free current present. This is equivalent to equating the potentials at $r = R$ $(\phi_1 = \phi_2)$; hence:

$$H_0 R + \frac{A}{R^2} = CR \qquad (S1.5)$$

2) At $r = R$, the normal $(\vec{i}_r)$ component of $\vec{B}$ is continuous:

$$-\mu_\circ H_0 + 2\mu_\circ\frac{A}{R^3} = -\mu C$$

$$-H_0 + 2\frac{A}{R^3} = -\frac{\mu}{\mu_\circ}C \qquad (S1.6)$$

From Eqs. S1.5 and S1.6, we can solve for the constants C and A:

$$C = \frac{3H_0\mu_\circ}{2\mu_\circ + \mu} \qquad (S1.7)$$

$$A = \frac{C}{3}\left(\frac{\mu_\circ - \mu}{\mu_\circ}\right)R^3 = H_0\left(\frac{\mu_\circ - \mu}{2\mu_\circ + \mu}\right)R^3 \qquad (S1.8)$$

Solution to Problem 2.1

$\vec{B}_1$ and $\vec{B}_2$ are thus:

$$\vec{B}_1 = \mu_0 H_0(-\cos\theta\,\vec{i}_r + \sin\theta\,\vec{i}_\theta)$$

$$+ \mu_0 \left(\frac{\mu_0 - \mu}{2\mu_0 + \mu}\right) H_0 \left(\frac{R}{r}\right)^3 (2\cos\theta\,\vec{i}_r + \sin\theta\,\vec{i}_\theta) \qquad (2.38a)$$

$$\vec{B}_2 = \frac{3\mu_0\mu H_0}{2\mu_0 + \mu}(-\cos\theta\,\vec{i}_r + \sin\theta\,\vec{i}_\theta) \qquad (2.38b)$$

Now let us consider the three special cases of μ/μ_0.

Case 1: $\mu/\mu_0 = 0$

With $\mu = 0$ inserted into Eqs. 2.38a and 2.38b, we obtain:

$$\vec{B}_1 = \mu_0 H_0(-\cos\theta\,\vec{i}_r + \sin\theta\,\vec{i}_\theta)$$

$$+ \frac{\mu_0 H_0}{2}\left(\frac{R}{r}\right)^3 (2\cos\theta\,\vec{i}_r + \sin\theta\,\vec{i}_\theta) \qquad (S1.9a)$$

$$\vec{B}_2 = 0 \qquad (S1.9b)$$

The sphere is like a superconductor; no magnetic flux density is allowed inside the sphere—the Meissner effect. As discussed in the next problem where a superconducting cylinder is considered, a discontinuity in the θ component of the $\vec{H}$ fields at $r = R$ requires a surface current (confined within a thin layer). Because this current, once set up, must flow persistently, it implies that the sphere's electrical conductivity, like the superconductor's, must be infinite. As discussed in Chapter 1, a perfectly diamagnetic material must at the same time be a perfect conductor; that is, such a material is automatically a superconductor.

Case 2: $\mu/\mu_0 = 1$

The problem reduces to the trivial case, equivalent to the absence of the sphere.

Case 3: $\mu/\mu_0 = \infty$

This is a case when the sphere is of ferromagnetic material such as iron. The magnetic field is *drawn* into the sphere. With $\mu = \infty$ inserted into Eqs. 2.38a and 2.38b, we obtain:

$$\vec{B}_1 = \mu_0 H_0(-\cos\theta\,\vec{i}_r + \sin\theta\,\vec{i}_\theta)$$

$$- \mu_0 H_0 \left(\frac{R}{r}\right)^3 (2\cos\theta\,\vec{i}_r + \sin\theta\,\vec{i}_\theta) \qquad (S1.10a)$$

$$\vec{B}_2 = 3\mu_0 H_0(-\cos\theta\,\vec{i}_r + \sin\theta\,\vec{i}_\theta) \qquad (S1.10b)$$

The important point to note is that the $\vec{B}$ field within the ferromagnetic sphere is 3 times that of the external $\vec{B}$ field. (Note that if the sphere's magnetization is saturated, μ would no longer be ∞.)

Solution to Problem 2.1

b) Field distributions for $\mu/\mu_o = 0.1$ and $\mu/\mu_o = 100$ are sketched below.

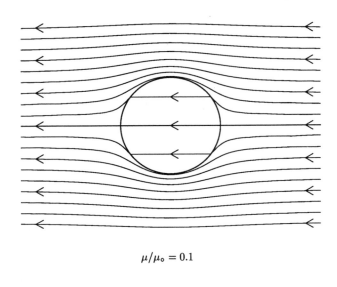

$$\mu/\mu_o = 0.1$$

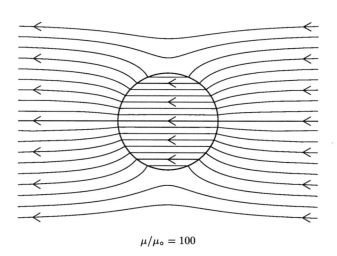

$$\mu/\mu_o = 100$$

Fig. 2.2 Field distributions inside and near the sphere for two values of μ/μ_o. For the case $\mu/\mu_o = 100$, the spacing between field lines inside the sphere is $\simeq \sqrt{3}$ times denser than that far from the sphere; the ratio would be $\simeq 2$ for a cylinder.

Problem 2.2: Type I superconducting rod in a uniform field

This problem deals with a Type I superconductor exhibiting the Meissner effect. The field solution is interpreted in terms of the classical London theory of superconductivity.

Figure 2.3 shows an infinitely long lead rod of circular cross section (radius R) subjected to a uniform external magnetic field perpendicular to its axis.

$$\vec{H}_\infty = H_0(-\cos\theta\,\vec{i}_r + \sin\theta\,\vec{i}_\theta) \tag{2.39}$$

where $\mu_o H_0 = 0.06\,\text{T}$. Initially the rod is at $4.2\,\text{K}$ and in the presence of this field it is in the *normal* state ($H_0 > H_c$). That is, the field given above is valid everywhere including inside the rod. The rod is then gradually cooled until it becomes superconducting.

a) Show that an expression for the field outside the superconducting rod ($\vec{H}_1$) after transient effects of the field change have subsided, is given by:

$$\vec{H}_1 = H_0(-\cos\theta\,\vec{i}_r + \sin\theta\,\vec{i}_\theta) + H_0\left(\frac{R}{r}\right)^2(\cos\theta\,\vec{i}_r + \sin\theta\,\vec{i}_\theta) \tag{2.40}$$

b) Show that an expression for the surface current density, $\vec{K}_f$ [A/m], flowing within a penetration depth $\lambda \ll R$, is given by:

$$\vec{K}_f = 2H_0\sin\theta\,\vec{i}_z \tag{2.41}$$

c) Convert the magnitude of the surface current density to that of current density, J_f [A/m²], and confirm that its numerical value is consistent with that for lead derivable from London's theory of superconductivity.

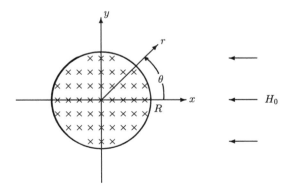

Fig. 2.3 Infinitely long, circular cross section superconducting rod in a uniform magnetic field.

Solution to Problem 2.2

a) The problem is divided into two regions, region 1 ($r \geq R$) and region 2 ($r \leq R$). Because we are dealing with a Type I superconductor, $\vec{B}_2 = 0$ ($\phi_2 = 0$) when it is in the superconducting state. The field in region 1 is derivable from appropriate potentials:

$$\phi_1 = H_0 r \cos\theta + \frac{A}{r}\cos\theta \qquad (S2.1)$$

Note that $\phi_1 \to H_0 r \cos\theta$ for $r \to \infty$, as required.

Using the ∇ operator for cylindrical coordinates (Eq. 2.35), we can derive $\vec{H}_1$:

$$\vec{H}_1 = -\frac{\partial}{\partial r}\left(H_0 r \cos\theta + \frac{A}{r}\cos\theta\right)\vec{\imath}_r - \frac{1}{r}\frac{\partial}{\partial\theta}\left(H_0 r \cos\theta + \frac{A}{r}\cos\theta\right)\vec{\imath}_\theta \quad (S2.2)$$

$$= -\left(H_0\cos\theta - \frac{A}{r^2}\cos\theta\right)\vec{\imath}_r - \left(-H_0\sin\theta - \frac{A}{r^2}\sin\theta\right)\vec{\imath}_\theta \qquad (S2.3)$$

Rearranging Eq. $S2.3$, we have:

$$\vec{H}_1 = H_0(-\cos\theta\,\vec{\imath}_r + \sin\theta\,\vec{\imath}_\theta) + \frac{A}{r^2}(\cos\theta\,\vec{\imath}_r + \sin\theta\,\vec{\imath}_\theta) \qquad (S2.4)$$

The boundary condition requires that $B_{r1} = B_{r2} = 0$ at $r = R$:

$$-H_0 + \frac{A}{R^2} = 0 \qquad (S2.5)$$

Solving Eq. $S2.5$ for A, we obtain:

$$A = R^2 H_0 \qquad (S2.6)$$

The field outside the superconducting rod (region 1) is thus given by:

$$\vec{H}_1 = H_0(-\cos\theta\,\vec{\imath}_r + \sin\theta\,\vec{\imath}_\theta) + H_0\left(\frac{R}{r}\right)^2(\cos\theta\,\vec{\imath}_r + \sin\theta\,\vec{\imath}_\theta) \qquad (2.40)$$

Note that at $r = R, \theta = 90°$, $|\vec{H}_1| = 2H_0$, or the field amplitude is twice the far-field amplitude.

b) Because of a discontinuity at $r = R$ of $2H_0\sin\theta$ in the tangential ($\vec{\imath}_\theta$) component of $\vec{H}$, there must be a surface current density $\vec{K}_f$ flowing in the rod, as given by Eq. 2.6. We thus have:

$$\vec{K}_f = \vec{\imath}_r \times (2H_0\sin\theta\vec{\imath}_\theta - 0)$$

$$= 2H_0\sin\theta\,\vec{\imath}_z \qquad (2.41)$$

Solution to Problem 2.2

c) According to the London theory of superconductivity, discussed briefly in Chapter 1, the critical current density J_c of Type I superconductors is given by $J_c = en_e v$, where e is the electronic charge (1.6×10^{-19} C), n_e is the electron density, and v is the speed of sound. n_e is given by Eq. 1.4:

$$n_e = \frac{2\varrho N_A}{W_A} \tag{1.4}$$

By inserting $\varrho = 11.4 \times 10^3 \, \text{kg/m}^3$, $N_A = 6.02 \times 10^{26}$ particle/kg-mole, and $W_A = 207.2$ kg/kg-mole for lead, we have:

$$n_e = \frac{(2)(11.4 \times 10^3 \, \text{kg/m}^3)(6.02 \times 10^{26} \, \text{particle/kg-mole})}{207.2 \text{kg/kg-mole}} \tag{S2.7}$$

$$= 6.62 \times 10^{28} \, \text{particle/m}^3 = 6.62 \times 10^{28} \, \text{electron/m}^3$$

Taking $v = 1200$ m/s for lead, we obtain:

$$J_c = en_e v \tag{S2.8}$$

$$= (1.6 \times 10^{-19} \, \text{C/electron})(6.62 \times 10^{28} \, \text{electron/m}^3)(1200 \, \text{m/s})$$

$$= 1.27 \times 10^{13} \, \text{A/m}^2$$

The current density required in the above lead cylinder must be less than J_c. The London theory also gives an expression (Eq. 1.3) for the penetration depth (λ) at the superconductor's surface within which superconducting current can flow. λ and n_e are related by:

$$\lambda = \sqrt{\frac{m}{\mu_o e^2 n_e}} \tag{1.3}$$

With $m = 9.1 \times 10^{-31}$ kg and appropriate values of e and n_e inserted into Eq. 1.3, we obtain:

$$\lambda = \sqrt{\frac{9.1 \times 10^{-31} \, \text{kg}}{(4\pi \times 10^{-7} \, \text{H/m})(1.6 \times 10^{-19} \, \text{C})^2(6.62 \times 10^{28} \, \text{electron/m}^3)}} \tag{S2.9}$$

$$= 2.1 \times 10^{-8} \, \text{m}$$

Because $K_f = J_f \lambda$:

$$J_f = \frac{2H_0}{\lambda} = \frac{2\mu_o H_0}{\mu_o \lambda} = \frac{2(0.06 \, \text{T})}{(4\pi \times 10^{-7} \, \text{H/m})(2.1 \times 10^{-8} \, \text{m})} \tag{S2.10}$$

$$\simeq 0.5 \times 10^{13} \, \text{A/m}^2$$

As required, $J_f < J_c$.

Problem 2.3: Magnetic shielding with a spherical shell

This problem deals with the essence of passive magnetic shielding, an important subject for MRI, maglev, and other systems where people and field-sensitive equipment might be exposed to a fringing field. The U.S. Food and Drug Administration limits the maximum fringing field in MRI systems to 5 gauss (0.5 mT).

Within a spherical region of space, a uniform magnetic field, $\vec{H}_\infty$, is to be shielded:

$$\vec{H}_\infty = H_0(-\cos\theta\,\vec{\imath}_r + \sin\theta\,\vec{\imath}_\theta) \qquad (2.37)$$

For passive shielding, a spherical shell of o.d. $2R$ and wall thickness $d/R \ll 1$ of highly permeable material ($\mu/\mu_o \gg 1$) may be used, as shown in Fig. 2.4.

a) Treating the problem as one of a magnetic spherical shell in a uniform external field, show that an expression for H_{ss}/H_0, where H_{ss} is the magnitude of the magnetic field in the spherical space ($r \le R - d$), is given by:

$$\frac{H_{ss}}{H_0} = \frac{9\mu\mu_o}{9\mu\mu_o + 2(\mu - \mu_o)^2\left[1 - \left(1 - \dfrac{d}{R}\right)^3\right]} \qquad (2.42)$$

b) Show that in the limits of $\mu/\mu_o \gg 1$ and $d/R \ll 1$, the ratio H_{ss}/H_0 given by Eq. 2.42 reduces to:

$$\frac{H_{ss}}{H_0} \simeq \frac{3}{2}\left(\frac{\mu_o}{\mu}\right)\left(\frac{R}{d}\right) \qquad (2.43)$$

c) Next, obtain Eq. 2.43 through a perturbation approach. First, solve the field in the shell ($R - d \le r \le R$) with $\mu = \infty$. Then use a perturbation approach for the case $\mu/\mu_o \gg 1$ and obtain Eq. 2.43.

d) In reality the magnetic flux in the shielding material must be kept below the material's saturation flux, $\mu_o M_{sa}$. Show that an expression for d/R to keep the shell unsaturated is given by:

$$\frac{d}{R} \ge \frac{3H_0}{2M_{sa}} \qquad (2.44)$$

e) Draw field lines for the case $\mu/\mu_o \gg 1$.

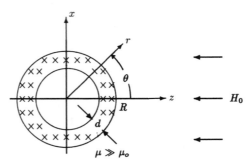

Fig. 2.4   Spherical magnetic shell in uniform magnetic field.

Solution to Problem 2.3

a) The problem is divided into three regions: region 1 ($r \geq R$), region 2 (the shell), and region 3 ($r \leq R-d$). The appropriate potential for each region is given below:

$$\phi_1 = H_0 r \cos\theta + \frac{A}{r^2}\cos\theta \tag{S3.1a}$$

$$\phi_2 = Cr\cos\theta + \frac{D}{r^2}\cos\theta \tag{S3.1b}$$

$$\phi_3 = H_{ss}r\cos\theta \tag{S3.1c}$$

Note that $\phi_1 \to H_0 r \cos\theta$ for $r \to \infty$ and that ϕ_3 remains finite as $r \to 0$.

Using the ∇ operator in spherical coordinates, we obtain:

$$\vec{H}_1 = H_0(-\cos\theta\,\vec{i}_r + \sin\theta\,\vec{i}_\theta) + \frac{A}{r^3}(2\cos\theta\,\vec{i}_r + \sin\theta\,\vec{i}_\theta) \tag{S3.2a}$$

$$\vec{H}_2 = C(-\cos\theta\,\vec{i}_r + \sin\theta\,\vec{i}_\theta) + \frac{D}{r^3}(2\cos\theta\,\vec{i}_r + \sin\theta\,\vec{i}_\theta) \tag{S3.2b}$$

$$\vec{H}_3 = H_{ss}(-\cos\theta\,\vec{i}_r + \sin\theta\,\vec{i}_\theta) \tag{S3.2c}$$

Boundary Conditions

1) At $r = R$, the tangential ($\vec{i}_\theta$) component of $\vec{H}$ (H_θ) is continuous: $\phi_1 = \phi_2$.

2) Similarly, at $r = R - d$, H_θ is continuous: $\phi_2 = \phi_3$.

3) At $r = R$, the normal ($\vec{i}_r$) component of $\vec{B}$ (B_r) is continuous.

4) Similarly, at $r = R - d$, B_r is continuous.

The above boundary conditions give rise to the following four equations:

$$H_0 R + \frac{A}{R^2} = CR + \frac{D}{R^2} \tag{S3.3a}$$

$$C(R-d) + \frac{D}{(R-d)^2} = H_{ss}(R-d) \tag{S3.3b}$$

$$\mu_0\left(-H_0 + \frac{2A}{R^3}\right) = \mu\left(-C + \frac{2D}{R^3}\right) \tag{S3.3c}$$

$$\mu\left[-C + \frac{2D}{(R-d)^3}\right] = -\mu_0 H_{ss} \tag{S3.3d}$$

Combining Eqs. S3.3a and S3.3b and eliminating C, we have:

$$\frac{A}{R^3} + D\left[\frac{1}{(R-d)^3} - \frac{1}{R^3}\right] - H_{ss} = -H_0 \tag{S3.4}$$

Solution to Problem 2.3

From Eqs. $S3.3b$ and $S3.3d$, we can obtain D in terms of H_{ss}:

$$D = \frac{\mu - \mu_o}{3\mu}(R - d)^3 H_{ss} \qquad (S3.5)$$

Combining Eqs. $S3.4$ and $S3.5$, we have A/R^3 in terms of H_{ss}:

$$\frac{A}{R^3} = H_{ss}\left\{1 - \frac{\mu - \mu_o}{3\mu}\left[1 - \left(1 - \frac{d}{R}\right)^3\right]\right\} - H_0 \qquad (S3.6)$$

From Eqs. $S3.3c$ and $S3.3d$, we obtain:

$$\frac{2A}{R^3} + 2\frac{\mu}{\mu_o}D\left[\frac{1}{(R - d)^3} - \frac{1}{R^3}\right] + H_{ss} = H_0 \qquad (S3.7)$$

Combining Eqs. $S3.4 \sim S3.7$ and expressing H_{ss} in terms of H_0, we obtain:

$$\frac{H_{ss}}{H_0} = \frac{9\mu\mu_o}{9\mu\mu_o + 2(\mu - \mu_o)^2\left[1 - \left(1 - \frac{d}{R}\right)^3\right]} \qquad (2.42)$$

b)   We may simplify Eq. 2.42 by dividing top and bottom by μ_o^2 and applying the limits $\mu/\mu_o \gg 1$ and $d/R \ll 1$:

$$\frac{H_{ss}}{H_0} \simeq \frac{9\mu/\mu_o}{9\left(\frac{\mu}{\mu_o}\right) + 2\left(\frac{\mu}{\mu_o}\right)^2\left[1 - \left(1 - 3\frac{d}{R}\right)\right]} \qquad (S3.8)$$

$$\simeq \frac{9}{9 + 6\left(\frac{\mu}{\mu_o}\right)\left(\frac{d}{R}\right)} \qquad (S3.9)$$

In the special case $\mu d/\mu_o R \gg 1$, Eq. $S3.9$ reduces to:

$$\frac{H_{ss}}{H_0} \simeq \frac{3}{2}\left(\frac{\mu_o}{\mu}\right)\left(\frac{R}{d}\right) \qquad (2.43)$$

c)   The same result given by Eq. 2.43 for H_{ss}/H_0 can be obtained directly by a perturbation approach for the case $\mu/\mu_o \gg 1$ and $d/R \ll 1$.

We proceed by assuming that μ of the shell material is infinite. We then find that the B lines enter and leave the shell at $r = R$ only normally. That is, $\vec{H}_1$ has only a radial ($\vec{\imath}_r$) component at $r = R$ because $\vec{H} = 0$ in the shell and H_θ is continuous at $r = R$. (This can be seen quite readily by noting that when $\mu = \infty$, $C = D = E = 0$.) From Eq. $S3.3a$, $A = -R^3 H_0$, and thus at $r = R$:

$$\vec{H}_1 = -3H_0\cos\theta\,\vec{\imath}_r \qquad (S3.10)$$

Solution to Problem 2.3

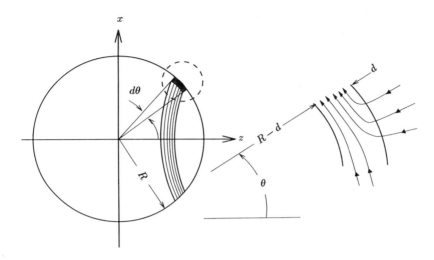

Fig. 2.5 Flux entering into the spherical shell over the surface bounded by $\pm\theta$.

The B lines stay inside the shell without "spilling" into region 3; that is, B inside the shell has only a $\vec{i}_\theta$ component. Let us now apply magnetic flux continuity, *i.e.* $\nabla \cdot \vec{B} = 0$, and solve for $\vec{B}_2$ when $\mu = \infty$. Once $\vec{B}_2$ is solved for this case, an approximate expression for $\vec{H}_3$ can be deduced for $\mu \neq \infty$ but $\mu/\mu_\circ \gg 1$.

First, we calculate the total magnetic flux Φ entering into the shell over the surface area bounded by $\pm\theta$ (Fig. 2.5). This surface area, as indicated in the figure, is given by a differential area (a ring of radius $R\sin\theta$ times $Rd\theta$) integrated from 0 to θ. Thus, we have:

$$\Phi = \mu_\circ \int_0^\theta \vec{H}_1 \cdot d\vec{A} = \mu_\circ \int_0^\theta 3H_0 \cos\theta \, 2\pi R^2 \sin\theta \, d\theta$$

$$= 3\pi\mu_\circ R^2 H_0 \sin^2\theta \qquad (S3.11)$$

This Φ must be equal to the total flux flowing in the θ-direction in the shell at θ. Because the shell's cross sectional area, A_2, at θ is given by the shell thickness d times the circumference of a ring of radius $R\sin\theta$, we have:

$$A_2 = d2\pi R\sin\theta \qquad (S3.12)$$

We thus have:

$$\Phi = 3\pi\mu_\circ R^2 H_0 \sin^2\theta$$

$$= B_2 A_2 = B_2 d2\pi R\sin\theta \qquad (S3.13)$$

Solution to Problem 2.3

Solving for B_2 from Eq. $S3.13$, we obtain:

$$\vec{B}_2 = \tfrac{3}{2}\mu_\circ \left(\frac{R}{d}\right) H_0 \sin\theta\,\vec{i}_\theta \qquad (S3.14)$$

Note that $\vec{B}_2$ is for $\mu = \infty$; we can now deduce an approximate solution for $\vec{H}_3$ because the $\vec{i}_\theta$-component of $\vec{H}$ must be continuous at $r = R - d$. Thus:

$$H_{\theta 3} \simeq \frac{B_{\theta 2}}{\mu} = \tfrac{3}{2}\left(\frac{\mu_\circ}{\mu}\right)\left(\frac{R}{d}\right) H_0 \sin\theta \qquad (S3.15)$$

Once $H_{\theta 3}$ is known, we have a complete expression for $\vec{H}_3$:

$$\vec{H}_3 \simeq \tfrac{3}{2}\left(\frac{\mu_\circ}{\mu}\right)\left(\frac{R}{d}\right) H_0(-\cos\theta\,\vec{i}_r + \sin\theta\,\vec{i}_\theta) \qquad (S3.16)$$

The ratio H_{ss}/H_0 deduced from Eq. $S3.16$ agrees with that given by Eq. 2.43. Note that in this perturbation approach the condition $\mu d/\mu_\circ R \gg 1$, required in the step going from Eq. $S3.9$ to Eq. 2.43, is unnecessary.

d) It is important to remember that d cannot be chosen arbitrarily small to satisfy the condition $d/R \ll 1$. The preceeding analysis is valid, in fact, only when:

$$\frac{\mu_\circ}{\mu} \ll \frac{d}{R} \ll 1 \qquad (S3.17)$$

In reality μ cannot be infinite and the shielding material will eventually saturate as the external field increases. Hence, the maximum magnetic flux inside the shell, which occurs at $\theta = 90°$, must be less than the saturation flux $\mu_\circ M_{sa}$ of the shell material. Thus:

$$\tfrac{3}{2}\left(\frac{R}{d}\right)\mu_\circ H_0 \le \mu_\circ M_{sa} \qquad (S3.18)$$

Solving Eq. $S3.18$ for d/R, we obtain:

$$\frac{d}{R} \ge \frac{3H_0}{2M_{sa}} \qquad (2.44)$$

Table 2.4 presents approximate values of *differential* $\mu/\mu_\circ$, defined as $(\mu/\mu_\circ)_{dif} \equiv \Delta M/\Delta H_0|_{\mu_\circ H_0}$, in the $\mu_\circ H_0$ range 5~1000 gauss (0.5~100 mT) and $\mu_\circ M_{sa}$ for annealed iron, as-cast iron, and as-cast steel. The data indicate that these materials are useful for magnetic shielding in external magnetic inductions up to ~200 gauss. The materials have $\mu_\circ M_{sa}$ values, respectively, of 2.15, 1.65, and 2.05 T.

Solution to Problem 2.3

Table 2.4: Approximate Values* of $(\mu/\mu_0)_{dif}$ and $\mu_0 M_{sa}$
For Iron Materials

$\mu_0 H_0$ [gauss]	$(\mu/\mu_0)_{dif} \equiv \Delta M/\Delta H_0\|_{\mu_0 H_0}$		
	Annealed Iron	*As-Cast Iron*	*As-Cast Steel*
5	250	610	1130
10	60	305	565
20	40	155	180
50	20	60	50
100	15	30	25
200	10	15	10
500	3	7	2
1000	1+	3	1+
$\mu_0 M_{sa}$	2.15 T	1.65 T	2.03 T

* Derived from M *vs* H plots given in *Permanent Magnet Manual*
(General Electric Company, 1963).

e)   Field lines for the case $\mu/\mu_0 = 100$ are shown in Fig. 2.6.

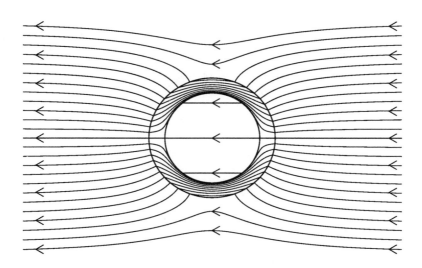

Fig. 2.6   Field distribution through a spherical magnetic shell in a uniform field.

Problem 2.4: Shielding with a cylindrical shell

Here is another problem of magnetic shielding, this time in the two-dimensional case with a cylindrical shell of o.d. $2R$ and wall thickness $d/R \ll 1$ of highly permeable material ($\mu/\mu_\circ \gg 1$) in a uniform external magnetic field, $\vec{H}_\infty$. Either the spherical shell model considered in the previous problem or the cylindrical shell model considered here is applicable in most practical cases requiring magnetic shielding. Which model to use must be considered on a case-by-case basis. Problem 3.11, for example, studies a case involving magnetic shielding of a computer exposed to a fringing field of a magnet located far from the computer; for this case, the spherical shell model is used.

In 2-D cylindrical coordinates $\vec{H}_\infty$ is given by:

$$\vec{H}_\infty = H_0(-\cos\theta\,\vec{i}_r + \sin\theta\,\vec{i}_\theta) \tag{2.39}$$

As in Fig. 2.4, which is a 2-D plot of 3-D spherical coordinates, θ is measured from the abscissa.

a) Using a perturbation technique similar to that studied in Problem 2.3, show that in the limits of $\mu/\mu_\circ \gg 1$ and $d/R \ll 1$, an expression for H_{cs}/H_0, where H_{cs} is the magnitude of a magnetic field in the cylindrical space ($r \leq R - d$) surrounded by the magnetic shell, is given by:

$$\frac{H_{cs}}{H_0} \simeq 2\left(\frac{\mu_\circ}{\mu}\right)\left(\frac{R}{d}\right) \tag{2.45}$$

As in the perturbation approach used in the spherical shell, the condition $\mu d/\mu_\circ R \gg 1$ is unnecessary to arrived at this equation.

b) Show that an expression of d/R for magnetic material with a saturation magnetization of M_{sa} is given by:

$$\frac{d}{R} \geq \frac{2H_\circ}{M_{sa}} \tag{2.46}$$

"... you don't know which way is straight up and which way is straight down."   —Ty Ty Walden

Solution to Problem 2.4

a) The problem is divided into three regions: region 1 ($r \geq R$), region 2 (shell), and region 3 ($r \leq R - d$). The appropriate potential for each region is:

$$\phi_1 = H_0 r \cos \theta + \frac{A}{r} \cos \theta \qquad (S4.1a)$$

$$\phi_2 = Cr \cos \theta + \frac{D}{r} \cos \theta \qquad (S4.1b)$$

$$\phi_3 = H_{cs} r \cos \theta \qquad (S4.1c)$$

We proceed by assuming that μ of the cylinder material is infinite. We then find that, as in the previous problem, the B lines enter and leave the cylinder at $r = R$ only normally. Thus at $r = R$ we have:

$$\vec{H}_1 = -2H_0 \cos \theta \, \vec{\imath}_\theta \qquad (S4.2)$$

B in the shell is θ-directed and because flux is continuous. For the case $d/R \ll 1$, the flux continuity requirement may be expressed by:

$$B_2 d = \int_0^\theta 2\mu_0 H_0 R \cos \theta \, d\theta = 2\mu_0 R H_0 \sin \theta \qquad (S4.3)$$

Thus:

$$\vec{B}_2 = 2\mu_0 \left(\frac{R}{d}\right) H_0 \sin \theta \, \vec{\imath}_\theta \qquad (S4.4)$$

Once $\vec{B}_2$ is known for $\mu = \infty$, we know $\vec{H}_2$ for $\mu/\mu_0 \gg 1$:

$$\vec{H}_2 = \frac{\vec{B}_2}{\mu} \simeq 2 \left(\frac{\mu_0}{\mu}\right) \left(\frac{R}{d}\right) H_0 \sin \theta \, \vec{\imath}_\theta \qquad (S4.5)$$

Because H_θ is continuous across a surface of discontinuity in the absence of surface current, the same H_θ must exist in regions 2 and 3: $H_{\theta 2} = H_{\theta 3}$. We thus have, at $r = R - d$:

$$H_{\theta 3} = H_{\theta 2} \simeq 2 \left(\frac{\mu_0}{\mu}\right) \left(\frac{R}{d}\right) H_0 \sin \theta \qquad (S4.6)$$

From Eq. $S4.6$, it follows that:

$$\vec{H}_3 \simeq 2 \left(\frac{\mu_0}{\mu}\right) \left(\frac{R}{d}\right) H_0 (-\cos \theta \, \vec{\imath}_r + \sin \theta \, \vec{\imath}_\theta) \qquad (S4.7)$$

$$\frac{H_{cs}}{H_0} \simeq 2 \left(\frac{\mu_0}{\mu}\right) \left(\frac{R}{d}\right) \qquad (2.45)$$

b) As in the spherical shell, the cylindrical shell cannot be arbitrarily thin; it must be thick enough to keep it from saturating:

$$\mu H_{cs} = 2\mu_0 H_0 \frac{R}{d} \leq \mu_0 M_{sa} \qquad (S4.8)$$

From Eq. $S4.8$, we obtain:

$$\frac{d}{R} \geq \frac{2H_0}{M_{sa}} \qquad (2.46)$$

Problem 2.5: The field far from a cluster of four dipoles

This problem considers the field far from a cluster of four dipoles arranged as shown in Fig. 2.7. The center-to-center distance between two opposing dipoles is $2\delta_d$. Each "ideal" jth dipole of zero winding thickness, diameter $2r_d$, and overall length ℓ_d can be modeled in the "far" field ($r_j \gg \ell_d$) as a spherical dipole, given by:

$$\vec{B}_j = \frac{r_d^2 \ell_d B_\circ}{2r_j^3}(\cos \vartheta_j \, \vec{i}_{r_j} + \tfrac{1}{2} \sin \vartheta_j \, \vec{i}_{\theta_j}) \tag{2.47}$$

where r_j is measured from the dipole center and ϑ_j is defined such that the field inside the winding points in the r_j-direction when $\vartheta_j = 0°$. Figure 2.7 indicates the direction of the field inside each dipole. Also defined in Fig. 2.7 are r-θ coordinates and z-x coordinates common to all the dipoles. Note that for $r \gg \delta_d$, we have $\vartheta_1 = \theta + 180°$, $\vartheta_2 = \theta - 90°$, $\vartheta_3 = \theta$, and $\vartheta_4 = \theta + 90°$.

Show that an approximate expression for the far field ($\vec{B}$ for $r/\delta_d \gg 1$) is given by:

$$\vec{B} \simeq \frac{6r_d^2 \ell_d B_\circ \delta_d}{2r^4}(-\sin 2\theta \, \vec{i}_r + \tfrac{1}{2} \cos 2\theta \, \vec{i}_\theta) \tag{2.48}$$

Neglect end effects of each dipole.

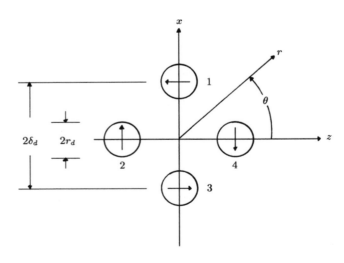

Fig. 2.7   Cross-sectional view of the four-dipole arrangement. The arrow in each dipole indicates the field direction inside the winding.

Solution to Problem 2.5

For $r \gg \delta_d$, r_j of each dipole may be given in terms of r and θ:

$$r_1 \simeq r - \delta_d \sin \theta \qquad (S5.1a)$$

$$r_2 \simeq r + \delta_d \cos \theta \qquad (S5.1b)$$

$$r_3 \simeq r + \delta_d \sin \theta \qquad (S5.1c)$$

$$r_4 \simeq r - \delta_d \cos \theta \qquad (S5.1d)$$

Inserting Eq. $S5.1$ into Eq. 2.47 for each dipole with ϑ_j expressed in terms of θ, we have:

$$\vec{B}_1 \simeq \frac{r_d^2 \ell_d B_\circ}{2(r - \delta_d \sin \theta)^3} (-\cos \theta \, \vec{\imath}_r - \tfrac{1}{2} \sin \theta \, \vec{\imath}_\theta) \qquad (S5.2a)$$

$$\vec{B}_2 \simeq \frac{r_d^2 \ell_d B_\circ}{2(r + \delta_d \cos \theta)^3} (\sin \theta \, \vec{\imath}_r - \tfrac{1}{2} \cos \theta \, \vec{\imath}_\theta) \qquad (S5.2b)$$

$$\vec{B}_3 \simeq \frac{r_d^2 \ell_d B_\circ}{2(r + \delta_d \sin \theta)^3} (\cos \theta \, \vec{\imath}_r + \tfrac{1}{2} \sin \theta \, \vec{\imath}_\theta) \qquad (S5.2c)$$

$$\vec{B}_4 \simeq \frac{r_d^2 \ell_d B_\circ}{2(r - \delta_d \cos \theta)^3} (-\sin \theta \, \vec{\imath}_r + \tfrac{1}{2} \cos \theta \, \vec{\imath}_\theta) \qquad (S5.2d)$$

For $r \gg \delta_d$ the denominator of each term may be expanded to the term containing δ_d/r (first order). Thus Eq. $S5.2$ may be written as:

$$\vec{B}_1 \simeq \frac{r_d^2 \ell_d B_\circ}{2r^3} \left[1 + 3 \left(\frac{\delta_d}{r} \right) \sin \theta \right] (-\cos \theta \, \vec{\imath}_r - \tfrac{1}{2} \sin \theta \, \vec{\imath}_\theta) \qquad (S5.3a)$$

$$\vec{B}_2 \simeq \frac{r_d^2 \ell_d B_\circ}{2r^3} \left[1 - 3 \left(\frac{\delta_d}{r} \right) \cos \theta \right] (\sin \theta \, \vec{\imath}_r - \tfrac{1}{2} \cos \theta \, \vec{\imath}_\theta) \qquad (S5.3b)$$

$$\vec{B}_3 \simeq \frac{r_d^2 \ell_d B_\circ}{2r^3} \left[1 - 3 \left(\frac{\delta_d}{r} \right) \sin \theta \right] (\cos \theta \, \vec{\imath}_r + \tfrac{1}{2} \sin \theta \, \vec{\imath}_\theta) \qquad (S5.3c)$$

$$\vec{B}_4 \simeq \frac{r_d^2 \ell_d B_\circ}{2r^3} \left[1 + 3 \left(\frac{\delta_d}{r} \right) \cos \theta \right] (-\sin \theta \, \vec{\imath}_r + \tfrac{1}{2} \cos \theta \, \vec{\imath}_\theta) \qquad (S5.3d)$$

Combining each field given by Eq. $S5.3$, we obtain:

$$\vec{B} = \vec{B}_1 + \vec{B}_2 + \vec{B}_3 + \vec{B}_4$$

$$\simeq \frac{6 r_d^2 \ell_d B_\circ \delta_d}{2r^4} (-\sin 2\theta \, \vec{\imath}_r + \tfrac{1}{2} \cos 2\theta \, \vec{\imath}_\theta) \qquad (2.48)$$

Note that $|\vec{B}|$ decreases $\propto 1/r^4$ rather than $\propto 1/r^3$, as would be the case with a single dipole.

Problem 2.6: Induction heating of a cylindrical shell

This problem deal with induction heating in a metallic (nonsuperconducting) cylindrical shell. It is a good example of a case involving sinusoidal electromagnetic fields, power flow (Poynting vector), and power dissipation. This and the next problem are the first examples of an AC loss, specifically an eddy-current loss, to be discussed further in Chapter 7. Although induction heating is widely used in electric furnaces to achieve high temperatures in conducting materials, it is sometimes used as a research tool in the study of the thermal behavior of superconducting windings. In superconducting magnet technology research, induction heating is most often used in the form of pulse fields to simulate transient disturbances that create small normal regions in otherwise superconducting windings.

Figure 2.8 shows a "long" metallic cylindrical shell of resistivity ρ_e, of o.d. $2R$, and of thickness $d \ll R$, placed in a sinusoidally time-varying magnetic field, which is within zeroth order, uniform and z-directed. Namely:

$$\vec{\mathcal{H}}_\infty(t) = \mathrm{Re}(\vec{H}_0 e^{j\omega t}) = \mathrm{Re}(H_0 e^{j\omega t})\vec{i}_z \tag{2.49}$$

where H_0 is a complex (real and imaginary parts) field amplitude.

We shall approach this problem first (Part 1) by solving for the appropriate fields by two methods and then (Part 2) solving for power dissipation in the cylinder by two methods.

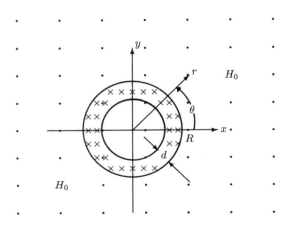

Fig. 2.8 Cylindrical metallic shell in a uniform sinusoidally time-varying magnetic field.

Problem 2.6: Induction heating—Part 1 (Field)

First, we shall solve appropriate fields using two methods, described below.

Method 1

a) Using the *integral* form of Maxwell's equations and neglecting end effects, show that expressions for the first-order electric field, $\vec{E}_1$, in the region $r \leq R$ and the first-order current density, $\vec{J}_1$, in the shell $(r \simeq R)$ are given by:

$$\vec{E}_1 = -\frac{j\omega\mu_o r H_0}{2} \, \vec{i}_\theta \tag{2.50}$$

$$\vec{J}_1 \simeq -\frac{j\omega\mu_o R H_0}{2\rho_e} \, \vec{i}_\theta \tag{2.51}$$

b) Show that the resulting first-order magnetic field, $\vec{H}_1$, in the region $r \leq R-d$ can be expressed by:

$$\vec{H}_1 = -\frac{j\omega\mu_o R d H_0}{2\rho_e} \, \vec{i}_z \tag{2.52}$$

c) Equations 2.50~2.52, derived using the quasi-static approximation, are valid only in the "low" frequency limit, or frequencies less than the "skin-depth" frequency, f_{sk}. Show that an expression for the skin-depth frequency is given by:

$$f_{sk} = \frac{\rho_e}{\pi\mu_o R d} \tag{2.53}$$

Method 2

$\vec{E}_1$, $\vec{J}_1$, and $\vec{H}_1$, derived by Method 1, each increasing with ω, are valid only for frequencies well below f_{sk}. We now demonstrate a new technique that enables us to derive the total field, $\vec{H}_T = (\vec{H}_0 + \vec{H}_R)$, in the bore, valid for the *entire range* of frequency. $\vec{H}_T$ is the total net field, $\vec{H}_0$ is the original field, $\vec{H}_R$ is the reaction field of the system in the bore. In this approach, first find the reaction field $\vec{H}_R$ in the bore by treating $\vec{H}_T = (\vec{H}_0 + \vec{H}_R)$ as a *zeroth order* field and solve for $\vec{H}_R$ as the usual 1st order magnetic field response.

d) Show that expressions for $\vec{H}_R$, $\vec{H}_T$, and $\vec{J}$ in the shell valid for $d \ll R$ are:

$$\vec{H}_R = -\frac{j\omega\mu_o R d H_0}{2\rho_e + j\omega\mu_o R d} \, \vec{i}_z \tag{2.54}$$

$$\vec{H}_T = \frac{2\rho_e H_0}{2\rho_e + j\omega\mu_o R d} \, \vec{i}_z \tag{2.55}$$

$$\vec{J} = -\frac{j\omega\mu_o R H_0}{2\rho_e + j\omega\mu_o R d} \, \vec{i}_\theta \tag{2.56}$$

Solution to Problem 2.6—Part 1

a) From symmetry in the θ-direction, $\vec{E}_1$ and $\vec{J}_1$ are constant in the θ-direction and only θ-directed. Thus:

$$\oint_C \vec{E}_1 \cdot d\vec{s} = -j\omega\mu_0 \int_S \vec{H}_0 \cdot d\vec{A} \tag{S6.1}$$

For $r \leq R$,

$$2\pi r E_{1\theta} = -j\omega\mu_0 \pi r^2 H_0 \tag{S6.2}$$

$$E_{1\theta} = -\frac{j\omega\mu_0 r H_0}{2} \tag{S6.3}$$

Thus:

$$\vec{E}_1 = -\frac{j\omega\mu_0 r H_0}{2} \, \vec{i}_\theta \tag{2.50}$$

The 1st order current flows only in the shell, and for $d/R \ll 1$:

$$\vec{J}_1 = \frac{\vec{E}_1}{\rho_e}$$

$$\simeq -\frac{j\omega\mu_0 R H_0}{2\rho_e} \, \vec{i}_\theta \tag{2.51}$$

Also for $d \ll R$, we may treat current as the 1st order surface current $\vec{K}_1$, by multiplying $\vec{J}_1$ with d:

$$\vec{K}_1 = -\frac{j\omega\mu_0 R d H_0}{2\rho_e} \, \vec{i}_\theta \tag{S6.4}$$

b) For $r > R$, $\vec{H}_1 = 0$. The discontinuity in $\vec{H}$ at $r = R$, from Eq. 2.6, is given by $\vec{K}_1$. That is:

$$\vec{K}_1 = \vec{i}_r \times [\vec{H}_0 - (\vec{H}_0 + \vec{H}_1)] = -\frac{j\omega\mu_0 R d H_0}{2\rho_e} \, \vec{i}_\theta$$

$$= \vec{i}_r \times -\vec{H}_1 = -\frac{j\omega\mu_0 R d H_0}{2\rho_e} \, \vec{i}_\theta \tag{S6.5}$$

Solving Eq. $S6.5$ for $\vec{H}_1$ ($r \leq R - d$) with $d \ll R$, we have:

$$\vec{H}_1 = -\frac{j\omega\mu_0 R d H_0}{2\rho_e} \, \vec{i}_z \tag{2.52}$$

Solution to Problem 2.6—Part 1

c) Equations 2.51, $S6.4$, and 2.52 show that $\vec{J}_1$, $\vec{K}_1$, and $\vec{H}_1$ all increase monotonically with frequency; this cannot be valid for the entire range of ω. Clearly these solutions are valid only in the "low" frequency limit, which is what the quasi-static approximation is based on. More precisely, Eq. 2.52 for $\vec{H}_1$ is valid only when $|\vec{H}_1| \ll |\vec{H}_0|$:

$$|\vec{H}_1| = \frac{\omega\mu_o Rd|H_0|}{2\rho_e} \ll |\vec{H}_0| \tag{S6.6}$$

From Eq. $S6.6$, we can obtain the frequency limit, f_{sk}, below which the above solutions are valid:

$$f_{sk} = \frac{\rho_e}{\pi\mu_o Rd} \tag{2.53}$$

d) In the second approach for computing the shell's reaction field, we have $\vec{H}_1 \equiv \vec{H}_R$, and substitute $\vec{H}_0 + \vec{H}_R$ for $\vec{H}_0$ in the expression for $\vec{H}_1$ given in Eq. 2.52:

$$\vec{H}_R = -\frac{j\omega\mu_o Rd(\vec{H}_0 + \vec{H}_R)}{2\rho_e} \tag{S6.7}$$

Solving Eq. $S6.7$ for $\vec{H}_R$, we obtain:

$$\vec{H}_R = -\frac{j\omega\mu_o RdH_0}{2\rho_e + j\omega\mu_o Rd}\vec{\imath}_z \tag{2.54}$$

Combining Eq. 2.54 and $\vec{H}_T = \vec{H}_0 + \vec{H}_R$, we have:

$$\vec{H}_T = \vec{H}_0 + \vec{H}_R = H_0\left(1 - \frac{j\omega\mu_o Rd}{2\rho_e + j\omega\mu_o Rd}\right)\vec{\imath}_z$$

$$= \frac{2\rho_e H_0}{2\rho_e + j\omega\mu_o Rd}\vec{\imath}_z \tag{2.55}$$

$\vec{J}$ and $\vec{H}_R$ are related by $\nabla \times \vec{H} = \vec{J}$, which, for $\vec{K} = \vec{J}d$, reduces to:

$$\vec{J} = \frac{1}{d}H_R\vec{\imath}_\theta \tag{S6.8}$$

$$= -\frac{j\omega\mu_o RH_0}{2\rho_e + j\omega\mu_o Rd}\vec{\imath}_\theta \tag{2.56}$$

Note that in the low frequency limit, $\vec{H}_R$ given by Eq. 2.54 reduces, as expected, to $\vec{H}_1$ given by Eq. 2.52. In the high frequency limit, $\vec{H}_R$ reduces, also as expected, to $-\vec{H}_0$ and $\vec{H}_T$ becomes 0. Similar comments may be made for $\vec{J}$.

Problem 2.6: Induction heating—Part 2 (Power Dissipation)

Now, we can solve for power dissipation in the cylinder; two methods are used.

Method 1

e) We may calculate the resistive power dissipated in the cylindrical shell by directly computing $<p>= \vec{E}\cdot\vec{J}^*/2 = \rho_e|J|^2/2$, where $\vec{J}$ is given by Eq. 2.56. Show that an expression for the time-averaged total power (per unit length) dissipated in the shell, $<P>$, for $d \ll R$ is given by:

$$<P> = 2\pi Rd <p>$$

$$= \frac{\pi\rho_e\omega^2\mu_o^2R^3d}{4\rho_e^2 + \omega^2\mu_o^2R^2d^2}|H_0|^2 \qquad (2.57)$$

Method 2

The same complex power supplied to the cylinder may also be viewed as a flow of complex power flux entering the cylinder at $r = R$ from a source located at $r > R$.

f) Show that an expression for the surface integral (per unit cylinder length) of the first-order complex Poynting vector $\vec{S}_1$ entering into the cylinder at $r = R$ is given by:

$$-\oint_S \vec{S}_1 \cdot d\mathcal{A} = \tfrac{1}{2}(2\pi R)E_{1\theta}H_0^*$$

$$= \frac{j\pi\rho_e\omega\mu_o R^2}{2\rho_e + j\omega\mu_o Rd}|H_0|^2 \qquad (2.58)$$

Note that $E_{1\theta} = \rho_e J_\theta$, where J_θ is given by Eq. 2.56.

g) Take the real part of the right-hand side on Eq. 2.58 and show that it is identical to the expression for $<P>$ given by Eq. 2.57.

h) Plot $<P>$ as a function of ρ_e. Because both a perfect conductor, i.e. $\rho_e = 0$, and a perfect insulator, i.e. $\rho_e = \infty$, obviously do not dissipate power, your $<P>$ vs ρ_e plot should start with $<P>= 0$ and $<P> \rightarrow 0$ as $\rho_e \rightarrow \infty$. $<P>$ given by Eq. 2.57 indeed indicates this behavior.

i) The behavior of the $<P>$ vs ρ_e plot described above suggests the existence of a critical resistivity, ρ_{e_c}, at which $<P>$ is maximum. Show that ρ_{e_c} is given by:

$$\rho_{e_c} = \frac{\omega\mu_o Rd}{2} \qquad (2.59)$$

j) Compute ρ_{e_c} for a copper tube 10-mm radius (R), 0.5-mm wall thickness (d), and $\rho_e = 2\times10^{-10}\,\Omega\,\mathrm{m}$. This resistivity corresponds roughly to copper's resistivity at liquid helium temperatures.

Solution to Problem 2.6—Part 2

e) In the sinusoidal case, the time-averaged total power (per unit length), $<p>$, is given by $\vec{E} \cdot \vec{J}^*/2 = \rho_e |J|^2/2$, where $\vec{J}$ is the complex current density (Eq. 2.56). We thus have:

$$<p> = \frac{\rho_e}{2}|J_\theta|^2 = \frac{\rho_e}{2}\left(\frac{\omega^2\mu_o^2 R^2}{4\rho_e^2 + \omega^2\mu_o^2 R^2 d^2}\right)|H_0|^2 \qquad (S6.9)$$

The time-averaged *total* power (per unit length) dissipated in the shell, $<P>$, is given by multiplying $<p>$ with the cross sectional area of the shell:

$$<P> = 2\pi Rd <p>$$

$$= \frac{\pi\rho_e\omega^2\mu_o^2 R^3 d}{4\rho_e^2 + \omega^2\mu_o^2 R^2 d^2}|H_0|^2 \qquad (2.57)$$

We now examine two limits of ρ_e:

$$\lim_{\rho_e \to 0} <P> = \frac{\pi\rho_e R}{d}|H_0|^2 \propto \rho_e \qquad \text{(perfect conductor)} \qquad (S6.10a)$$

$$\lim_{\rho_e \to \infty} <P> = \frac{\pi\omega^2\mu_o^2 R^3 d}{4\rho_e}|H_0|^2 \propto \frac{1}{\rho_e} \qquad \text{(perfect insulator)} \qquad (S6.10b)$$

Note that in both limits, $<P> \to 0$, as expected.

f) The complex Poynting vector, $\vec{S}$, expanded to the first order fields, is:

$$\vec{S}_1 = \frac{1}{2}\left(\vec{E}_0 \times \vec{H}_0^* + \vec{E}_0 \times \vec{H}_1^* + \vec{E}_1 \times \vec{H}_0^*\right) \qquad (S6.11)$$

In this particular case, we have $\vec{E}_0 = 0$; thus Eq. $S6.11$ simplifies to:

$$\vec{S}_1 = \frac{1}{2}\left(\vec{E}_1 \times \vec{H}_0^*\right) \qquad (S6.12)$$

where from Eq. 2.56, $\vec{E}_1$ is given by:

$$\vec{E}_1 = \rho_e\vec{J} = -\frac{j\rho_e\omega\mu_o RH_0}{2\rho_e + j\omega\mu_o Rd}\vec{i}_\theta \qquad (S6.13)$$

We thus have:

$$-\oint_S \vec{S}_1 \cdot d\mathcal{A} = \frac{1}{2}(2\pi R)E_{1\theta}H_0^*$$

$$= \frac{j\pi\rho_e\omega\mu_o R^2}{2\rho_e + j\omega\mu_o Rd}|H_0|^2 \qquad (2.58)$$

Solution to Problem 2.6—Part 2

g) The real part of Eq. 2.58 is given by:

$$<P> = \frac{(\pi\rho_e\omega\mu_o R^2)(\omega\mu_o Rd)}{4\rho_e^2 + \omega^2\mu_o^2 R^2 d^2}|H_0|^2 = \frac{\pi\rho_e\omega^2\mu_o^2 R^3 d}{4\rho_e^2 + \omega^2\mu_o^2 R^2 d^2}|H_0|^2 \qquad (S6.14)$$

$<P>$ derived by Method 2 (Eq. $S6.14$) is identical to Eq. 2.57 (Method 1).

h) Figure 2.9 shows a plot of $<P>$ vs ρ_e.

i) We differentiate $<P>$ with respect to ρ_e and set that equal to 0 at ρ_{e_c}:

$$\frac{d<P>}{d\rho_e}\bigg|_{\rho_{e_c}} = \left[\frac{\pi\omega^2\mu_o^2 R^3 d}{4\rho_{e_c}^2 + \omega^2\mu_o^2 R^2 d^2} - \frac{8\pi\rho_{e_c}^2\omega^2\mu_o^2 R^3 d}{(4\rho_{e_c}^2 + \omega^2\mu_o^2 R^2 d^2)^2}\right] = 0 \qquad (S6.15)$$

Solving Eq. $S6.15$ for ρ_{e_c}, we have:

$$\rho_{e_c} = \frac{\omega\mu_o Rd}{2} \qquad (2.59)$$

Equation 2.59 is important in induction heating. A uniform, sinusoidally time-varying magnetic field is applied to a conducting sample to heat it by eddy currents. For a given material resistivity and sample dimensions, the critical frequency, derivable from Eq. 2.59 and identical to f_{sk}, is selected to maximize the heating.

j) For a copper cylinder of $R = 1\,\text{cm}$, $d = 0.5\,\text{mm}$, $\rho_e = 2\times10^{-10}\,\Omega\,\text{m}$ (corresponding to copper's resistivity at $\sim4\,\text{K}$), we obtain:

$$f_{sk} = \frac{\omega_{sk}}{2\pi} = \frac{\rho_{e_c}}{\pi\mu_o Rd} \qquad (2.53)$$

$$= \frac{2\times10^{-10}\,\Omega\,\text{m}}{\pi(4\pi\times10^{-7}\,\text{H/m})(10\times10^{-3}\,\text{m})(0.5\,\text{m})} \simeq 10\,\text{Hz}$$

Another way of interpreting f_{sk} is, as discussed earlier, that the induced magnetic field ($\vec{H}_1$) is negligible compared with $\vec{H}_0$ for $f \ll f_{sk}$.

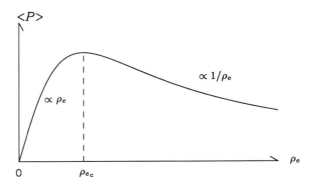

Fig. 2.9   Power dissipation vs resistivity for induction heated cylindrical shell.

Problem 2.7: Eddy-current loss in a metallic strip

In this problem an expression for eddy-current loss in a metallic strip subjected to a time-varying magnetic field is derived. It is useful in computing eddy-current heating in copper matrix superconductor strips. (When induced-current heating is beneficial, it is usually called induction heating; when it is not, it is often called eddy-current loss.)

One relevant example where eddy-current heating is a nuisance as a source of extraneous heat is in high-field, low-temperature (millikelvin range) experiments. A typical 20-T Bitter magnet at the FBNML contains an 84-Hz ripple field of an amplitude of $\sim 2\,\mathrm{mT}$; eddy currents induced in a test sample by the ripple fields could prevent it from reaching temperatures below $\sim 100\,\mathrm{mK}$. For these millikelvin experiments, ripple fields need to be less than $\sim 0.2\,\mathrm{mT}$.

Figure 2.10 shows a "long" (in the x-direction) metallic strip of electrical resistivity ρ_e, width b (in the y-direction), and thickness a (in the z-direction) placed in a time-varying external magnetic induction, $dB_0/dt = \dot{B}_0$, which is, within zeroth order, uniform and z-directed.

a) Show that the 1st order electric field, $\vec{E}_1$, can be expressed by:

$$E_{1x} = y\dot{B}_0 \qquad (2.60)$$

b) Show that the *spatially-averaged* power dissipation density $\tilde{p}$ (over the unit strip volume) can be expressed by:

$$\tilde{p} = \frac{(b\dot{B}_0)^2}{12\rho_e} \qquad (2.61)$$

c) When the external magnetic induction is sinusoidally varying in time with radial frequency ω, $B(t) = B_0 \sin \omega t$, show that an expression for the *time-averaged* $\tilde{p}$, $<\tilde{p}>$, is given by:

$$<\tilde{p}> = \frac{(b\omega B_0)^2}{24\rho_e} \qquad (2.62)$$

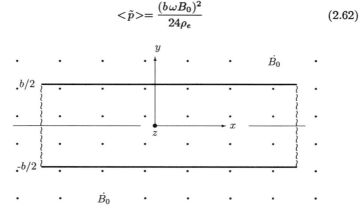

Fig. 2.10   Metallic strip in a time-varying magnetic field.

Solution to Problem 2.7

a)   Because the system is independent of x and $\vec{B}_0$ is uniform, $\vec{E}_1$ points only in the x-direction and depends only on y. That is, $\nabla \times \vec{E}_1 = -\partial \vec{B}_0/\partial t$ reduces to:

$$-\frac{dE_{1x}}{dy} = -\frac{dB_0}{dt} = -\dot{B}_0 \qquad (S7.1)$$

From symmetry, $E_{1x}(y = 0) = 0$, and we have from Eq. $S7.1$:

$$E_{1x} = y\dot{B}_0 \qquad (2.60)$$

b)   The *local* power density dissipated in the strip, $p(y)$, is given by $\vec{E}_1 \cdot \vec{J}_1$. The total power dissipation (per unit strip length), P, is thus given by integrating $p(y)$ over the strip width:

$$P = a \int_{-b/2}^{b/2} p(y)\,dy = \frac{2a(\dot{B}_0)^2}{\rho_e} \int_0^{b/2} y^2\,dy = \frac{ab(b\dot{B}_0)^2}{12\rho_e} \qquad (S7.2)$$

Equation $S7.2$ is valid for "reasonably" slow variation of B_0 and "reasonably" resistive materials. That is, it is valid only when the induced (1st order) magnetic induction generated by $\vec{J}_1$ is small compared with B_0.

The *spatially-averaged* dissipation density, $\tilde{p}$, is P (given by Eq. $S7.2$) divided by the strip cross section:

$$\tilde{p} = \frac{P}{ab} = \frac{(b\dot{B}_0)^2}{12\rho_e} \qquad (2.61)$$

c)   Under sinusoidal excitations the *time-averaged* total power dissipation density, $<p>$, is given by:

$$<p> = \tfrac{1}{2}E_{1x}J_{1x}^* \qquad (S7.3)$$

With $E_{1x} = j\omega y B_0$ and $J_{1x} = E_{1x}/\rho_e$ and $<p>$ now averaged *spatially* over the strip volume, we have:

$$<\tilde{p}> = \frac{2a(\omega B_0)^2}{2\rho_e(ab)} \int_0^{b/2} y^2\,dy = \frac{(b\omega B_0)^2}{24\rho_e} \qquad (2.62)$$

Note that $\tilde{p}$ and $<\tilde{p}>$ are proportional, respectively, to $(b\dot{B}_0)^2$ and $(b\omega B_0)^2$, *i.e.* both depend not only on the square of time rate of change of magnetic induction but also on the square of *conductor width*.

Lamination to Reduce Eddy-Current Loss

Suppose the strip is cut into two strips, each having a total width of $b/2$. Then from Eq. 2.61 we see that both $\tilde{p}$ and the total power dissipation (over the two narrow strips) will be 1/4 and 1/2 the original values, respectively. Thus, it is possible to reduce eddy-current power dissipation to an arbitrarily small value by dividing a conducting strip into many narrow strips. This lamination technique is used most widely in transformers where iron yokes are made up of many iron sheets. Similarly, as we shall see in Chapter 5, superconductors also benefit from subdivision, leading to the universal use of multifilamentary conductors.

CHAPTER 3
MAGNETS, FIELDS, AND FORCES

3.1 Introduction

In this chapter we study magnets, fields, and forces. Magnets discussed in this chapter, all in the Problem Section, include: 1) solenoid; 2) Helmholtz coil; 3) ideal dipole; 4) ideal quadrupole; 5) racetrack; and 6) ideal torus. The two important solenoidal magnets for generation of high magnetic fields (Bitter and hybrid) are also described in the Problem Section.

At the present time, field and force computations are generally performed with computer codes that, for a given magnet configuration, give accurate numerical solutions at any location. If the magnet is composed of separate coils, these codes can also compute the inductance matrix and forces between the coils [3.1]. Analytical approaches derived in the Problem Section can give field values only at restricted locations, usually at the magnet center; however, they elucidate subtle relationships among fields, forces, and magnet parameters.

In this introductory section, we shall present the law of Biot and Savart and apply it to derive a simple expression for the magnetic field of a current-carrying loop. The expression is useful for solving field problems in solenoidal magnets. Using a long solenoid, we also introduce the concepts of magnetic force and pressure.

3.2 Law of Biot and Savart

The differential magnetic field $d\vec{H}$ produced at point P by a differential current element $I\,d\vec{s}$ located at point O, a distance r away from P, is given by:

$$d\vec{H} = \frac{(I\,d\vec{s} \times \vec{r})}{4\pi r^3} \tag{3.1}$$

Equation 3.1 is known as the law of Biot and Savart. It indicates that the magnitude of $d\vec{H}$ at any location decreases inversely as the square of the distance from the differential current element: $|d\vec{H}| \propto 1/r^2$. At a fixed radius, $|d\vec{H}|$ varies as $\sin\theta$, where θ is the angle between the $\vec{s}$ and $\vec{r}$ vectors, with $\theta = 0$ defined when the two vectors point in the same direction. Most field computation codes are based on Eq. 3.1. We apply Eq. 3.1 to derive an expression for the central ($r = 0$) field as a function of axial (z) position created by a loop of radius a located at $z = 0$ and carrying current I, as illustrated in Fig. 3.1. The loop's axis defines the z axis; θ, measured from the $z = 0$ plane, is defined in Fig. 3.1.

As seen from Fig. 3.1, the r-component of $\vec{H}$ (H_r) at each axial (z) location cancels, leaving only the z-component, H_z. For this particular case, the $(I\,d\vec{s}\times r)$ term in Eq. 3.1 simplifies to $I(2\pi a)r\sin\theta$, leading to:

$$H_z = \frac{I(2\pi a)r\sin\theta}{4\pi r^3} = \frac{Ia\sin\theta}{2r^2} \tag{3.2}$$

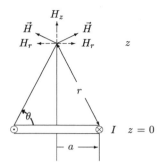

Fig. 3.1   Loop of radius a carrying current I.

With $\sin\theta = a/r$ and $r^2 = a^2 + z^2$, Eq. 3.2 becomes:

$$H_z = \frac{a^2 I}{2r^3} = \frac{a^2 I}{2(a^2 + z^2)^{3/2}} \qquad (3.3)$$

Equation 3.3 is one of the most useful expressions for field computation. It can be used to derive an expression for the H_z-component along the axis of a solenoid of arbitrary winding cross section having current distributions that are invariant in the θ-direction. Problems 3.1 and 3.2 are good examples.

3.3 Lorentz Force and Magnetic Pressure

In the presence of magnetic induction $\vec{B}$, an electric charge q in motion with velocity $\vec{v}$ is acted on by a force $\vec{F}_L$, called the Lorentz force: $\vec{F}_L = q\vec{v} \times \vec{B}$. For a conductor element carrying current density $\vec{J}$ in the presence of $\vec{B}$, the Lorentz force density, $\vec{f}_L$, is given by:

$$\vec{f}_L = \vec{J} \times \vec{B} \qquad (3.4)$$

Equation 3.4 is the basic expression for computing magnetic forces and stresses in magnets. As stated at the beginning of Chapter 1, whether superconducting and operating at 4.2 K or 77 K, or resistive and operating at room temperature, magnets producing the same field must deal with essentially the same stress level; its ultimate field is dictated by the strengths of its structural elements that include the current-carrying conductor. Thus a 50-T superconducting magnet—if it is ever to be built—and a 50-T resistive magnet must both withstand tremendous Lorentz stresses. As illustrated below, a 50-T magnetic induction corresponds to an equivalent magnetic pressure of 10 GPa (10,000 atm).

Let us consider an infinitely long, "thin-walled" solenoid (thickness δ) of an average diameter $2a$ carrying a uniformly distributed current, approximated by the surface current density K_θ [A/m]. The B_z-component of magnetic induction at the center (0,0) is given by a modified expression of Eq. 3.3:

$$B_z = \frac{\mu_o a^2 K_\theta}{2} \int_{-\infty}^{\infty} \frac{dz}{(a^2 + z^2)^{3/2}} = \mu_o K_\theta \equiv B_o \qquad (3.5)$$

The integration is taken over the entire range of z because of the presence of current "rings" that extend from $z = -\infty$ to $z = \infty$. The application of Ampere's law gives that for this infinitely long solenoid, $\vec{B}$ outside the solenoid ($r > a$) is zero and that within the solenoid bore ($r < a$) is B_o, uniform in both z- and r-directions. (From symmetry of current distribution, the field is also uniform in the θ-direction. Also note that Eq. 3.3 is valid only for such cases.) That is, the magnetic induction within the bore of an infinitely long solenoid is completely uniform and directed only in the z-direction.

The B_z field just inside the winding is B_o and that just outside the winding is zero, decreasing linearly with r over the winding thickness δ. The average induction acting on the current element in the winding, $\tilde{B}_z$, is thus $B_o/2$, resulting in an r-directed average Lorentz force density $f_{L_r}\, \vec{i}_r$ acting on the winding given by:

$$f_{L_r}\, \vec{i}_r = \frac{K_\theta}{\delta}\, \tilde{B}_z\, \vec{i}_r = \frac{K_\theta B_o}{2\delta}\, \vec{i}_r \qquad (3.6)$$

An r-directed Lorentz force, $F_{L_r}\, \vec{i}_r$ acting on the winding volume element, defined in Fig. 3.2, is equivalent to a magnetic pressure, $p_m\, \vec{i}_r$, acting on the winding surface element, also defined in Fig. 3.2. Thus:

$$F_{L_r}\, \vec{i}_r = f_{L_r}[(a\,\Delta\theta)\times\delta\times\Delta z]\,\vec{i}_r = p_m[(a\,\Delta\theta)\times\Delta z]\,\vec{i}_r \qquad (3.7)$$

Combining Eqs. 3.6 and 3.7 and solving for p_m, we obtain:

$$p_m = \frac{B_o^2}{2\mu_o} \qquad (3.8)$$

That is, the magnetic pressure is equivalent to magnetic energy density. For B_o equal to 1 T, Eq. 3.8 gives a magnetic pressure of 3.98×10^5 Pa or ~4 atm, from which it follows that for $B_o = 50$ T, a magnetic pressure of ~10 GPa is reached.

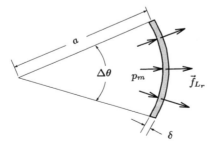

Fig. 3.2   Top view of a differential element for a "thin-walled" solenoid (thickness δ) of an average diameter $2a$. The element is Δz high in the z-direction (out of the paper).

Problem 3.1: Uniform-current-density solenoids

This problem deals with the axial field generated at the midplane of a uniform-current-density solenoid of winding i.d. $2a_1$, winding o.d. $2a_2$, and total winding length $2b$. Figure 3.3 defines the winding cross section and important parameters. The differential magnetic field at the center, $dH_z(0,0)$, generated by a current-carrying ring of differential cross section dA located at (r, z) is given by a modified form of Eq. 3.3:

$$dH_z(0,0) = \frac{r^2 \lambda J \, dA}{2(r^2 + z^2)^{3/2}} \tag{3.9}$$

where λJ is the overall current density within the differential cross section. λ, called the space factor, acknowledges that not all the winding cross section is occupied by current-carrying conductors. Note that λJ in this model is uniform over the winding cross section.

a) By integrating Eq. 3.9 from $r = a_1$ to $r = a_2$ and from $z = -b$ to $z = b$, show that an expression for $H_z(0,0)$ for the solenoid is given by:

$$H_z(0,0) = \lambda J a_1 F(\alpha, \beta) \tag{3.10}$$

$$F(\alpha, \beta) = \beta \ln \left(\frac{\alpha + \sqrt{\alpha^2 + \beta^2}}{1 + \sqrt{1 + \beta^2}} \right) \tag{3.11}$$

where $\alpha = a_2/a_1$ and $\beta = b/a_1$. $F(\alpha, \beta)$ is the "field factor" for a uniform-current-density coil [3.2].

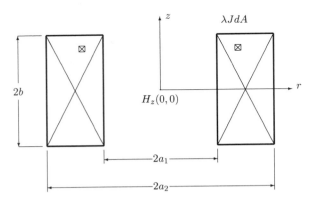

Fig. 3.3   Cross section of a uniform-current-density solenoid.

Problem 3.1: Uniform-current-density solenoids

b) By substituting $NI/[2b(a_2 - a_1)]$—the total ampere-turns (N is the total number of turns and I is the current) divided by the solenoid overall cross section—for λJ, we can also express $H_z(0,0)$ by:

$$H_z(0,0) = \frac{NI}{2a_1(\alpha - 1)} \ln\left(\frac{\alpha + \sqrt{\alpha^2 + \beta^2}}{1 + \sqrt{1 + \beta^2}}\right) \tag{3.12}$$

Equation 3.12 states that for a given set of α and β, $H_z(0,0)$ is proportional to NI and inversely proportional to a_1.

"Filamentary" Ring Simplify Eq. 3.12 for a "filamentary" ring of radius a_1 carrying total current NI and reduce it to:

$$H_z(0,0) = \frac{NI}{2a_1} \tag{3.13}$$

c) **"Long" Solenoid**:   Simplify Eq. 3.12 for the case of a solenoid that is much longer than its winding thickness and reduce it to:

$$H_z(0,0) = \frac{NI}{2b} \tag{3.14}$$

Equation 3.14 is essentially equivalent to Eq. 3.5, because $NI/2b = K_\theta$.

d) **"Pancake" Coil**:   Simplify Eq. 3.12 this time for the case of a winding that is very short compared with its winding thickness and reduce it to:

$$H_z(0,0) = \frac{NI}{2a_1}\left(\frac{\ln \alpha}{\alpha - 1}\right) \tag{3.15}$$

Equation 3.15 is valid for a "pancake" coil. More about pancake coils later.

e) **Field vs Power**:   If the solenoid is layer-wound with copper conductor (resistivity ρ_{cu}), it will dissipate Joule heating power, P. Because the current density J is constant through the solenoid, P may be given simply by: $P = \rho_{cu} J^2 \times$ <total conductor volume>. Expressed in terms of $\rho_{cu} J^2$ and solenoid parameters, a_1, α, and β, show that P may be given by:

$$P = \rho_{cu} J^2 \lambda a_1^3 2\pi \beta(\alpha^2 - 1) \tag{3.16}$$

f) Combining Eqs. 3.10, 3.11, and 3.16, show that an expression for the central magnetic field $H_z(0,0)$ in a resistive solenoid is given by:

$$H_z(0,0) = G(\alpha, \beta)\sqrt{\frac{\lambda P}{\rho_{cu} a_1}} \tag{3.17}$$

$$G(\alpha, \beta) = \sqrt{\frac{\beta}{2\pi(\alpha^2 - 1)}} \ln\left(\frac{\alpha + \sqrt{\alpha^2 + \beta^2}}{1 + \sqrt{1 + \beta^2}}\right) \tag{3.18}$$

$G(\alpha, \beta)$ is known as the G factor for a uniform-current-density coil [3.2].

Solution to Problem 3.1

a) By integrating Eq. 3.9 over the appropriate limits for $z-$ and r-coordinates, we have:

$$H_z(0,0) = \frac{\lambda J}{2} \int_{a_1}^{a_2} \int_{-b}^{b} \frac{r^2 \, dz \, dr}{(r^2 + z^2)^{3/2}}$$

$$= \lambda J \int_{a_1}^{a_2} \int_{0}^{b} \frac{r^2 \, dz \, dr}{(r^2 + z^2)^{3/2}} \qquad (S1.1)$$

$$\int_{0}^{b} \frac{r^2 \, dz}{(r^2 + z^2)^{3/2}} = \left[\frac{r^2 z}{r^2 \sqrt{r^2 + z^2}} \right]_{0}^{b} = \frac{b}{\sqrt{r^2 + b^2}} \qquad (S1.2)$$

From Eqs. $S1.1$ and $S1.2$, we have:

$$H_z(0,0) = \lambda J \int_{a_1}^{a_2} \frac{b \, dr}{\sqrt{r^2 + b^2}}$$

$$= \lambda J b \left[\ln \left(r + \sqrt{r^2 + b^2} \right) \right]_{a_1}^{a_2}$$

$$= \lambda J b \left[\ln \left(a_2 + \sqrt{a_2^2 + b^2} \right) - \ln \left(a_1 + \sqrt{a_1^2 + b^2} \right) \right]$$

$$= \lambda J b \ln \left(\frac{a_2 + \sqrt{a_2^2 + b^2}}{a_1 + \sqrt{a_1^2 + b^2}} \right) \qquad (S1.3)$$

By defining $a_2/a_1 = \alpha$ and $b/a_1 = \beta$, we can express Eq. $S1.3$ as:

$$H_z(0,0) = \lambda J a_1 \left(\frac{b}{a_1} \right) \ln \left[\frac{a_2/a_1 + \sqrt{(a_2/a_1)^2 + (b/a_1)^2}}{(a_1/a_1) + \sqrt{(a_1/a_1)^2 + (b/a_1)^2}} \right] \qquad (S1.4)$$

$$H_z(0,0) = \lambda J a_1 F(\alpha, \beta) \qquad (3.10)$$

$$F(\alpha, \beta) = \beta \ln \left(\frac{\alpha + \sqrt{\alpha^2 + \beta^2}}{1 + \sqrt{1 + \beta^2}} \right) \qquad (3.11)$$

b) We shall first take the limit as $\beta \to 0$ for the quantity within the log:

$$\lim_{\beta \to 0} \ln \left(\frac{\alpha + \sqrt{\alpha^2 + \beta^2}}{1 + \sqrt{1 + \beta^2}} \right) = \ln \alpha \qquad (S1.5)$$

Next, we let $\alpha = 1 + \epsilon$: $\alpha \to 1$, $\epsilon \to 0$, and $\ln(1 + \epsilon) \to \epsilon = \alpha - 1$. Hence, $\ln \alpha \to \alpha - 1$ as $\alpha \to 1$. We thus have:

Solution to Problem 3.1

$$H_z(0,0) = \frac{NI}{2a_1(\alpha-1)}(\alpha-1) = \frac{NI}{2a_1} \qquad (3.13)$$

Note that Eq. 3.13 is equivalent to a simple expression for the central field of a ring of radius a_1 carrying a total current NI. Note also that Eq. 3.13 is equivalent to Eq. 3.3 with $z = 0$ and a and I replaced respectively with a_1 and NI.

c)   We take another limit for which $\beta \gg \alpha$:

$$\lim_{\beta \gg \alpha} \ln\left(\frac{\alpha + \sqrt{\alpha^2 + \beta^2}}{1 + \sqrt{1 + \beta^2}}\right) = \ln\left[\frac{\alpha/\beta + \sqrt{(\alpha/\beta)^2 + (\beta/\beta)^2}}{1/\beta + \sqrt{(1/\beta)^2 + (\beta/\beta)^2}}\right]$$

$$= \ln\left(\frac{\alpha/\beta + 1}{1/\beta + 1}\right) \qquad (S1.6)$$

Again, using the approximation $\ln(1+x) \simeq x$ for $x \ll 1$, we have:

$$\lim_{\beta \to \infty} \ln\left(\frac{\alpha/\beta + 1}{1/\beta + 1}\right) = \ln(\alpha/\beta + 1) - \ln(1/\beta + 1)$$

$$\simeq \frac{\alpha}{\beta} - \frac{1}{\beta} = \frac{\alpha - 1}{\beta} \qquad (S1.7)$$

We thus obtain:

$$\lim_{\beta \to \infty} H_z(0,0) = \frac{NI}{2a_1(\alpha-1)} \frac{(\alpha-1)}{\beta} = \frac{NI}{2b} \qquad (3.14)$$

Note that Eq. 3.14 is a simple expression for the field inside a "long" solenoid of length $2b$ carrying a total current of NI. As remarked in the problem statement, $NI/2b$ may be considered as ampere-turns per unit coil length or surface current density K_θ (Eq. 3.5).

d)   For a "pancake" coil, an appropriate limit is $\beta \to 0$:

$$\lim_{\beta \to 0} \ln\left(\frac{\alpha + \sqrt{\alpha^2 + \beta^2}}{1 + \sqrt{1 + \beta^2}}\right) = \ln\left(\frac{2\alpha}{2}\right) = \ln\alpha \qquad (S1.8)$$

$$H_z(0,0) = \frac{NI}{2a_1}\left(\frac{\ln\alpha}{\alpha-1}\right) \qquad (3.15)$$

Note that the central field of a pancake coil is equal to that of a "ring" coil times a factor $\ln\alpha/(\alpha-1)$; as $\alpha \to 1$, Eq. 3.15 reduces to Eq. 3.13.

Solution to Problem 3.1

e) Total conductor volume is equal to $\lambda \times$ <winding volume>. Winding volume is given by:

$$< \text{winding volume} > = 2b\pi(a_2^2 - a_1^2) = a_1^3 2\pi\beta(\alpha^2 - 1) \tag{S1.9}$$

And thus:

$$P = \rho_{cu} J^2 \lambda a_1^3 2\pi\beta(\alpha^2 - 1) \tag{3.16}$$

where J is current density in conductor only.

f) From Eq. 3.16, we can solve for J in terms of P and other parameters:

$$J = \sqrt{\frac{P}{\rho_{cu}\lambda a_1}} \left[\frac{1}{a_1\sqrt{2\pi\beta(\alpha^2 - 1)}} \right] \tag{S1.10}$$

Combining Eqs. 3.10, 3.11, and S1.10, we get:

$$\begin{aligned}
H_z(0,0) &= \lambda\sqrt{\frac{P}{\rho_{cu}\lambda a_1}} \left[\frac{1}{a_1\sqrt{2\pi\beta(\alpha^2 - 1)}} \right] a_1\beta\ln\left(\frac{\alpha + \sqrt{\alpha^2 + \beta^2}}{1 + \sqrt{1 + \beta^2}} \right) \\
&= \sqrt{\frac{\lambda P}{\rho_{cu} a_1}} \sqrt{\frac{\beta}{2\pi(\alpha^2 - 1)}} \ln\left(\frac{\alpha + \sqrt{\alpha^2 + \beta^2}}{1 + \sqrt{1 + \beta^2}} \right) \\
&= G(\alpha, \beta)\sqrt{\frac{\lambda P}{\rho_{cu} a_1}}
\end{aligned} \tag{3.17}$$

with

$$G(\alpha, \beta) = \sqrt{\frac{\beta}{2\pi(\alpha^2 - 1)}} \ln\left(\frac{\alpha + \sqrt{\alpha^2 + \beta^2}}{1 + \sqrt{1 + \beta^2}} \right) \tag{3.18}$$

Note that Eq. 3.17 implies that in resistive solenoids the required power P increases, with α and β constant, quadratically with the central magnetic field:

$$P = \frac{\rho_{cu} a_1 H_z^2(0,0)}{\lambda G^2(\alpha, \beta)} \tag{S1.11}$$

A field efficiency (η_f) may be defined as $\eta_f \equiv B_z^2(0,0)/P$ and from Eq. S1.11, we have:

$$\eta_f \equiv \frac{H_z^2(0,0)}{P} = \frac{\lambda G^2(\alpha, \beta)}{\rho_{cu} a_1} \tag{S1.12}$$

Note that η_f is proportional to λ and $G^2(\alpha, \beta)$, and is inversely proportional to ρ_{cu} and a_1. A further discussion of η_f follows Problem 3.2.

Bitter Magnet

Although this book deals superconducting magnets, here we shall devote one problem to the resistive magnet, specifically the Bitter magnet. We may call the development of iron-free electromagnets that Francis Bitter initiated at MIT in the 1930s as the basis for modern magnet technology. Bitter successfully built and operated the first 10-T Bitter magnet (1.7 MW), having a room-temperature magnet-bore of 50 mm, in the period 1937~38. His design employed a conductor in the form of a stack of annular plates, each separated by a thin sheet of insulation except over a segment. The bare segment makes pressure contact to the next plate's bare segment. Each plate, punched with hundreds of cooling holes, has a radial slit (Fig. 3.4). The slit forces the current to commutate from one plate to the next in a helical path as it flows from one end of the stack to the other. These plates with cooling holes are called Bitter plates. (Plates with etched radial cooling channels are also called Bitter plates.) To produce the required field, tens of thousands of amperes of current are pushed through the electrically resistive stack, consuming megawatts of electrical power, which the stack converts into heat. This heat is removed by water forced through the cooling holes at high velocity.

A key feature of the Bitter magnet construction is that it is modular, consisting of many similar plates. Plate thickness, mechanical properties, and electrical properties can be adjusted along the axial position to optimize magnet performance.

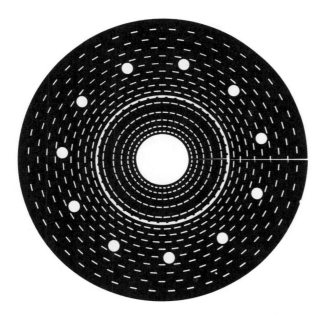

Fig. 3.4   Silhouette of two nested Bitter plates. The shape and distribution of cooling holes of these plates have been modified by R.J. Weggel from those of Bitter's original design. The outer plate's o.d. is 211 mm.

Problem 3.2: Bitter magnet

As evident from results of Problem 3.1, the smaller the inner winding radius, a_1, the higher is the field at the bore, $H_z(0,0)$, for a given total ampere turn, NI. This can be seen readily from a simple expression such as Eq. 3.13. Another advantage of smaller a_1 is that for the same power consumption, P, higher $H_z(0,0)$ is generated, as seen from Eq. 3.17. What these results point to is that for production of highest field for a given power, it is obviously best to place ampere-turns closest to the magnet bore. Since in practice, they cannot all be placed at the innermost winding radius, they must be distributed over the winding volume. As we shall study here, in Bitter magnets more ampere-turns are placed in the innermost winding radius than in uniform-current-density magnet, making Bitter magnet design superior to uniform-current-density magnet design. Additional comments on current distributions are presented in p. 57.

a) Show that, ignoring the effect of the cooling holes, the current density in each plate in the θ-direction, J_θ, is inversely proportional to r:

$$J_\theta = J_0 \frac{a_1}{r} \qquad (3.19)$$

where $J_0 = V_B/(2\pi\rho_{cu}a_1)$ and V_B is the voltage across the slit.

b) Starting from Eq. 3.9 of Problem 3.1, show that the axial field at the coil center in the Bitter magnet, $[H_z(0,0)]_B$, is given by:

$$[H_z(0,0)]_B = \lambda_B J_0 a_1 [F(\alpha,\beta)]_B \qquad (3.20)$$

$$[F(\alpha,\beta)]_B = \ln\left(\alpha \frac{\beta + \sqrt{1+\beta^2}}{\beta + \sqrt{\alpha^2+\beta^2}}\right) \qquad (3.21)$$

c) By deriving first an expression for the power P_B for Bitter coils similar to Eq. 3.16 for uniform-current-density solenoids, show that expressions for $[H_z(0,0)]_B$ and $[G(\alpha,\beta)]_B$ for the Bitter magnet are given by:

$$[H_z(0,0)]_B = [G(\alpha,\beta)]_B \sqrt{\frac{\lambda_B P_B}{\rho_{cu}a_1}} \qquad (3.22)$$

$$[G(\alpha,\beta)]_B = \frac{1}{\sqrt{4\pi\beta\ln\alpha}} \ln\left(\alpha \frac{\beta + \sqrt{1+\beta^2}}{\beta + \sqrt{\alpha^2+\beta^2}}\right) \qquad (3.23)$$

d) Using Eqs. 3.22 and 3.23, compute the power required to generate a central magnetic induction of $20\,\text{T}$ in a Bitter magnet having $2a_1 = 6\,\text{cm}$, $2a_2 = 40\,\text{cm}$, $2b = 22\,\text{cm}$, $\lambda_B = 0.8$, and $\rho_{cu} = 2\times10^{-6}\,\Omega\,\text{cm}$. (The power computed here is typical of the 20-T Bitter magnets operated at the FBNML. Note that your answer should be less than the FBNML's power capacity of $9\,\text{MW}$.)

Solution to Problem 3.2

a) Consider a single disk, in which current, by geometry, is constrained to flow in the θ-direction. The resulting E field gives potential V_B that varies with θ only:

$$\int_C \vec{E} \cdot d\vec{s} = V_B \tag{S2.1}$$

Because $\vec{E}$ is only θ-directed (E_θ) and it is constant at a given r, we have, from Eq. $S2.1$:

$$E_\theta 2\pi r = V_B \tag{S2.2}$$

That is, E_θ varies as $\propto 1/r$. Noting that $\vec{J} = \vec{E}/\rho_{cu}$, we have:

$$J_\theta = \frac{E_\theta}{\rho_{cu}} = \frac{V_B}{2\pi\rho_{cu}r} \tag{S2.3}$$

Defining $J_0 = J(r = a_1) = V_B/(2\pi\rho_{cu}a_1)$, we can express Eq. $S2.3$:

$$J_\theta = J_0 \frac{a_1}{r} \tag{3.19}$$

b) Equation 3.9 may be integrated over appropriate limits:

$$[H_z(0,0)]_B = \lambda_B J_0 a_1 \int_{a_1}^{a_2} \int_0^b \frac{r\, dr\, dz}{2(r^2 + z^2)^{3/2}} \tag{S2.4}$$

From a *Table of Integrals*, we have:

$$\int_0^b \frac{r\, dz}{2(r^2 + z^2)^{3/2}} = \left[\frac{rz}{r^2\sqrt{r^2 + z^2}}\right]_0^b = \frac{b}{r\sqrt{r^2 + b^2}} \tag{S2.5}$$

Therefore:

$$[H_z(0,0)]_B = \lambda_B J_0 a_1 b \int_{a_1}^{a_2} \frac{dr}{r\sqrt{r^2 + b^2}} \tag{S2.6}$$

$$= \lambda_B J_0 a_1 b \left[-\frac{1}{b} \ln\left(\frac{b + \sqrt{b^2 + r^2}}{r}\right)\right]_{a_1}^{a_2}$$

$$= \lambda_B J_0 a_1 \ln\left(\frac{a_2(b + \sqrt{b^2 + a_1^2})}{a_1(b + \sqrt{b^2 + a_2^2})}\right)$$

$$= \lambda_B J_0 a_1 \ln\left(\alpha\frac{\beta + \sqrt{1 + \beta^2}}{\beta + \sqrt{\alpha^2 + \beta^2}}\right) \tag{S2.7}$$

We may express Eq. $S2.7$ as:

$$[H_z(0,0)]_B = \lambda_B J_0 a_1 [F(\alpha, \beta)]_B \tag{3.20}$$

$$[F(\alpha, \beta)]_B = \ln\left(\alpha\frac{\beta + \sqrt{1 + \beta^2}}{\beta + \sqrt{\alpha^2 + \beta^2}}\right) \tag{3.21}$$

Note that there is a subtle difference between Eqs. 3.11 and 3.21.

Solution to Problem 3.2

c) Here, power density $\rho_{cu} J^2(r)$ varies as $\propto 1/r^2$ and it must be integrated over the entire winding volume:

$$P_B = \int_{a_1}^{a_2} \int_{-b}^{b} \rho_{cu} \lambda_B \left(J_0 \frac{a_1}{r} \right)^2 2\pi r \, dz \, dr = J_0^2 \rho_{cu} \lambda_B a_1^3 (4\pi b \ln \alpha) \qquad (S2.8)$$

From Eq. $S2.8$, we obtain:

$$J_0 = \frac{1}{a_1 \sqrt{4\pi b \ln \alpha}} \sqrt{\frac{P_B}{\rho_{cu} \lambda_B a_1}} \qquad (S2.9)$$

Combining Eqs. 3.20, 3.21, and $S2.9$, we have:

$$[H_z(0,0)]_B = [G(\alpha, \beta)]_B \sqrt{\frac{\lambda_B P_B}{\rho_{cu} a_1}} \qquad (3.22)$$

$$[G(\alpha, \beta)]_B = \frac{1}{\sqrt{4\pi \beta \ln \alpha}} \ln \left(\alpha \frac{\beta + \sqrt{1 + \beta^2}}{\beta + \sqrt{\alpha^2 + \beta^2}} \right) \qquad (3.23)$$

As in uniform-current-density solenoids, $[H_z(0,0)]_B$ increases as the square root of P_B; P_B is a quadratic function of field. The maximum of $[G(\alpha, \beta)]_B$ occurs when $\alpha \simeq 6.4$ and $\beta \simeq 2.1$: $[G(6.4, 2.1)]_B \simeq 0.17$. $[G(\alpha, \beta)]_B$ stays within 99% of its peak value for $5 \leq \alpha \leq 8$ and $1.8 \leq \beta \leq 2.3$. That is, P_B for a given field is minimal, and roughly constant within this range of α and β. However, because field homogeneity for a given value of a_1 improves with $2b$ or β, most of Bitter magnets used at FBNML have β values greater than 2.3 as in the magnet considered in **d)** below.

d) We have: $\alpha = (2a_2/2a_1) = 40/6 = 6.67$; $\beta = (2b/2a_1) = 22/6 = 3.67$. Thus:

$$[G(6.67, 3.67)]_B = \frac{1}{\sqrt{4\pi 3.67 \ln(6.67)}} \ln \left[6.67 \frac{3.67 + \sqrt{1 + (3.67)^2}}{3.67 + \sqrt{(6.67)^2 + (3.67)^2}} \right]$$

$$\simeq 0.159$$

We have an expression of P_B, similar to Eq. $S1.11$, in terms of $[H_z(0,0)]_B$ and other parameters:

$$P_B = \frac{\rho_{cu} a_1 [H_z(0,0)]_B^2}{\lambda_B [G(\alpha, \beta)]_B^2} = \frac{\rho_{cu} a_1 \{\mu_o [H_z(0,0)]_B\}^2}{\lambda_B \mu_o^2 [G(\alpha, \beta)]_B^2} \qquad (S2.10)$$

$$= \frac{(2 \times 10^{-8} \, \Omega \, \text{m})(3 \times 10^{-2} \, \text{m})(20 \, \text{T})^2}{(0.8)(4\pi \times 10^{-7} \, \text{H/m})^2 (0.159)^2} \simeq 7.5 \, \text{MW}$$

This power requirement is roughly equal to that needed for a typical 20-T Bitter magnet.

Additional Comments on Water-Cooled Magnets

One important parameter in water-cooled magnets is field efficiency, defined as the square of the field at the magnet center divided by the total power input to the magnet: $[H_z(0,0)]_B^2/P_B$. As with η_f (Eq. $S1.12$, p. 52) for uniform-current-density magnets, the field efficiency for Bitter magnets is proportional to λ_B and $[G(\alpha, \beta)]_B^2$ and inversely proportional to a_1 and conductor resistivity ρ_{cu}.

Current Density Distributions

We have considered two current density distributions so far: 1) uniform or $J(r,z) = J_o$ and 2) Bitter or $J(r,z) \propto 1/r$. The uniform distribution means J is independent of r or z. Superconducting magnets wound with "graded" conductors have $J(r)$ that changes with r in discrete steps. Those consisting of many nested coils, each of which is wound with a different conductor, also have $J(r)$ distributions that change in discrete steps.

We describe here three other current distributions of interest for water-cooled magnets [3.3].

Kelvin Coil: The current density that gives the best field efficiency is known as the Kelvin distribution, $J_K(r,z)$:

$$J_K(r,z) \propto \frac{r}{(r^2+z^2)^{3/2}} \qquad (3.24)$$

Its unique feature is that every portion of the Kelvin coil produces the same field per unit power. In comparison with a Kelvin coil, a uniform-current-density coil for the same total power produces 66% of field at the magnet center; for a Bitter coil, the efficiency is 79%. In practice it is not possible to fabricate coils having the Kelvin current density.

Gaume: The Gaume distribution, $J_G(r,z)$, also gives a good field efficiency:

$$J_G(r,z) \propto \frac{1}{r}\left(\frac{1}{\sqrt{a_1^2+z^2}} - \frac{1}{\sqrt{a_2^2+z^2}} \right) \qquad (3.25)$$

The Gaume coils make each turn produce the same field per unit power as every other turn. A Gaume coil produces 89% of the field of a Kelvin coil. The current distribution of Bitter coils designed and built at the FBNML approximates to a degree the Gaume distribution. This is achieved by using thicker Bitter plates axially away from the magnet midplane: $J_B(r,z) \propto 1/r(z+z_o)$.

Polyhelix: A polyhelix coil consists of many nested single-layer coils, in which the current density of each layer is adjusted to maximize the field efficiency and/or to match the stress in each layer to its conductor strength:

$$J_P(r,z) \propto \frac{1}{r^2} \qquad (3.26)$$

A polyhelix coil is 92% as efficient as a Kelvin coil. In practice, because of the need to have many electrodes at both ends, polyhelix coils are considered more difficult to manufacture than Bitter coils.

Hybrid Magnet

The hybrid magnet was conceived in the late 1960s by Montgomery and others as a means to achieve DC fields higher than $\sim$25 T, the limit of Bitter magnets at the FBNML [3.4]. The name hybrid is used to denote a magnet system combining both a water-cooled magnet and a superconducting magnet. Figure 3.5 shows a drawing of Hybrid III, a 35-T system at FBNML. Hybrid III follows a series of hybrids built at FBNML since the early 1970s [3.5$\sim$3.8]. There are three other active 30-plus tesla hybrid systems elsewhere: Tohoku University in Sendai, Japan [3.9]; the High Field Magnet Laboratory, Nijmegen, The Netherlands [3.6, 3.10]; and the High Magnetic Field Laboratory in Grenoble, France [3.11]. In addition, a 40-T hybrid at the National Research Institute for Metals in Tsukuba, Japan [3.12] begins operation in 1995 and a 45-T hybrid is under construction at the National High Magnetic Field Laboratory (NHMFL), Tallahassee, Florida [3.13].

The Hybrid III's footprint occupies a 4-m^2 area and its overall height, excluding service chimneys, is $\sim$3 m. The water-cooled insert (the two inner coils in Fig. 3.5) resides in the room-temperature bore of the cryostat that houses the superconducting magnet (two outer coils). The 9-MW power access to the insert is from below, while the 5000-liter/min cooling water access is at both top and bottom.

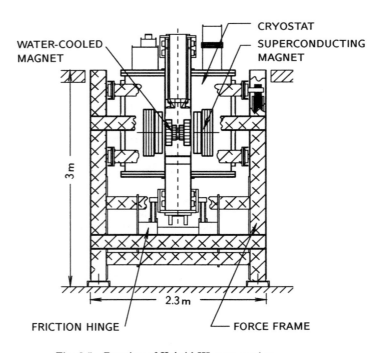

Fig. 3.5 Drawing of Hybrid III cross section.

Parameters of Hybrid III Superconducting Magnet (SCM)

Because Hybrid III's superconducting magnet (SCM) is used as a vehicle to discuss important issues of cryogenics in Chapter 4, of stability in Chapter 6, of dissipation in Chapter 7, and of protection in Chapter 8, its basic parameters are presented here. Specific problems will present other relevant parameters.

The Hybrid III SCM is comprised of two sections, an inner coil wound with two grades of Nb_3Sn conductor and an outer wound with two grades of Nb-Ti conductor. In both coils the grading is between a high-field (HF) conductor inside and a low-field (LF) conductor outside. The coils are housed in a single cryostat and, at a nominal temperature of 1.8 K, generate a field of 12.3 T at the magnet center with an operating current I_{op} of 2100 A. Table 3.1 gives coil parameters. (For the reader unfamiliar with these conductors, articles by Suenaga [3.14] and Larbalestier [3.15] and a treatise on titanium alloys by Collings [3.16] are recommended.)

The Nb_3Sn coil is layer wound and contains a total of 18 layers. The first 8 layers of the HF conductor were wound with a single piece of conductor. At the end of the coil between layers 8 and 9, the HF conductor was spliced to the LF conductor. Between layers 14 and 15, the LF conductor was spliced again because the low-field conductor was not obtainable as a single length.

The Nb-Ti coil is an assembly of 32 "double pancakes." A double pancake is wound flat, consisting of a continuous piece of conductor that spirals in from the outside of one pancake and then spirals back out in the other. The transition turn at the inside diameter is managed without any space-consuming splicing or possible loss of strength. Figure 3.6 presents a pictorial view of a double-pancake coil. As stated above, each double pancake in the Nb-Ti coil contains two grades of Nb-Ti conductor, a splice occurring at a diameter of 756 mm (Table 3.1). Problem 7.9 in Chapter 7 examines Joule dissipation generated in the splices in the Nb-Ti coil.

Table 3.1: Hybrid III SCM Parameters

Parameter		Nb_3Sn		Nb-Ti	
		HF	LF	HF	LF
$2a_1$, winding i.d.	[mm]	432	516	658	756
$2a_2$, winding o.d.	[mm]	516	621	756	907
$2b$, winding length	[mm]	550	551	640	640
# of layers		8	10	—	
# of double pancakes		—		32	
# turns		454	593	1120	2160
λJ @ I_{op}	[MA/m^2]	41	43	75	94
B_{peak} @ I_{op}	[T]	13.2	11.5	9.4	7.5

Problem 3.3: Hybrid magnet

a) Explain why the hybrid magnet configuration invariably consists of a super-conducting magnet surrounding a water-cooled resistive insert and not the other way around.

b) *Optional* Discuss the design and operational issues that are unique to hybrid magnet systems and those which are relatively inconsequential or nonexistent in either a Bitter magnet or a superconducting magnet alone.

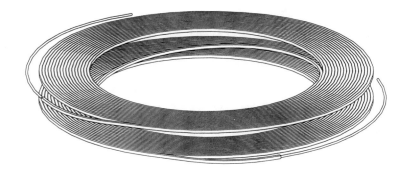

Fig. 3.6 Pictorial view of a double-pancake coil with the two individual pancakes separated axially. The pancakes in this drawing are wound with a CIC (cable-in-conduit) conductor with its conduit removed at each end over ~90° arc for splicing.

> *"We have learned that 'paper' hybrid magnets:*
> *are always built with perfect superconductors;*
> *can be built in perfect confidence with utmost materials;*
> *are not subject to fatigue; never have shorts . . .;*
> *have cryostats closed with zippers and are always vacuum tight;*
> *provide unlimited experimental access; operate themselves;*
> *. . . are always on schedule and within budget."* —Mathias J. Leupold

Solution to Problem 3.3

a) The unique features of water-cooled magnets may be summarized as follows:

- As expressed in Eq. $S2.10$ (Problem 3.2), the power requirement P_B for Bitter magnets is proportional to a_1 and $[B_z(0,0)]_B^2$, with P_B typically in the 6~30 MW range. Bitter magnets are "power hungry," and it is best to minimize their overall volume;

- Unlike superconductors which have practical field limits of 15~20 T, water-cooled magnets have no "intrinsic" limit—those made of stronger materials generally require greater power (because they are more resistive) and thus require more cooling than those of weaker materials, but there are no clear-cut limits above which the magnets cannot be operated;

- Field efficiency decreases with magnet size, again requiring the magnets to be as compact as feasible.

The unique features of superconducting magnets, on the other hand, are:

- Superconducting materials have fairly well defined upper field limits above which they cannot operate;

- The total energy storage increases with magnet size, but the power required remains insignificant. A 100-MJ magnet does not require a 100-MW power supply; typically 10~100 kW supplies suffice.

This combination of features makes it natural for hybrid magnets to consist of a water-cooled insert surrounded by a superconducting magnet.

b) A unique and demanding aspect of a hybrid magnet system arises from inter-active forces between the two magnets. If the two magnets are perfectly aligned axially and radially, they exert no force on each other. However, relative displacements of their field centers result in forces of increasing magnitude. Axial displacements produce axial restoring forces; the magnets tend to center themselves. Radial displacements of the field centers result in forces of the same sign, *i.e.* instability. Normally, forces are modest; careful design and construction can cope with them relatively easily. However, the failure of a necessarily high-performance water-cooled insert must be accepted as inevitable. In such a failure, large forces can suddenly develop from a field displacement created when part of the insert winding becomes shorted and ceases to produce field. Design problems related to this insert fault mode are discussed in more detail in Problems 3.14~3.16.

Although less demanding than the structural requirements to contain fault forces, magnet monitoring for electrical protection is also complicated because of magnetic coupling (mutual inductance) of the two systems. Obviously each magnet and its power system must have electrical protection of some sort to prevent damage or injury if something goes wrong, but there are also strong electrical interactions between the two magnets which would not exist were they separate. Problems 8.3 and 8.4 in Chapter 8 discuss coil monitoring in more detail for magnets in general and for hybrid magnets in particular.

Problem 3.4: Helmholtz coil

Highly uniform magnetic fields are desirable in many applications. An arrangement known as the "Helmholtz coil" achieves a high uniformity of the field over a limited region of space by simple means. It uses two identical coils spaced coaxially a distance d apart (Fig. 3.7a) in the magnet axis (z-direction); the top coil is located at $z = d/2$ and the bottom coil is located at $z = -d/2$. The spacing d is adjusted to satisfy, at the magnet center ($r = 0, z = 0$):

$$\left. \frac{d^2 H_z(0,z)}{dz^2} \right|_{z=0} = 0 \tag{3.27}$$

a) Idealizing the two coils by two single-wire loops each of radius a, show that when $d = a$, $dH_z^2(0,z)/dz^2 = 0$ at the magnet center. The solid curve in Fig. 3.7b gives $H_z(0,z)$ of a Helmholtz coil for $d = 2a$ (not optimized).

b) Show that if the current polarity of the bottom coil is reversed, a gradient field is generated at the magnet center. Evaluate this dH_z/dz at $z = 0$. (Note that $d^3 H_z(0,z)/dz^3 \neq 0$ at $z = 0$ even when $d = a$.) The dotted curve in Fig. 3.7b gives $H_z(z)$ of a gradient coil for $d = 2a$ (not optimized).

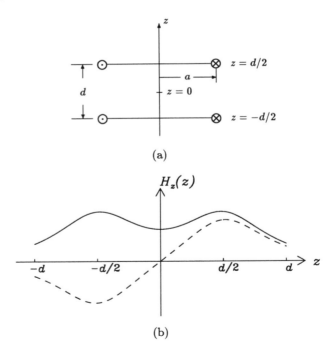

Fig. 3.7   (a) Ideal Helmholtz coil arrangement; (b) $H_z(0,z)$ for a "uniform" field case (solid) and $H_z(0,z)$ for a "gradient" field case (dotted).

Solution to Problem 3.4

a) The z-component of magnetic field along the axis ($r = 0$) due to the bottom loop, located at $z = -d/2$, may be given by Eq. 3.3:

$$H_z(0, z) = \frac{a^2 I}{2[(z + d/2)^2 + a^2]^{3/2}} \qquad (S4.1)$$

Thus, with the top coil located at $z = d/2$, we have:

$$H_z(0, z) = \frac{a^2 I}{2} \left\{ \frac{1}{[(z + d/2)^2 + a^2]^{3/2}} + \frac{1}{[a^2 + (z - d/2)^2]^{3/2}} \right\} \qquad (S4.2)$$

Differentiating $B_z(z)$ given by Eq. $S4.2$ with z, we have:

$$\frac{dH_z(0, z)}{dz} = \frac{3a^2 I}{2} \left\{ -\frac{(z + d/2)}{[a^2 + (z + d/2)^2]^{5/2}} - \frac{(z - d/2)}{[a^2 + (z - d/2)^2]^{5/2}} \right\} \qquad (S4.3)$$

Note that from symmetry $dH_z(0, z)/dz = 0$ at $z = 0$ for any value of a.

The second derivative of Eq. $S4.2$ is given by:

$$\frac{d^2 H_z(0, z)}{dz^2} = \frac{3a^2 I}{2} \left\{ -\frac{a^2 - 4(z + d/2)^2}{[a^2 + (z + d/2)^2]^{7/2}} - \frac{a^2 - 4(z - d/2)^2}{[a^2 + (z - d/2)^2]^{7/2}} \right\} \qquad (S4.4)$$

The second derivative is also zero at $z = 0$ but only if $d = a$. This technique of locating two identical coils with axial spacing equal to the coil radius to produce a region of field homogeneity is the basic principle used in MRI and other magnets requiring a high spatial field homogeneity.

b) For this system with the current polarity of the bottom coil reversed, we have:

$$H_z(0, z) = \frac{a^2 I}{2} \left\{ -\frac{1}{[a^2 + (z + d/2)^2]^{3/2}} + \frac{1}{[a^2 + (z - d/2)^2]^{3/2}} \right\} \qquad (S4.5)$$

$B_z(z)$, as expected, is zero at $z = 0$. We have:

$$\frac{dH_z(0, z)}{dz} = \frac{3a^2 I}{2} \left\{ \frac{(z + d/2)}{[a^2 + (z + d/2)^2]^{5/2}} - \frac{(z - d/2)}{[a^2 + (z - d/2)^2]^{5/2}} \right\} \qquad (S4.6)$$

Evaluating Eq. $S4.6$ at $z = 0$, we have:

$$\left. \frac{dH_z(0, z)}{dz} \right|_0 = \frac{3a^2 I d}{2[a^2 + (d/2)^2]^{5/2}} \qquad (S4.7)$$

This method of locating two identical coils having opposite currents to achieve a field gradient is the basic principle used in magnets requiring a gradient field at the midplane. A pulsed magnet used in an MRI system to produce a gradient field (to extract spatial information for imaging) is one example.

Problem 3.5: Spatially homogeneous fields

In spherical coordinates (r,θ,φ) defined in Fig. 3.8, the magnetic field in the z-direction, H_z, in a source-free space generated by a solenoidal magnet consisting of nested solenoidal coils can be expressed by:

$$H_z(r,\theta,\varphi) = \sum_{n=1}^{\infty} \sum_{m=0}^{n-1} r^{n-1}(n+m)P_{n-1}^m(u)(A_n^m \cos m\varphi + B_n^m \sin m\varphi) \quad (3.28)$$

You may recognize Eq. 3.28 as same product solution from Eq. 2.32. $P_{n-1}^m(u)$ is the set of Legendre functions with $u = \cos\theta$ (Table 2.2) and associated Legendre functions (Table 2.3):

Legendre functions: $P_0^0 = 1 \quad P_1^0 = u \quad P_2^0 = \frac{1}{2}(3u^2 - 1)$

Associate functions: $P_1^1 = s \quad P_2^1 = 3us \quad P_2^2 = 3s^2$

where $s \equiv \sin\theta$. A_n^m and B_n^m are constants, which, except for A_1^1 and B_1^1, are to be minimized, because they contribute to field nonuniformity. A_n^m and B_n^m may be minimized by adjusting the parameters of each coil comprising the magnet [3.17]. These parameters include winding i.d. $(2a_1)$, winding o.d. $(2a_2)$, winding length $(2b)$, coil midplane location relative to the magnet center, current density (λJ), and number of turns (N).

Show that expressions for $H_z(r,\theta,\varphi)$ in Cartesian coordinates, $H_z(x,y,z)$, for n's up to 1, 2, and 3 are given by:

$$n = 1 \quad H_z(x,y,z) = A_1^0 \tag{3.29a}$$

$$n = 2 \quad H_z(x,y,z) = A_1^0 + 2zA_2^0 + z(A_2^1 x + B_2^1 y) \tag{3.29b}$$

$$n = 3 \quad H_z(x,y,z) = A_1^0 + 2zA_2^0 + z(A_2^1 x + B_2^1 y)$$

$$+ \tfrac{3}{2}A_3^0(2z^2 - x^2 - y^2) + 12z(A_3^1 x + B_3^1 y)$$

$$+ 15[A_3^2(x^2 - y^2) + 2B_3^2 xy] \tag{3.29c}$$

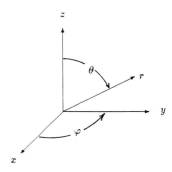

Fig. 3.8   Spherical coordinates.

Solution to Problem 3.5

The spherical coordinate parameters, r, $u = \cos\theta$, $s = \sin\theta$, $\sin\varphi$, and $\cos\varphi$ are given in terms of x, y, and z:

$$r = \sqrt{x^2 + y^2 + z^2} \tag{S5.1a}$$

$$u = \cos\theta = \frac{z}{\sqrt{x^2 + y^2 + z^2}} \tag{S5.1b}$$

$$s = \sin\theta = \frac{\sqrt{x^2 + y^2}}{\sqrt{x^2 + y^2 + z^2}} \tag{S5.1c}$$

$$\sin\varphi = \frac{y}{\sqrt{x^2 + y^2}} \tag{S5.1d}$$

$$\cos\varphi = \frac{x}{\sqrt{x^2 + y^2}} \tag{S5.1e}$$

Equations 3.28 and $S5.1a{\sim}S5.1e$ are combined for cases $n = 1$, $n = 2$, and $n = 3$.

$n = 1$:

$$H_z(x,y,z) = \sum_{m=0}^{0} r^0(1+0)P_0^0(u)(A_1^0 \cos 0 + B_1^0 \sin 0) \tag{S5.2a}$$

$$= (1)(1)(1)(A_1^0) \tag{S5.2b}$$

For n up to 1, we thus have:

$$H_z(x,y,z) = A_1^0 \tag{3.29a}$$

Note that A_1^0 represents field at the magnet center $(0,0,0)$.

$n = 2$:

$$H_z(x,y,z) = \sum_{m=0}^{1} r^1(2+m)P_1^m(A_2^m \cos m\varphi + B_2^m \sin m\varphi) \tag{S5.3a}$$

$$= r^1(2+0)P_1^0(A_2^0) + r^1(2+1)P_1^1(A_2^1 \cos\varphi + B_2^1 \sin\varphi)$$

$$= 2ruA_2^0 + 3rs(A_2^1 \cos\varphi + B_2^1 \sin\varphi)$$

$$= 2\sqrt{x^2 + y^2 + z^2}\,\frac{z}{\sqrt{x^2 + y^2 + z^2}}A_2^0$$

$$+ 3\sqrt{x^2 + y^2 + z^2}\,\frac{\sqrt{x^2 + y^2}}{\sqrt{x^2 + y^2 + z^2}}\left(\frac{A_2^1 x + B_2^1 y}{\sqrt{x^2 + y^2}}\right)$$

$$= 2zA_2^0 + z(A_2^1 x + B_2^1 y) \tag{S5.3b}$$

Solution to Problem 3.5

Thus for n up to 2, we have:

$$H_z(x,y,z) = A_1^0 + 2zA_2^0 + z(A_2^1 x + B_2^1 y) \tag{3.29b}$$

Note that $H_z(x,y,z)$ contains terms up to those varying as z, zx, and zy.

$n = 3$:

$$H_z(x,y,z) = \sum_{m=0}^{2} r^2(3+m)P_2^m(A_3^m \cos m\varphi + B_3^m \sin m\varphi) \tag{S5.4a}$$

$$= r^2(3+0)P_2^0(A_3^0) + r^2(3+1)P_2^1(A_3^1 \cos\varphi + B_3^1 \sin\varphi)$$

$$+ r^2(3+2)P_2^2(A_3^2 \cos 2\varphi + B_3^2 \sin 2\varphi)$$

$$H_z(x,y,z) = 3(x^2+y^2+z^2)\tfrac{1}{2}\left(\frac{3z^2-x^2-y^2}{x^2+y^2+z^2}\right)A_3^0$$

$$+ 4(x^2+y^2+z^2)\frac{3z\sqrt{x^2+y^2}}{x^2+y^2+z^2}\left(A_3^1\frac{x}{\sqrt{z^2+y^2}} + B_3^1\frac{y}{\sqrt{z^2+y^2}}\right)$$

$$+ 5(x^2+y^2+z^2)\frac{3(x^2+y^2)}{x^2+y^2+z^2}\left[A_3^2\left(\frac{2x^2}{x^2+y^2}-1\right) + B_3^2\frac{2xy}{x^2+y^2}\right]$$

$$= \tfrac{3}{2}A_3^0(2z^2-x^2-y^2)$$

$$+ 12z(A_3^1 x + B_3^1 y) + 15[A_3^2(x^2-y^2) + 2B_3^2 xy] \tag{S5.4b}$$

Summing Eqs. 5.2b, S5.3b, and S5.4b, we have, for n up to 3:

$$H_z(x,y,z) = A_1^0 + 2zA_2^0 + z(A_2^1 x + B_2^1 y)$$

$$+ \tfrac{3}{2}A_3^0(2z^2-x^2-y^2) + 12z(A_3^1 x + B_3^1 y)$$

$$+ 15[A_3^2(x^2-y^2) + 2B_3^2 xy] \tag{3.29c}$$

Note that $H_z(x,y,z)$ contains terms up to those varying as z^2, x^2, y^2, zx, zy, and xy.

> *"Ignorance is like a delicate exotic fruit; touch it and the bloom is gone."*
> —Lady Bracknell

Problem 3.6: Notched solenoid

The principle of the Helmholtz coil—to place current-carrying elements symmetrically about the solenoid center to create a spatially homogeneous field in the central zone—is the basis for notched solenoids. Notched solenoids are indispensable components in MRI and NMR magnets.

For a simple solenoid, with winding inner radius a_1, winding outer radius a_2, total winding length $2b$, and overall current density λJ, recall that the axial magnetic field at the center, $H_z(0,0)$, is given by Eqs. 3.10 and 3.11:

$$H_z(0,0) = \lambda J a_1 F(\alpha, \beta) \qquad (3.10)$$

$$F(\alpha, \beta) = \beta \ln \left(\frac{\alpha + \sqrt{\alpha^2 + \beta^2}}{1 + \sqrt{1 + \beta^2}} \right) \qquad (3.11)$$

where $\alpha = a_2/a_1$ and $\beta = b/a_1$.

Using symmetry considerations and superposition, show that an expression for $H_z(0, z_1)$ of the notched solenoid having a uniform current density λJ shown in Fig. 3.9 is given by:

$$H_z(0, z_1) = \tfrac{1}{2} \lambda J a_1 \left[F(\alpha_1, \beta_1 + \gamma_1) + F(\alpha_1, \beta_1 - \gamma_1) \right]$$
$$- \tfrac{1}{2} \lambda J a_3 \left[F(\alpha_2, \beta_2 + \gamma_2) + F(\alpha_2, \beta_2 - \gamma_2) \right] \qquad (3.30)$$

where $\alpha_1 = a_2/a_1$, $\beta_1 = b_1/a_1$, $\gamma_1 = z_1/a_1$, $\alpha_2 = a_2/a_3$, $\beta_2 = b_2/a_3$, and $\gamma_2 = z_1/a_3$. The coil parameters, a_1, a_2, a_3, b_1, and b_2 are defined in Fig. 3.9. Note also that Eq. 3.30 is valid for $0 \le z_1 \le b_2$.

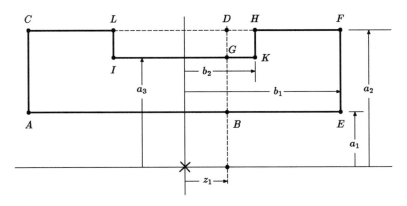

Fig. 3.9   Geometrical arrangement of a notched solenoid.

Solution to Problem 3.6

To solve $H_z(0, z_1)$, we may divide the solenoid into four single solenoids whose cross sections are designated by corner points as follows:

Solenoid 1: $ABDC$ (with $\alpha_1 = a_2/a_1$, $\beta_1 = b_1/a_1$, $\gamma_1 = z_1/a_1$);

Solenoid 2: $BEFD$ (with $\alpha_1 = a_2/a_1$, $\beta_1 = b_1/a_1$, $\gamma_1 = z_1/a_1$);

Solenoid 3: $IGDL$ (with $\alpha_2 = a_2/a_3$, $\beta_2 = b_2/a_3$, $\gamma_2 = z_1/a_3$);

Solenoid 4: $GKHD$ (with $\alpha_2 = a_2/a_3$, $\beta_2 = b_2/a_3$, $\gamma_2 = z_1/a_3$).

Note that in each solenoid J is equal, but in Solenoids 3 and 4, it is in the opposite direction from that in Solenoids 1 and 2. Also note that each solenoid is notchless.

Field from Solenoid 1: $H_z(0, z_1)$ from Solenoid 1 ($2b = b_1 + z_1$) is one half of the center field of a solenoid $2b = 2(b_1 + z_1)$ long having the same values of a_1, a_2, and λJ. This may be best seen by noting that the center field $H_z(0,0)$ of a notchless solenoid $2b$ long is the sum of the field generated by one half of the solenoid (from 0 to $z = b$) and that generated by the other half (from $z = -b$ to 0). That is, each half of the solenoid generates 50% of the total field $H_z(0,0)$. Thus:

$$H_z(0, z_1)|_1 = \tfrac{1}{2}\lambda J a_1 F(\alpha_1, \beta_1 + \gamma_1) \qquad (S6.1)$$

Field from Solenoid 2: At $(0, z_1)$, H_z from Solenoid 2 ($2b = b_1 - z_1$) is one half of the center field of a solenoid $2b = 2(b_1 - z_1)$ long, both solenoids having the same values of a_1, a_2, and λJ.

$$H_z(0, z_1)|_2 = \tfrac{1}{2}\lambda J a_1 F(\alpha_1, \beta_1 - \gamma_1) \qquad (S6.2)$$

Field from Solenoid 3: At $(0, z_1)$, H_z from Solenoid 3 ($2b = b_2 + z_1$) is one half of the center field of a solenoid having $2b = 2(b_2 + z_1)$, both solenoids with the same values of a_3, a_2, and λJ. Because J is directed opposite to that of Solenoids 1 and 2, we have:

$$H_z(0, z_1)|_3 = -\tfrac{1}{2}\lambda J a_3 F(\alpha_2, \beta_2 + \gamma_2) \qquad (S6.3)$$

Field from Solenoid 4: At $(0, z_1)$, H_z from Solenoid 4 ($2b = b_2 - z_1$) is one half of the center field of a solenoid $2b = 2(b_2 - z_1)$ long.

$$H_z(0, z_1)|_4 = -\tfrac{1}{2}\lambda J a_3 F(\alpha_2, \beta_2 - \gamma_2) \qquad (S6.4)$$

Field from the Notched Solenoid

$H_z(0, z_1)$ from the original notched solenoid is given by the sum of Eqs. $S6.1 \sim S6.4$:

$$\begin{aligned}
H_z(0, z_1) &= H_z(0, z_1)|_1 + H_z(0, z_1)|_2 + H_z(0, z_1)|_3 + H_z(0, z_1)|_4 \\
&= \tfrac{1}{2}\lambda J a_1 \Big[F(\alpha_1, \beta_1 + \gamma_1) + F(\alpha_1, \beta_1 - \gamma_1) \Big] \\
&\quad - \tfrac{1}{2}\lambda J a_3 \Big[F(\alpha_2, \beta_2 + \gamma_2) + F(\alpha_2, \beta_2 - \gamma_2) \Big]
\end{aligned} \qquad (3.30)$$

Problem 3.7: Ideal dipole magnet

This problem studies an ideal dipole magnet, which is infinitely long (no end effects) and whose fields, directed normal to the dipole axis, are generated by a longitudinal surface current having zero winding thickness. Field and force solutions for real dipole magnets are more complex than those treated in this problem; nevertheless the ideal dipole magnet, except for complications at the ends, illustrates most of the key aspects. Dipole magnets are used in systems that require a uniform field directed transverse to the magnet axis; superconducting versions have been used in high-energy particle accelerators [3.18, 3.19], MHD power generators [3.20, 3.21], and electric generators [3.22, 3.23].

A long (two-dimensional) dipole magnet of radius R and of "zero" winding thickness is energized by a surface current flowing in the z-direction at the dipole shell $(r = R)$. The magnetic field within the bore $(r \le R)$, $\vec{H}_{d1}$, and that outside the shell $(r \ge R)$, $\vec{H}_{d2}$, are given by:

$$\vec{H}_{d1} = H_0(\sin\theta\,\vec{i}_r + \cos\theta\,\vec{i}_\theta) \tag{3.31a}$$

$$\vec{H}_{d2} = H_0\left(\frac{R}{r}\right)^2(\sin\theta\,\vec{i}_r - \cos\theta\,\vec{i}_\theta) \tag{3.31b}$$

The 2-D coordinates are defined in Fig. 3.10. Note that the $+z$-direction is out of the paper. In answering the following questions, neglect end effects.

a) Draw neatly the field profile of the dipole for both regions, $r < R$ and $r > R$.

b) Show that an expression for the surface current $\vec{K}_{fd}$ at $r = R$ is given by:

$$\vec{K}_{fd} = -2H_0\cos\theta\,\vec{i}_z \tag{3.32}$$

Indicate on a sketch its direction either with circles (o) where $\vec{K}_{fd}$ is $+z$-directed (out of the paper) or with crosses ($\times$) where $\vec{K}_{fd}$ is $-z$-directed.

c) Show that an expression for the Lorentz force flux, $\vec{f}_{Ld}$ [N/m^2], acting on a current-carrying element of the shell (per unit length), is given by:

$$\vec{f}_{Ld} = -\mu_o H_0^2 \sin 2\theta\,\vec{i}_\theta \tag{3.33}$$

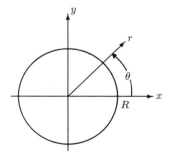

Fig. 3.10   Two-dimensional cylindrical coordinates.

Problem 3.7: Ideal dipole magnet

d) Show that the net x-directed Lorentz force (per unit dipole length), F_{Ldx} [N/m], acting on the right-hand segment $(-90° < \theta < 90°)$ is given by:

$$F_{Ldx} = \frac{4R\mu_o H_0^2}{3} \qquad (3.34)$$

e) Show that an expression for the total magnetic energy stored (per unit dipole length), E_m [J/m], is given by:

$$E_m = \frac{\pi R^2 B_0^2}{\mu_o} \qquad (3.35)$$

Compute E_m for $\mu_o H_0 = 5\,\text{T}$ and $R = 20\,\text{mm}$. Also compute the inductance, L_d, of a 10-m long dipole with an operating current I_{op} of 5000 A.

To reduce the field outside the dipole, an iron yoke $(\mu = \infty)$ of radial thickness d is placed outside the dipole, as shown in Fig. 3.11.

f) Show that the new $\vec{K}_{fd1}$ needed to generate the *same* $\vec{H}$ field inside the dipole is exactly one half that given by Eq. 3.32. Explain this current reduction.

g) In reality, the iron yoke cannot maintain its high μ for an unlimited value of H_0. Show that an expression for the minimum d_m to keep the yoke unsaturated is given by:

$$d_m = R\left(\frac{H_0}{M_{sa}}\right) \qquad (3.36)$$

where M_{sa} is the yoke material's saturation magnetization. Compute d_m for the following set: $\mu_o H_o = 5\ \text{T}$; $\mu_o M_{sa} = 1.2\,\text{T}$; $R = 20\,\text{mm}$.

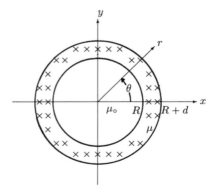

Fig. 3.11   Ideal dipole with an iron yoke of thickness d.

Solution to Problem 3.7

a) The field lines are sketched in Fig. 3.12a for both regions. Note that the normal (r-directed) component of the field is continuous at the boundary ($r = R$).

b) The discontinuity in the tangential (θ-directed) component of the field at $r = R$ is equal to the surface current density ($\vec{K}_{fd}$) flowing there. From Eq. 2.6:

$$\vec{K}_{fd} = \vec{i}_r \times (\vec{H}_{d2} - \vec{H}_{d1}) = \vec{i}_r \times -2H_0 \cos\theta\,\vec{i}_\theta$$
$$= -2H_0 \cos\theta\,\vec{i}_z \qquad (3.32)$$

As indicated in Fig. 3.12b, $\vec{K}_{fd}$ in the $-90° < \theta < 90°$ segment points into the $-z$-direction, while that in the $90° < \theta < 270°$ segment points into the $+z$-direction.

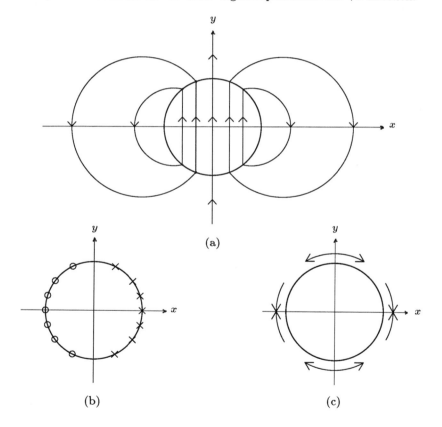

(a)

(b)   (c)

Fig. 3.12   a) Dipole fields for inside and outside the magnet; b) surface current density vectors on the magnet; c) force vectors on the magnet (solution for **c**).

Solution to Problem 3.7

c) $\vec{f}_{Ld}$ is given by the cross-product of $\vec{K}_{fd}$ and $\vec{B}_d$, where $\vec{B}_d = (\vec{B}_{d1} + \vec{B}_{d2})/2$. Thus:

$$\vec{f}_{Ld} = \vec{K}_{fd} \times \mu_o H_0 \sin\theta \, \vec{i}_r \qquad (S7.1)$$

$$= -2\mu_o H_0^2 \cos\theta \sin\theta \, \vec{i}_\theta$$

$$= -\mu_o H_0^2 \sin 2\theta \, \vec{i}_\theta \qquad (3.33)$$

Note that $\vec{f}_{Ld}$ has no r-component; it only has the θ-component (Fig. 3.12c). Note that the force density is maximum at $\theta = \pi/4 + n\pi/2$ and zero at $\theta = 0 + n\pi/2$ ($n = 0, 1, 2, 3$).

d) It is clear from Fig. 3.12c that the net Lorentz force acting on the right-hand segment of the shell is $+x$-directed. Thus, F_{Ldx} [N/m] is given by:

$$F_{Ldx} = \int \vec{f}_{Ld} \cdot d\vec{x} = -R \int_{-\pi/2}^{\pi/2} f_{Ld\theta} \sin\theta \, d\theta \qquad (S7.2a)$$

$$= -2R \int_0^{\pi/2} f_{Ld\theta} \sin\theta \, d\theta = 4R\mu_o H_0^2 \int_0^{\pi/2} \cos\theta \sin^2\theta \, d\theta \qquad (S7.2b)$$

From Eq. $S7.2b$, we obtain:

$$F_{Ldx} = \frac{4R\mu_o H_0^2}{3} \qquad (3.34)$$

The net Lorentz force on the left-hand segment has the same magnitude as that on the right-hand segment except it is $-x$-directed. That is, there is a large force trying to pull the two halves of the dipole apart. In fact structural support to contain these forces is a key design issues for dipole magnets.

e) E_m [J/m] may be computed by integrating $\mu_o |H(r,\theta)|^2/2$, the magnetic energy density, over the entire surface, from $r = 0$ to $r = \infty$ and $\theta = 0$ to $\theta = 2\pi$, transverse to the dipole axis.

$$E_m = \frac{\mu_o}{2} \int_0^R |H_{d1}|^2 2\pi r \, dr + \frac{\mu_o}{2} \int_R^\infty |H_{d2}|^2 2\pi r \, dr \qquad (S7.3a)$$

$$= \frac{\mu_o}{2} H_0^2 \pi R^2 + \frac{\mu_o}{2} H_0^2 \pi R^2 \qquad (S7.3b)$$

$$= \mu_o \pi R^2 H_0^2 = \frac{\pi R^2 B_0^2}{\mu_o} \qquad (3.35)$$

From Eq. $S7.3b$ it is clear that the total stored magnetic energy is divided equally inside and outside the dipole shell. This implies, as it will become even clearer in f), that one half of the current flowing in the dipole is used to create $\vec{H}_{d1}$ and the other half is used to create $\vec{H}_{d2}$. Inserting $\mu_o H_0 = B_0 = 5\,\mathrm{T}$ and $R = 0.02\,\mathrm{m}$ into the above expression, we obtain:

$$E_m = \frac{\pi (2 \times 10^{-2}\,\mathrm{m})^2 (5\,\mathrm{T})^2}{4\pi \times 10^{-7}\,\mathrm{H/m}} = 25\,\mathrm{kJ/m} \qquad (S7.4)$$

Solution to Problem 3.7

For a 10-m, 5-T long dipole, the total magnetic energy becomes 250 kJ. This total energy may be equated to the dipole's total inductive energy:

$$\frac{1}{2} L_d I_{op}^2 = 250 \text{ kJ} \tag{$S7.5$}$$

Solving Eq. $S7.5$ for L_d with $I_{op} = 5000$ A, we have:

$$L_d = \frac{2(250 \times 10^3 \text{ J})}{(5000 \text{ A})^2} \tag{$S7.6$}$$

$$= 20 \text{ mH}$$

Note that if the dipole's operating current is, for example, 1000 A, then the dipole must have an inductance of 0.5 H or it must have about five times more winding turns than the 20-mH dipole.

f) Because $\mu = \infty$, $\vec{H}_{d2} = 0$ for $R < r < R + d$ and $\vec{H}_{d2} = 0$ also for $r > R + d$. We still have $\vec{H}_{d1}$ as before. Clearly:

$$\vec{K}_{fd1} = -H_0 \cos \theta \, \vec{i}_\theta \tag{$S7.7$}$$

which is exactly one half of $\vec{K}_{fd}$ given by Eq. 3.32.

Considering the surface current requirements for both cases, with and without the iron yoke, we might think of $\vec{K}_{fd1} = -H_0 \cos \theta \, \vec{i}_\theta$ as the source for the field *inside* and the $K_{fd2} = -H_0 \cos \theta \, \vec{i}_\theta$ as the source for the field *outside* the dipole. In the presence of an iron yoke of $\mu = \infty$, there is no need to create the field outside the shell; thus the current requirement is halved.

g) All the flux entering the yoke of radial thickness d between 0 and $\theta = 90°$ must be equal to or less than $\mu_o M_{sa} d$. That is:

$$R\mu_o H_0 \int_0^{\pi/2} \sin \theta \, d\theta = R\mu_o H_0 \leq \mu_o M_{sa} d \tag{$S7.8$}$$

The minimum yoke thickness d_m is thus given by:

$$d_m = R \left(\frac{H_0}{M_{sa}} \right) \tag{3.36}$$

With $R = 20$ mm, $\mu_o H_0 = 5$ T, and $\mu_o M_{sa} = 1.2$ T, we obtain:

$$d_m = (20 \text{ mm}) \frac{5 \text{ T}}{1.2 \text{ T}} = 83 \text{ mm}$$

Problem 3.8: Ideal quadrupole magnet

This problem studies an ideal quadrupole magnet, which is infinitely long (no end effects) and whose fields, directed normal to the magnet axis, are generated by a longitudinal surface current having zero winding thickness.

A long quadrupole magnet of radius R and of "zero" winding thickness is energized by a surface current flowing in the z-direction at the quadrupole shell $(r = R)$. The magnetic field within the bore $(r \leq R)$, $\vec{H}_1$, and that outside the shell $(r \geq R)$, $\vec{H}_2$, are given by:

$$\vec{H}_{q1} = H_0 \left(\frac{r}{R}\right)(\sin 2\theta \, \vec{i}_r + \cos 2\theta \, \vec{i}_\theta) \qquad (3.37a)$$

$$\vec{H}_{q2} = H_0 \left(\frac{R}{r}\right)^2 (\sin 2\theta \, \vec{i}_r - \cos 2\theta \, \vec{i}_\theta) \qquad (3.37b)$$

In answering the following questions, neglect end effects.

a) Draw *neatly* the field profiles of the quadrupole for both regions.

b) Show that an expression for the surface current $\vec{K}_{fq}$ at $r = R$ is given by:

$$\vec{K}_{fq} = -2H_0 \cos 2\theta \, \vec{i}_z \qquad (3.38)$$

Indicate neatly on a sketch its direction either with circles (o) where $\vec{K}_{fq}$ is $+z$-directed (out of the paper) or with crosses ($\times$) where $\vec{K}_{fq}$ is $-z$-directed.

c) Show that an expression for the Lorentz force flux, $\vec{f}_{Lq}$, acting on a current-carrying element of the shell is given by:

$$\vec{f}_{Lq} = -\mu_\circ H_0^2 \sin 4\theta \, \vec{i}_\theta \qquad (3.39)$$

d) Show that an expression for the "magnetic spring constant," k_{Lqx}, in the x-direction for a proton travelling in the $+z$-direction along the center of the magnet with a speed nearly equal to that of light, c, is given by:

$$k_{Lqx} \simeq \frac{qc\mu_\circ H_0}{R} \qquad (3.40)$$

e) Similarly, show that an expression for the "magnetic spring constant," k_{Lqy}, in the y-direction for a proton travelling in the $+z$-direction along the center of the magnet with a speed nearly equal to that of light, c, is given by:

$$k_{Lqy} \simeq -\frac{qc\mu_\circ H_0}{R} \qquad (3.41)$$

f) By stating whether k_{Lqx} and k_{Lqy} are unstable or restoring, describe the function of quadrupoles for proton (and electron) accelerators.

Solution to Problem 3.8

a) The field lines are sketched in Fig. 3.13a. Note that as with the ideal dipole, the r-component of the field is continuous at $r = R$.

b) The discontinuity in the θ-component of the field at the boundary is equal to the surface current density $(\vec{K}_{fq})$ flowing at $r = R$. Thus:

$$\vec{K}_{fq} = \vec{\imath}_r \times (\vec{H}_2 - \vec{H}_1) = \vec{\imath}_r \times -2H_0 \cos 2\theta \, \vec{\imath}_\theta$$

$$= -2H_0 \cos 2\theta \, \vec{\imath}_z \tag{3.38}$$

The $\vec{K}_{fq}$ vectors change directions four times around the magnet shell (Fig. 3.13b).

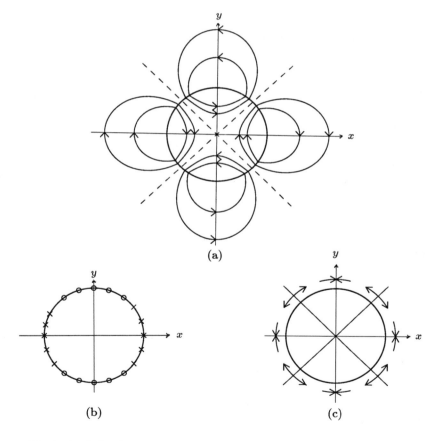

(a)

(b)   (c)

Fig. 3.13   a) Quadrupole fields for inside and outside the magnet; b) surface current density vectors on the magnet; c) force vectors on the magnet.

Solution to Problem 3.8

c) The Lorentz force flux, $\vec{f}_{Lq}$, is given by the cross product of $\vec{K}_{fq}$ and $\mu_o\vec{H}$ at $r = R$. At $r = R$, we have $\vec{H} = H_0 \sin 2\theta\, \vec{i}_r$, because the θ-component changes sign from $+$ to $-$ as we approach from $r < R$. Noting that $\vec{K}_{fq} = -2H_0 \cos 2\theta\, \vec{i}_z$, we have:

$$\vec{f}_{Lq} = -2\mu_o H_0^2 \sin 2\theta \cos 2\theta\, \vec{i}_\theta$$

$$= -\mu_o H_0^2 \sin 4\theta\, \vec{i}_\theta \qquad (3.39)$$

The $\vec{f}_{Lq}$ distribution is sketched in Fig. 3.13c.

d) We may define the magnetic spring constant in the x-direction as:

$$k_{Lqx} = -\frac{\partial F_{Lqx}}{\partial x} \qquad (S8.1)$$

F_{Lqx}, the Lorentz force in the x-direction acting on the proton of electric charge q travelling in the z-direction with velocity c, is given by:

$$F_{Lqx} \simeq [q(c\,\vec{i}_z) \times \mu_o H_1\, \vec{i}_\theta]_{\theta=0^\circ}$$

$$\simeq -qc\mu_o H_0 \left(\frac{r}{R}\right)\vec{i}_x \qquad (S8.2)$$

k_{Lqx} is thus given by:

$$k_{Lqx} = -\frac{\partial F_{Lqx}}{\partial x} = -\frac{\partial F_{Lqx}}{\partial r}$$

$$\simeq \frac{qc\mu_o H_0}{R} \qquad (3.40)$$

e) In the y-direction (r-direction at $\theta = 90^\circ$), the magnetic force F_{Lqy} is given by:

$$F_{Lqy} \simeq [q(c\,\vec{i}_z) \times \mu_o H_1\, \vec{i}_\theta]_{\theta=90^\circ}$$

$$\simeq qc\mu_o H_0 \left(\frac{r}{R}\right)\vec{i}_y \qquad (S8.3)$$

k_{Lqy} is thus given by:

$$k_{Lqy} = -\frac{\partial F_{Lqy}}{\partial y} = -\frac{\partial F_{Lqy}}{\partial r}$$

$$\simeq -\frac{qc\mu_o H_0}{R} \qquad (3.41)$$

f) The plus sign for k_{Lqx} indicates that F_{Lqx} is restoring, while the minus sign for k_{Lqy} indicates that F_{Lqy} is unstable, tending to diverge the beam in the y-direction. In accelerator rings, quadrupole magnets are thus used in pairs, one that focuses the beam in x-direction followed by another that focuses the beam in the y-direction.

Problem 3.9: Magnet comprised of two ideal "racetracks"

This problem deals with a magnet comprised of two infinitely long ideal "race-track" coils, placed parallel at a distance of $2c$. The name "racetrack" comes about because in real magnets the axial extent is not infinite, and at each end the conductor loops around $180°$, like the end of a race track. Because racetrack coils are flat, they are suitable for maglev [3.24~3.27].

Figure 3.14 shows a cross sectional view of the winding configuration of a magnet comprised of two infinitely long ideal racetracks. Two sets of very long racetrack coils, placed parallel to each other as shown, can sometimes substitute for a dipole magnet. For example, if a long length of a superconductor must be tested in a uniform field directed transverse to its major axis, this magnet configuration could be used; it has the advantage of requiring simpler winding rigs than a dipole magnet. This configuration also has been used as an MHD magnet [3.28]. As indicated in Fig. 3.14, each racetrack has a winding outer width of $2a_2$, inner width of $2a_1$, and contains N turns. The two coils are placed a distance $2c$ apart.

The direction of current in the right-hand side of each racetrack coil is in the $+z$-direction (out of the paper), while the direction of current in the left-hand side is in the $-z$-direction (into the paper). In answering the following questions, neglect end effects.

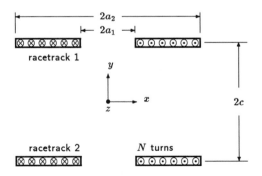

Fig. 3.14   Cross section of a magnet comprised of two ideal racetrack coils.

Problem 3.9: Magnet comprised of two ideal "racetracks"

a) Show that when each racetrack is energized with current I, an expression for the y-component of magnetic field, $H_y(x,y)$, in the region ($|x| \leq a_1$, $|y| \leq c$) and that at the magnet center $(0,0)$, are given, respectively, by:

$$H_y(x,y) = -\frac{K}{4\pi}\left\{\ln\left[\frac{(a_2-x)^2+(c-y)^2}{(a_1-x)^2+(c-y)^2}\right]+\ln\left[\frac{(a_2+x)^2+(c-y)^2}{(a_1+x)^2+(c-y)^2}\right]\right.$$

$$\left.+\ln\left[\frac{(a_2-x)^2+(c+y)^2}{(a_1-x)^2+(c+y)^2}\right]+\ln\left[\frac{(a_2+x)^2+(c+y)^2}{(a_1+x)^2+(c+y)^2}\right]\right\} \quad (3.42a)$$

and

$$H_y(0,0) = -\frac{K}{\pi}\ln\left(\frac{a_2^2+c^2}{a_1^2+c^2}\right) \quad (3.42b)$$

where $K = NI/(a_2 - a_1)$.

b) Show that near the origin $H_y(x,y)$ given by Eq. 3.42 may be expressed as the sum of $H_y(0,0)$ and a term containing (x^2+y^2). Namely:

$$H_y(x,y) \simeq H_y(0,0) + \frac{K(a_2^2-a_1^2)}{\pi(a_2^2+c^2)(a_1^2+c^2)}(x^2+y^2) \quad (3.43)$$

c) Current distribution in the magnet may be approximated by four current elements, 1, 2, 3, and 4, as shown in Fig. 3.15. Each element carries a net current of NI in the direction indicated in the figure. A dot indicates it is in the $+z$-direction and a cross in the $-z$-direction. Under this approximation, show that expressions for $H_y(0,0)$ and $H_y(x,y)$ near the magnet center may be given, respectively, by:

$$H_y(0,0) = -\frac{2aNI}{\pi(a^2+c^2)} \quad (3.44a)$$

$$H_y(x,y) \simeq H_y(0,0)\left(1-\frac{x^2+y^2}{a^2+c^2}\right) \quad (3.44b)$$

d) The same 4-current-element model may be used to compute the Lorentz forces (per unit axial length) acting on current element 1, $\vec{F}_1$, as the sum of Lorentz interactions on current element 1 by current elements 2, 3, and 4:

$$\vec{F}_1 = \vec{F}_1|_2 + \vec{F}_1|_3 + \vec{F}_1|_4 \quad (3.45)$$

where $\vec{F}_1|_2$, $\vec{F}_1|_3$, and $\vec{F}_1|_4$ are the force vectors on element 1 by element 2, element 3, and element 4, respectively.

Problem 3.9: Magnet comprised of two ideal "racetracks"

Neglecting end effects, show that expressions for $\vec{F}_1$ (per unit magnet axial length) in the x-direction, F_{1x}, and in the y-direction, F_{1y}, are given, respectively, by:

$$F_{1x} = \frac{\mu_o N^2 I^2}{4\pi}\left(\frac{1}{a} + \frac{a}{a^2 + c^2}\right) \tag{3.46a}$$

$$F_{1y} = -\frac{\mu_o N^2 I^2}{4\pi}\left(\frac{1}{c} - \frac{c}{a^2 + c^2}\right) \tag{3.46b}$$

e) Generalize $\vec{F}_1$ to the remaining current elements, and describe interaction forces within each racetrack and between the racetracks.

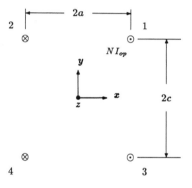

Fig. 3.15   Current distribution model for force calculation.

"Sometimes there is as much magic as science in the explanations of the force. Yet what is a magician but a practicing theorist?"   —Obi Wan Kenobi

Solution to Problem 3.9

a)   The differential field at (x, y), $d\vec{H}_{1+}$, from a differential surface current, $K\,d\xi$, located at ξ in the right-hand side of racetrack 1 (with + current) as shown in Fig. 3.16, is given by:

$$d\vec{H}_{1+} = \frac{K\,d\xi}{2\pi r_1}\,\vec{i} \tag{S9.1}$$

where the field direction is shown in the figure and $K = NI/(a_2 - a_1)$. Note that Eq. $S9.1$ is an expression of the law of Biot and Savart. The y-component of field contributed by the entire surface current, from $\xi = a_1$ to $\xi = a_2$, is given by:

$$H_{y1+} = -\frac{K}{2\pi} \int_{a_1}^{a_2} \frac{\cos\theta_1\,d\xi}{r_1} \tag{S9.2}$$

By inserting $r_1 = \sqrt{(\xi - x)^2 + (c - y)^2}$ and $\cos\theta_1 = (\xi - x)/r_1$ into Eq. $S9.2$, we obtain:

$$H_{y1+} = -\frac{K}{2\pi} \int_{a_1}^{a_2} \frac{(\xi - x)\,d\xi}{(\xi - x)^2 + (c - y)^2} = -\frac{K}{4\pi} \ln\left[\frac{(a_2 - x)^2 + (c - y)^2}{(a_1 - x)^2 + (c - y)^2}\right] \tag{S9.3a}$$

Similarly, the contribution from each of the three remaining current sheets is given by:

$$H_{y1-} = -\frac{K}{2\pi} \int_{a_1}^{a_2} \frac{(\xi + x)\,d\xi}{(\xi + x)^2 + (c - y)^2} = -\frac{K}{4\pi} \ln\left[\frac{(a_2 + x)^2 + (c - y)^2}{(a_1 + x)^2 + (c - y)^2}\right] \tag{S9.3b}$$

$$H_{y2+} = -\frac{K}{2\pi} \int_{a_1}^{a_2} \frac{(\xi - x)\,d\xi}{(\xi - x)^2 + (c + y)^2} = -\frac{K}{4\pi} \ln\left[\frac{(a_2 - x)^2 + (c + y)^2}{(a_1 - x)^2 + (c + y)^2}\right] \tag{S9.3c}$$

$$H_{y2-} = -\frac{K}{2\pi} \int_{a_1}^{a_2} \frac{(\xi + x)\,d\xi}{(\xi + x)^2 + (c + y)^2} = -\frac{K}{4\pi} \ln\left[\frac{(a_2 + x)^2 + (c + y)^2}{(a_1 + x)^2 + (c + y)^2}\right] \tag{S9.3d}$$

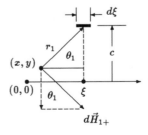

Fig. 3.16   Field produced by a differential current element.

Solution to Problem 3.9

Combining these four contributions, we have:

$$H_y(x,y) = -\frac{K}{4\pi}\left\{\ln\left[\frac{(a_2-x)^2+(c-y)^2}{(a_1-x)^2+(c-y)^2}\right] + \ln\left[\frac{(a_2+x)^2+(c-y)^2}{(a_1+x)^2+(c-y)^2}\right]\right.$$

$$\left. + \ln\left[\frac{(a_2-x)^2+(c+y)^2}{(a_1-x)^2+(c+y)^2}\right] + \ln\left[\frac{(a_2+x)^2+(c+y)^2}{(a_1+x)^2+(c+y)^2}\right]\right\} \quad (3.42a)$$

By inserting $x = 0$ and $y = 0$ into Eq. 3.42a, we obtain:

$$H_y(0,0) = -\frac{K}{\pi}\ln\left(\frac{a_2^2+c^2}{a_1^2+c^2}\right) \quad (3.42b)$$

b) H_{y1+} derived above (Eq. $S9.3a$) may be expressed as:

$$H_{y1+} = -\frac{K}{4\pi}\ln\left[\frac{(a_2-x)^2+(c-y)^2}{(a_1-x)^2+(c-y)^2}\right]$$

$$= -\frac{K}{4\pi}\left\{\ln[(a_2-x)^2+(c-y)^2] - \ln[(a_1-x)^2+(c-y)^2]\right\} \quad (S9.4)$$

The term $\ln[(a_2-x)^2+(c-y)^2]$ may be given as:

$$\ln[(a_2-x)^2+(c-y)^2] = \ln\left[(a_2^2+c^2)\left(1+\frac{x^2+y^2-2a_2x-2cy}{a_2^2+c^2}\right)\right]$$

$$= \ln(a_2^2+c^2) + \ln\left(1+\frac{x^2+y^2-2a_2x-2cy}{a_2^2+c^2}\right) \quad (S9.5)$$

By using $\ln(1+x) \simeq x$ for $x \ll 1$, we have:

$$\ln[(a_2-x)^2+(c-y)^2] \simeq \ln(a_2^2+c^2) + \frac{x^2+y^2-2a_2x-2cy}{a_2^2+c^2} \quad (S9.6)$$

Similarly,

$$\ln[(a_1-x)^2+(c-y)^2] \simeq \ln(a_1^2+c^2) + \frac{x^2+y^2-2a_1x-2cy}{a_1^2+c^2} \quad (S9.7)$$

Thus:

$$H_{y1+}(x,y) \simeq -\frac{K}{4\pi}\left[\ln\left(\frac{a_2^2+c^2}{a_1^2+c^2}\right)\right.$$

$$\left. + \frac{x^2+y^2-2a_2x-2cy}{a_2^2+c^2} - \frac{x^2+y^2-2a_1x-2cy}{a_1^2+c^2}\right] \quad (S9.8a)$$

Solution to Problem 3.9

Similarly, H_{y1-}, H_{y2+}, and H_{y2-} may be expressed as:

$$H_{y1-}(x,y) \simeq -\frac{K}{4\pi}\left[\ln\left(\frac{a_2^2+c^2}{a_1^2+c^2}\right)\right.$$
$$\left.+\frac{x^2+y^2+2a_2x-2cy}{a_2^2+c^2}-\frac{x^2+y^2+2a_1x-2cy}{a_1^2+c^2}\right] \quad (S9.8b)$$

$$H_{y2+}(x,y) \simeq -\frac{K}{4\pi}\left[\ln\left(\frac{a_2^2+c^2}{a_1^2+c^2}\right)\right.$$
$$\left.+\frac{x^2+y^2-2a_2x+2cy}{a_2^2+c^2}-\frac{x^2+y^2-2a_1x+2cy}{a_1^2+c^2}\right] \quad (S9.8c)$$

$$H_{y2-}(x,y) \simeq -\frac{K}{4\pi}\left[\ln\left(\frac{a_2^2+c^2}{a_1^2+c^2}\right)\right.$$
$$\left.+\frac{x^2+y^2+2a_2x+2cy}{a_2^2+c^2}-\frac{x^2+y^2+2a_1x+2cy}{a_1^2+c^2}\right] \quad (S9.8d)$$

Combining each term, near $(0,0)$ we have:

$$H_y(x,y) \simeq -\frac{K}{\pi}\left[\ln\left(\frac{a_2^2+c^2}{a_1^2+c^2}\right)-\frac{a_2^2-a_1^2}{(a_2^2+c^2)(a_1^2+c^2)}(x^2+y^2)\right]$$
$$\simeq H_y(0,0)+\frac{K(a_2^2-a_1^2)}{\pi(a_2^2+c^2)(a_1^2+c^2)}(x^2+y^2) \quad (3.43)$$

Note that $H_y(x,y)$ has zero slope with respect to both x and y at $(0,0)$. In other words, the magnetic field is fairly uniform at the magnet center.

c)   We can further simplify the expression for the magnetic field near the center of the racetrack magnet by approximating each of the four current sheets as a current element carrying NI, illustrated in Fig. 3.15. In this case, we let $a_1 = a$, $a_2 = a+\epsilon$, and $K\epsilon = K(a_2-a) = NI$. Substituting these parameters into Eq. 3.42 and noting $\ln x = x$ for $x \ll 1$, we have:

$$H_y(0,0) = -\frac{K}{\pi}\ln\left(\frac{a_2^2+c^2}{a_1^2+c^2}\right)$$
$$= -\frac{K}{\pi}\ln\left(\frac{a^2+c^2+2a\epsilon}{a^2+c^2}\right) = -\frac{K2a\epsilon}{\pi(a^2+c^2)}$$
$$= -\frac{2aNI}{\pi(a^2+c^2)} \quad (3.44a)$$

Solution to Problem 3.9

The quantity $K(a_2^2 - a_1^2) = K(a_2 + a_1)(a_2 - a_1)$ in the second term of the right-hand side of Eq. 3.43 becomes $2aNI$. Thus:

$$\frac{K(a_2^2 - a_1^2)}{\pi(a_2^2 + c^2)(a_1^2 + c^2)}(x^2 + y^2) = \frac{2aNI}{\pi(a^2 + c^2)^2}(x^2 + y^2)$$

$$= -H_y(0,0)\left(\frac{x^2 + y^2}{a^2 + c^2}\right) \qquad (S9.9)$$

Combining Eqs. 3.43 and S9.9, we obtain:

$$H_y(x,y) \simeq H_y(0,0)\left(1 - \frac{x^2 + y^2}{a^2 + c^2}\right) \qquad (3.44b)$$

d) The force on element 1 by element 2, $\vec{F}_1|_2$, is $+x$-directed and given by:

$$\vec{F}_1|_2 = \frac{\mu_o I_1 I_2}{4\pi a}\vec{i}_x = \frac{\mu_o N^2 I^2}{4\pi a}\vec{i}_x \qquad (S9.10)$$

Similarly, the force on element 1 by element 3, $\vec{F}_1|_3$, is $-y$-directed and given by:

$$\vec{F}_1|_3 = -\frac{\mu_o I_1 I_3}{4\pi c}\vec{i}_y = -\frac{\mu_o N^2 I^2}{4\pi c}\vec{i}_y \qquad (S9.11)$$

The force on element 1 by element 4, $\vec{F}_1|_4$, is directed along both x and y axes and given by:

$$\vec{F}_1|_4 = \frac{\mu_o I_1 I_4}{4\pi\sqrt{a^2 + c^2}}\left(\frac{a}{\sqrt{a^2 + c^2}}\vec{i}_x + \frac{c}{\sqrt{a^2 + c^2}}\vec{i}_y\right)$$

$$= \frac{\mu_o N^2 I^2}{4\pi}\left(\frac{a}{a^2 + c^2}\vec{i}_x + \frac{c}{a^2 + c^2}\vec{i}_y\right) \qquad (S9.12)$$

The x- and y-components of the electromagnetic force on element 1 from the three other current elements, F_{1x} and F_{1y}, are thus given, respectively, by:

$$F_{1x} = \frac{\mu_o N^2 I^2}{4\pi}\left(\frac{1}{a} + \frac{a}{a^2 + c^2}\right) \qquad (3.46a)$$

$$F_{1y} = -\frac{\mu_o N^2 I^2}{4\pi}\left(\frac{1}{c} - \frac{c}{a^2 + c^2}\right) \qquad (3.46b)$$

e) Because $c^2 < a^2 + c^2$, F_{1y} points in the $-y$-direction. The net force between elements 1 and 2, within racetrack 1, is repulsive because their current polarities are opposite. Similarly, that between elements 3 and 4 in racetrack 2 is also repulsive. This means that in the absence of external restraint, each racetrack seeks a circular geometry.

The net force between elements 1 and 3, because their polarities are the same, is attractive. Similarly, that between elements 2 and 4 is attractive. As indicated by Eq. 3.46b, the net force between the two racetracks is attractive.

Problem 3.10: Ideal toroidal magnet

This problem deals with an ideal toroidal magnet, illustrating key features of toroidal magnets.

An ideal circular-cross-section torus of major radius R and minor radius a is energized with a surface current sheet with equivalent total ampere turns NI (Fig. 3.17). Assume the surface current to occupy zero thickness and to flow around the torus in the plane purely perpendicular to the toroidal direction.

a) Show that an expression for the toroidal magnetic induction, B_φ, within the torus is given by:

$$B_\varphi = \frac{\mu_\circ NI}{2\pi r} \tag{3.47}$$

Also show that B_φ outside the torus is zero.

b) Assuming that the torus consists of N coils, each carrying a current of I, show that an expression for the net radial Lorentz force F_{L+} acting on a single coil is given by:

$$F_{L+} = \mu_\circ NI^2 \left(1 - \frac{R}{\sqrt{R^2 - a^2}}\right) \tag{3.48}$$

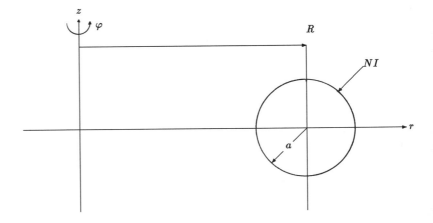

Fig. 3.17   Ideal toroidal magnet.

Solution to Problem 3.10

a) Noting that H_φ, from symmetry, is constant in the φ-direction, we can apply Ampere's integral law within the torus:

$$\int_0^{2\pi} H_\varphi r\, d\varphi = 2\pi r H_\varphi = NI \qquad (S10.1)$$

$$B_\varphi = \mu_0 H_\varphi = \frac{\mu_0 NI}{2\pi r} \qquad (3.47)$$

Outside the torus, no net current is enclosed when the above integral is performed over the entire circumference. Therefore $H_\varphi = 0$ and $B_\varphi = 0$.

b) Figure 3.18 shows a single coil in which differential force $d\vec{F}_L$ acting on differential element $d\vec{s}$ with differential force dF_{Lr} in the r-direction. $d\vec{F}_L$ is given by:

$$d\vec{F}_L = -I\, ds\, \vec{i}_\theta \times B_\varphi(r)\vec{i}_\varphi = \frac{\mu_0 NI^2 ds}{2\pi r}\, \vec{i}_\xi \qquad (S10.2)$$

where the vector $\vec{i}_\xi$ points in the direction of $\vec{F}_L$ (Fig. 3.18). The r-component of this differential force is given by:

$$dF_{Lr} = \frac{\mu_0 NI^2 \cos\theta\, ds}{2\pi r} \qquad (S10.3)$$

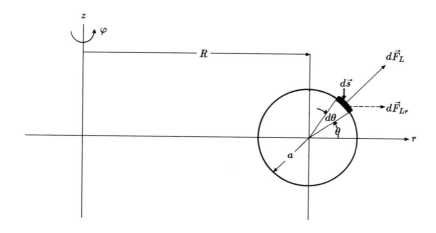

Fig. 3.18   Differential force acting on a single coil.

Solution to Problem 3.10

Because $ds = a\,d\theta$ and $r = R + a\cos\theta$, we can write Eq. $S10.3$ for dF_{Lr} as:

$$dF_{Lr} = \frac{\mu_o N I^2 a \cos\theta\,d\theta}{2\pi(R + a\cos\theta)} \qquad (S10.4)$$

By integrating Eq. $S10.4$ for the entire minor circle, we have:

$$F_{Lr} = \frac{\mu_o N I^2 a}{2\pi}\int_0^{2\pi}\frac{\cos\theta\,d\theta}{R + a\cos\theta} = \frac{\mu_o N I^2 a}{\pi}\int_0^{\pi}\frac{\cos\theta\,d\theta}{R + a\cos\theta} \qquad (S10.5)$$

Using a *Tables of Integral*, we obtain:

$$F_{Lr} = \frac{\mu_o N I^2 a}{\pi}\left(\left.\frac{\theta}{a}\right|_0^{\pi} - \frac{R}{a}\int_0^{\pi}\frac{d\theta}{R + a\cos\theta}\right) \qquad (S10.6)$$

$$= \frac{\mu_o N I^2 a}{\pi}\left\{\frac{\pi}{a} - \frac{2R}{a\sqrt{R^2 - a^2}}\tan^{-1}\left[\frac{(R-a)\tan\frac{\theta}{2}}{\sqrt{R^2 - a^2}}\right]_0^{\pi}\right\} \qquad (S10.7)$$

$$F_{Lr} = \mu_o N I^2\left(1 - \frac{R}{\sqrt{R^2 - a^2}}\right) \qquad (3.48)$$

Note that as $R \to \infty$, the torus becomes a straight solenoid of diameter $2a$ and, as expected, $F_{Lr} \to 0$.

Nuclear Fusion and Magnetic Confinement

If nuclei of light elements are confined and heated to a very high temperature ($\sim$100 MK), they fuse. Because the total mass of the fusion products, M_f, is less than the total mass of the original nuclei, M_n, a net energy $E_n = (M_n - M_f)c^2$ is released by the reaction. (c is the speed of light.) The sun generates energy through this process. A controlled thermonuclear fusion reactor is a miniature man-made sun. The sun, through its enormous size and mass, loses relatively little energy by radiation and confines unstable hot plasma gravitationally. Magnetic pressure can substitute for the gravitational pressure; the technique of using magnetic fields to stabilize hot plasma is known as magnetic confinement.

Power-generating fusion reactors will most likely use the Tokamak, a toroidal-shaped machine configuration that uses magnetic confinement. The tokamak was conceived in the 1950s by L.A. Artsimovich and A.D. Sakharov of the Kurchatov Institute of Atomic Energy, Moscow. Recently, a program known as the International Thermonuclear Experimental Reactor (ITER) was launched by the European Union, Japan, Russia, and the United States. ITER's ultimate goal is to construct a break-even Tokamak based on superconducting magnets. The ITER's toroidal magnet, not circular as studied above but D-shaped, will have a major radius (R) of $\sim$8 m and be $\sim$12 m tall ($2a$ in the z-direction); its toroidal magnetic induction (B_φ) is $\sim$6 T, with a peak induction at the conductor of $\sim$13 T.

Problem 3.11: Fringing field

This problem deals with the fringing field—an unwanted field outside a magnet system. The fringing field is important because it presents a safety hazard to those working on the system; it may also disrupt or distort field-sensitive equipment. For computing the fringing field, $\vec{H}_f$, at locations *far* from the magnet, the magnet can be modeled as a spherical dipole with an effective radius R_e:

$$\vec{H}_f = H_0 \left(\frac{R_e}{r}\right)^3 (\cos\theta\,\vec{i}_r + \tfrac{1}{2}\sin\theta\,\vec{i}_\theta) \tag{3.49}$$

where $\mu_0 H_0$, the induction at the center, is 12.2 T for Hybrid III at $I_{op} = 2100\,\text{A}$. We may compute R_e by noting that the dipole's far field $(r \gg R_e)$ along the z-axis (r-direction at $\theta = 0$), given by Eq. 3.49, and that $(z \gg a)$ of a ring of current I and radius a, given by Eq. 3.3, are equal and thus:

$$H_0 R_e^3 = \tfrac{1}{2}a^2 I \tag{3.50}$$

For a winding having a total ampere-turn of NI_{op}, i.d. of $2a_1$ and o.d. of $2a_2$, Eq. 3.50 may be given by:

$$H_0 R_e^3 = \tfrac{1}{4}(a_2^2 + a_1^2)NI_{op} \tag{3.51a}$$

Note that R_e is independent of magnet length, $2b$, and because H_0 is proportional to NI_{op}, it is also independent of NI_{op}; it depends purely on magnet geometry. For the Hybrid III SCM comprised of four winding sections, we apply Eq. 3.51a:

$$H_0 R_e^3 = \tfrac{1}{4}I_{op}\sum_{j=1}^{4}(a_{2_j}^2 + a_{1_j}^2)N_j \tag{3.51b}$$

By inserting parameter values given in Table 3.1 and solving Eq. 3.51b for R_e, we obtain: $R_e = 0.4\,\text{m}$. (Because the effective radius R_e for the insert in Hybrid III is much smaller than that of the superconducting magnet, the insert contributes little to the fringing fields and is neglected in the computation of R_e.)

a) The computers used for monitoring and controlling Hybrid III are located at approximately $x = 8.8\,\text{m}$, $y = 0\,\text{m}$, and $z = 2.0\,\text{m}$, with the origin $(0,0,0)$ located at the magnet center. Compute the magnitude $|\vec{H}_f|$ at the computers for $\mu_0 H_0 = 12.2\,\text{T}$ and $R_e = 0.4\,\text{m}$. Note that the θ-direction is measured from the z-axis; that is, the magnet midplane is at $\theta = 90°$ (see Fig. 3.8).

b) A computer monitor display begins to be distorted when the monitor is exposed to a field as low as $\sim$2 oersted (equivalent to a magnetic induction of 0.2 mT). Compute an approximate value of I at which the Hybrid III monitors will begin to distort.

c) If the computer is to be shielded against this fringing field for currents up to 2100 A, give an estimate of the iron (as-cast) sheet that will surround the computer over five of its six surfaces. Use the magnetic properties of as-cast iron given in Table 2.4 (p. 31).

Solution to Problem 3.11

a)   First, we must find appropriate values of r and θ corresponding to $(x, y, z) = (x = 8.8\,\text{m}, y = 0, z = 2.0\,\text{m})$:

$$r = \sqrt{x^2 + y^2 + z^2} \qquad\qquad (S11.1)$$

$$= \sqrt{(8.8\,\text{m})^2 + 0 + (2.0\,\text{m})^2}$$

$$= 9.0\,\text{m}$$

$$\theta = \tan^{-1}\left(\frac{8.8}{2.0}\right) = 77.2° \qquad\qquad (S11.2)$$

Thus for $I_{op} = 2100$ A, we have:

$$\mu_0 \vec{H}_f = \vec{B}_f = (12.2\,\text{T})\left(\frac{0.4\,\text{m}}{9.0\,\text{m}}\right)^3 \left[\cos(77.2°)\,\vec{\imath}_r + \tfrac{1}{2}\sin(77.2°)\,\vec{\imath}_\theta\right] \quad (S11.3)$$

$$\simeq (10.7\times10^{-4}\,\text{T})(0.222\,\vec{\imath}_r + 0.488\,\vec{\imath}_\theta)$$

$$|\vec{B}_f| = (10.7\times10^{-4}\,\text{T})\sqrt{(0.222)^2 + (0.488)^2} \simeq 5.7\times10^{-4}\,\text{T} \ (S11.4)$$

$$= 0.57\,\text{mT} \ (5.7\,\text{gauss})$$

b)   Because $I_{op} = 2100$ A gives rise to a magnetic field of 5.7 oersted (or a magnetic induction of 5.7 gauss), a field of $\sim$2 oersted is reached at $\sim$740 A. The screens of the Hybrid III computers indeed begin to be distorted at around 700 A.

c)   As computed in a), a fringing field reaching the computer at 2100 A will be 5.7 oersted (or an induction of 5.7 gauss). At this field level, from Table 2.4, as-cast iron is still quite effective as shielding material.

We approximate this shielding to be of spherical shape of radius R of 25 cm. From Eq. 2.44 (p. 26):

$$\frac{d}{R} \geq \frac{3H_0}{2M_{sa}} \qquad\qquad (2.44)$$

Solving Eq. 2.44 for d, we can compute a minimum thickness d_{mn} for the sheet ($\mu_0 M_{sa} = 1.65$ T from Table 2.4):

$$d_{mn} = \frac{3\mu_0 H_0}{2\mu_0 M_{sa}}R$$

$$= \frac{3(5.7\times10^{-4}\,\text{T})}{2(1.65\,\text{T})}(0.25\,\text{m}) = 1.3\times10^{-4}\,\text{m}$$

$$\simeq 0.1\,\text{mm}$$

The as-cast iron sheets of $\sim$1-mm thickness often used in the laboratory to shield computers and oscilloscopes are thus quite adequate for external magnetic inductions up to $\sim$100 gauss—at this field the material is still effective $(\mu/\mu_0)_{dif} = 30$.

Problem 3.12: Circulating proton in an accelerator

The Superconducting Supercollider (SSC), terminated in 1993, was to have been the largest (super) "atom smasher" (collider) for high-energy physics research. The prefix superconducting was used because its nearly 10,000 dipole and quadrupole magnets were to be superconducting. (The SSC would have been the largest single consumer of superconductor in history—nearly a thousand tons of Nb-Ti multifilamentary conductor.)

The SSC's main ring was to have two counter-circulating beams of protons, each accelerated to an energy (E_p) of 20 TeV.

a) Assuming that the main ring is perfectly circular, compute the ring's radius R_a for a proton circulating with an energy E_p of 20 TeV in the presence of a vertical magnetic induction (B_z) of 5 T. You may also assume that the proton speed is equal to the speed of light. Note: $1\,\text{eV} = 1.6 \times 10^{-19}\,\text{J}$ (Appendix I).

b) Show that the proton speed at an energy of 20 TeV is indeed very close to the speed of light.

Particle Accelerators

The simple principle that an electric field $(\vec{E})$ accelerates charged particles is the basis for particle accelerators. Early machines of Cockroft-Walton (1928) and Van de Graaff (1930) were linear, accelerating particles along a straight path over a potential $(\int \vec{E} \cdot d\vec{s})$. A large potential is always required to produce highly energetic particles. A linear accelerator thus requires either a large $\vec{E}$ field, a long distance, or a combination of both. The Stanford Linear Accelerator ($\sim$20 GeV) has a beam distance of 2 miles (3.2 km).

E.O. Lawrence in the 1930s developed the cyclotron, a circular accelerator. Modern circular accelerators are variations of Lawrence's cyclotron. In a circular accelerator, charged particles are accelerated by a modest potential each time they revolve around the machine; by circulating them many times it is possible to energize them to energy levels well beyond those achievable by linear accelerators. One essential component of a circular accelerator is a set of magnets that supplies a magnetic field (usually in the vertical direction) to bend the particles into a circular trajectory; modern machines use dipole magnets, while Lawrence's first 1.2-MeV cyclotron used magnet polepieces, which sandwiched the beam trajectory.

As shown in the next page, the particle energy (E_p) in circular accelerators is proportional to the beam trajectory's radius (R_a), beam velocity, and vertical magnetic induction (B_z). For a particle energy of 20 TeV proposed for the SSC, it meant—also shown in the next page—a machine radius exceeding 10 km! Compare this with $\sim$0.1 m, the radius of Lawrence's first cyclotron. If SSC were designed to use $\sim$1 T, the strength of B_z in Lawrence's first cyclotron, the factor of $\sim$10^5 increase in radius and increase in beam velocity would still bring E_p to only $\sim$4 TeV. For SSC, the designers proposed to gain another factor of $\sim$5 needed to reach the energy level of 20 TeV through increased field strength, a goal achievable only by the use of superconducting dipole magnets.

Solution to Problem 3.12

a)   The centripetal force, $\vec{F}_{cp}$, on a circulating proton (mass M_p) is balanced by the Lorentz force, $\vec{F}_L$. The direction of B_z is chosen to make F_L points radially inward because F_{cp} always points radially outward. The two forces are given by:

$$\vec{F}_{cp} = \frac{M_p v^2}{R_a} \vec{i}_r \simeq \frac{M_p c^2}{R_a} \vec{i}_r \qquad (S12.1)$$

$$= \frac{E_p}{R_a} \vec{i}_r \qquad (S12.2)$$

$$\vec{F}_L = -qcB_z \vec{i}_r \qquad (S12.3)$$

Solving for R_a from $\vec{F}_{cp} + \vec{F}_L = 0$, we obtain:

$$R_a = \frac{E_p}{qcB_z} \qquad (S12.4)$$

From Eq. $S12.4$, we have:

$$R_a = \frac{(1.6 \times 10^{-19}\,\text{J/eV})(20 \times 10^{12}\,\text{eV})}{(1.6 \times 10^{-19}\,\text{C})(3 \times 10^8\,\text{m/s})(5\,\text{T})} \simeq 1.33 \times 10^4\,\text{m} \qquad (S12.5)$$

$$\simeq 13.3\,\text{km}$$

The value of $26.6\,\text{km}$ ($2R_a$) is close to the diameter of the SSC's main ring. Also note that by making $B_m = 10\,\text{T}$, the ring diameter can be halved; 10-T superconducting dipole magnets are not out of the question [3.29].

b)   The proton mass M_p, travelling at speed v, is related to its rest mass, M_{p_o} (1.67×10^{-27} kg), by:

$$M_p = \frac{M_{p_o}}{\sqrt{1 - \left(\dfrac{v}{c}\right)^2}} \qquad (S12.6)$$

$$= \frac{E_p}{c^2} \qquad (S12.7)$$

Solving for v/c from Eqs. $S12.6$ and $S12.7$, we have:

$$1 - \left(\frac{v}{c}\right)^2 = \frac{M_{p_o}^2 c^4}{E_p^2} \qquad (S12.8)$$

$$\frac{v}{c} = \sqrt{1 - \frac{M_{p_o}^2 c^4}{E_p^2}} \qquad (S12.9)$$

Because v/c is very close to 1, Eq. $S12.9$ may be approximated by:

$$\frac{v}{c} \simeq 1 - \frac{M_{p_o}^2 c^4}{2E_p^2} = 1 - \frac{(1.67 \times 10^{-27}\,\text{kg})^2 (3 \times 10^8\,\text{m/s})^4}{2(1.6 \times 10^{-19}\,\text{eV})^2 (20 \times 10^{12}\,\text{J/eV})^2} \qquad (S12.10)$$

$$\simeq 1 - 1 \times 10^{-9}$$

That is, the proton velocity is within one part per billion of the speed of light.

Problem 3.13: Magnetic force on an iron sphere

For safety considerations, it is extremely important to keep ferromagnetic objects away from a large magnet. This problem presents a quantitative approach to computing the magnetic force on an iron sphere placed *far* from a solenoidal magnet.

As discussed in Problem 3.11, the fringing (*far*) field of a solenoidal magnet is given approximately by a dipole field:

$$\vec{H}_f = H_0 \left(\frac{R_e}{r}\right)^3 (\cos\theta\, \vec{i}_r + \tfrac{1}{2}\sin\theta\, \vec{i}_\theta) \tag{3.49}$$

$$= H_{f_r}\, \vec{i}_r + H_{f_\theta}\, \vec{i}_\theta \tag{3.52}$$

where, as in Problem 3.11, for Hybrid III $\mu_o H_0 = 12.2\,\text{T}$ at $I_{op} = 2100\,\text{A}$ and $R_e = 0.4\,\text{m}$.

When a magnetic object, such as an iron sphere, is placed in a magnetic field that is spatially varying, the object will be subjected to a net magnetic force density, $\vec{f}_m$, given by:

$$\vec{f}_m(r,\theta) = \nabla e_m \tag{3.53}$$

where ∇ is the grad operator in spherical coordinates and e_m is the magnetic energy density stored in the iron due to its magnetization. We also know from Problem 2.1 that for a ferromagnetic sphere with $\mu/\mu_o \gg 1$, the magnetic induction inside the sphere, $\vec{B}_{sp}$, is three times that of the "uniform" applied magnetic induction: $\vec{B}_{sp} \simeq 3\mu_o\vec{H}_f = 3\vec{B}_f$. For a sphere whose diameter is much smaller than the distance from the magnet center to the sphere, we may assume $\vec{B}_f$ to be uniform over the sphere. Thus:

$$e_m = \frac{\vec{B}_{sp} \cdot \vec{B}_f}{2\mu_o} = \frac{3|\vec{B}_f|^2}{2\mu_o} \tag{3.54}$$

When the iron is saturated with magnetization $\vec{M}_{sa}$, its magnetic induction, because $\vec{M}_{sa} \gg \vec{H}_f$, is approximately equal to $\vec{B}_{sa}$ ($= \mu_o\vec{M}_{sa}$), which is constant and aligned with $\vec{B}_f$. Its energy density e_{ms} is thus given by:

$$e_{ms} \simeq \frac{\vec{B}_{sa} \cdot \vec{B}_f}{2\mu_o} \tag{3.55}$$

In driving Eqs. 3.54 and 3.55, we have assumed that the impinging field is "uniform" for energy density computation, retaining the spatial variation of $\vec{B}_f$ for force density computation, which comes next.

a) Show that $\vec{f}_m(r,\theta)$ acting on the unsaturated iron sphere is given by:

$$\vec{f}_m(r,\theta) = -\frac{9\mu_o H_0^2}{4R_e}\left(\frac{R_e}{r}\right)^7 \left[(1 + 3\cos^2\theta)\vec{i}_r + \sin\theta\cos\theta\,\vec{i}_\theta\right] \tag{3.56}$$

Problem 3.13: Magnetic force on an iron sphere

Note that $\vec{f}_m(r,\theta)$ varies as $1/r^7$ and as the minus sign indicates, as expected for any ferromagnetic object, the r-component of $\vec{f}_m(r,\theta)$ is directed *towards* the magnet center.

b) Show that $\vec{f}_{ms}(r,\theta)$ acting on the saturated iron sphere is given by:

$$\vec{f}_{ms}(r,\theta) = -\frac{3\mu_\circ M_{sa} H_0}{R_e}\left(\frac{R_e}{r}\right)^4$$

$$\times\left(\sqrt{\cos^2\theta + \tfrac{1}{4}\sin^2\theta}\,\vec{i}_r + \frac{\sin\theta\cos\theta}{4\sqrt{\cos^2\theta + \tfrac{1}{4}\sin^2\theta}}\,\vec{i}_\theta\right) \quad (3.57)$$

Note that $\vec{f}_{ms}$ varies as $1/r^4$ and its r-component is also directed towards the magnet center.

c) The Hybrid III platform, installed for use by experimenters, is located at $z = 1\,\text{m}$, measured from the magnet center. Compute $y_{0.1g}$, the distance from the Hybrid III center line on the platform, at which $|f_{my}|$, the magnitude of the y-component of the magnetic force at $z = 1\,\text{m}$ acting on the iron sphere of density ϱ, is $0.1\varrho g$, the equivalent force for 0.1 "gee" (Fig. 3.19). Assume the iron sphere to be unsaturated at this location. Use the following values: $\mu_\circ H_0 = 12.2\,\text{T}$ (at $I_{op} = 2100\,\text{A}$); $R_e = 0.4\,\text{m}$; $\varrho = 8000\,\text{kg/m}^3$ (iron); $g = 9.8\,\text{m/s}^2$.

d) Show that at $y_{0.1g}$ computed above for the iron sphere is still unsaturated. Assume $(\mu/\mu_\circ)_{dif} \sim 10$ for iron at this location.

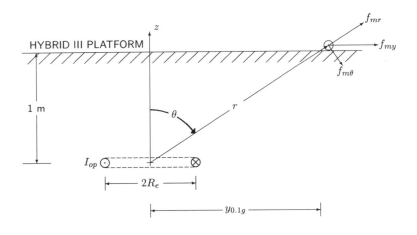

Fig. 3.19   Iron sphere placed on the Hybrid III platform.

Solution to Problem 3.13

a) We first compute $|\vec{B}_f|^2$ from Eq. 3.49:

$$|\vec{B}_f|^2 = \mu_\circ^2 H_0^2 \left(\frac{R_e}{r}\right)^6 (\cos^2\theta + \tfrac{1}{4}\sin^2\theta) \qquad (S13.1)$$

Combining Eqs. $S13.1$, 3.53, and 3.54 and using the grad operator in spherical coordinates (Eq. 2.36), we obtain:

$$\vec{f}_m(r,\theta) = \frac{3\mu_\circ H_0^2}{2}\left[(\cos^2\theta + \tfrac{1}{4}\sin^2\theta)\frac{\partial}{\partial r}\left(\frac{R_e}{r}\right)^6 \vec{i}_r\right.$$
$$\left.+ \frac{1}{r}\left(\frac{R_e}{r}\right)^6\frac{\partial}{\partial\theta}\left(\cos^2\theta + \tfrac{1}{4}\sin^2\theta\right)\vec{i}_\theta\right] \qquad (S13.2a)$$

$$= \frac{3\mu_\circ H_0^2}{2R_e}\left(\frac{R_e}{r}\right)^7\left[-\tfrac{6}{4}(4\cos^2\theta + \sin^2\theta)\vec{i}_r\right.$$
$$\left.+ \tfrac{1}{4}(-8\cos\theta\sin\theta + 2\sin\theta\cos\theta)\vec{i}_\theta\right] \qquad (S13.2b)$$

We can simplify Eq. $S13.2b$ to:

$$\vec{f}_m(r,\theta) = -\frac{9\mu_\circ H_0^2}{4R_e}\left(\frac{R_e}{r}\right)^7[(1 + 3\cos^2\theta)\vec{i}_r + \sin\theta\cos\theta\,\vec{i}_\theta] \qquad (3.56)$$

As stated above, for any θ, $\vec{f}_m(r,\theta)$ always has a $-r$-directed component, or ferromagnetic objects will be attracted towards the magnet center.

b) The magnetic energy density of Eq. 3.55 is given by:

$$e_{ms} = \mu_\circ M_{sa} H_0 \left(\frac{R_e}{r}\right)^3\sqrt{\cos^2\theta + \tfrac{1}{4}\sin^2\theta} \qquad (S13.3)$$

Performing a similar grad operation on e_m, we obtain:

$$\vec{f}_{ms}(r,\theta) = -\frac{3\mu_\circ M_{sa} H_0}{R_e}\left(\frac{R_e}{r}\right)^4$$
$$\times \left(\sqrt{\cos^2\theta + \tfrac{1}{4}\sin^2\theta}\,\vec{i}_r + \frac{\sin\theta\cos\theta}{4\sqrt{\cos^2\theta + \tfrac{1}{4}\sin^2\theta}}\vec{i}_\theta\right) \qquad (3.57)$$

The magnetic force thus varies as $1/r^4$ when the iron sphere is saturated. Note also that because it is $-r$-directed, as in the unsaturated case, the iron sphere is attracted to the magnet center.

Solution to Problem 3.13

c) f_{my}, the y-component of $\vec{f}_m$ at $(x = 0, y_{0.1g}, z)$, is given by:

$$f_{my} = f_{mr} \sin\theta + f_{m\theta} \cos\theta \qquad (S13.4)$$

where f_{mr} and $f_{m\theta}$ are, respectively, the r- and θ-components of the magnetic force. Combining Eq. 3.56 and $S13.4$, we have:

$$f_{my} = \frac{9\mu_o H_0^2}{4R_e}\left(\frac{R_e}{r}\right)^7 [-(1+3\cos^2\theta)\sin\theta - \sin\theta\cos^2\theta] \qquad (S13.5a)$$

$$= -\frac{9\mu_o H_0^2}{4R_e}\left(\frac{R_e}{r}\right)^7 (1+4\cos^2\theta)\sin\theta \qquad (S13.5b)$$

The minus sign in Eq. $S13.5b$ indicates f_{my} actually points in the direction opposite from that indicated in Fig. 3.19. r, $\sin\theta$, and $\cos\theta$ for the case $x = 0$ are given by:

$$r = \sqrt{y_{0.1g}^2 + z^2} \qquad \sin\theta = \frac{y_{0.1g}}{\sqrt{y_{0.1g}^2 + z^2}} \qquad \cos\theta = \frac{z}{\sqrt{y_{0.1g}^2 + z^2}} \qquad (S13.6)$$

Combining Eq. $S13.5b$ and $S13.6$ and equating f_{my} to $0.1\varrho g$, the equivalent force for 0.1 "gee," we have:

$$0.1\varrho g = \frac{9\mu_o H_0^2 R_e^6}{4}\left(\frac{1}{y_{0.1g}^2 + z^2}\right)^{3.5}\left(1+\frac{4z^2}{y_{0.1g}^2 + z^2}\right)\frac{y_{0.1g}}{\sqrt{y_{0.1g}^2 + z^2}} \qquad (S13.7)$$

By inserting appropriate values into Eq. $S13.7$, we have:

$$0.1(8000\,\mathrm{kg/m^3})(9.81\,\mathrm{m/s^2}) = \frac{9(12.2\,\mathrm{T})^2(0.4\,\mathrm{m})^6}{4(4\pi\times10^{-7}\,\mathrm{H/m})}\left[\frac{1}{y_{0.1g}^2 + (1.0\,\mathrm{m})^2}\right]^{3.5}$$

$$\times\left[1+\frac{4(1.0\,\mathrm{m})^2}{y_{0.1g}^2 + (1.0\,\mathrm{m})^2}\right]\frac{y_{0.1g}}{\sqrt{y_{0.1g}^2 + (1.0\,\mathrm{m})^2}} \qquad (S13.8)$$

Solving Eq. $S13.8$ for $y_{0.1g}$, we find: $y \simeq 1.93\,\mathrm{m}$.

d) With $r_{0.1g} = \sqrt{(1.93\,\mathrm{m})^2 + (1.0\,\mathrm{m})^2} \simeq 2.17\,\mathrm{m}$ and $\theta = \tan^{-1}(1.92\,\mathrm{m}/1\,\mathrm{m}) = 62.6°$ substituted into Eq. 3.49, we have:

$$|\mu_o\vec{H}_f| = (12.2\,\mathrm{T})\left(\frac{0.4\,\mathrm{m}}{1.93\,\mathrm{m}}\right)^3\sqrt{\cos^2 62.6° + \tfrac{1}{4}\sin^2 62.6°} \qquad (S13.9)$$

$$= (12.2\,\mathrm{T})(8.9\times10^{-3})(0.64)$$

$$= 0.070\,\mathrm{T}$$

At $\mu_o H_0 = 0.07\,\mathrm{T}$, as-cast iron has a $(\mu/\mu_o)_{dif}$ value of less than 7 (Table 2.4) or the analysis based on the assumption of unsaturated magnetization performed here is barely valid. This means, at $y_{0.1g} = 1.93\,\mathrm{m}$ from the magnet center line, the actual magnetic force is slightly less than 0.1 gee. Nevertheless, during Hybrid III operation no equipment, ferromagnetic or nonferromagnetic, is allowed on the platform within a circular boundary defined by this distance.

Problem 3.14: Fault condition in hybrid magnets

1. Fault-mode forces

Of the interaction problems between the water-cooled insert and the supercon-ducting magnet in a hybrid system, the potentially violent consequences of an insert failure have the greatest impact upon the design. As discussed in Problem 3.2, Bitter magnets are operated at very high power densities, in the range 1 to 10 GW/m^3, all of which must be balanced by cooling provided by high-velocity ($\sim$10 m/s) water flow. Moreover, Bitter plates are highly stressed, reaching stresses in the range from 300 to 600 MPa.

The designer is challenged to provide strength and cooling in a limited space which he would like to fill completely with current-carrying copper. Thus, the magnets are by no means conservative and, as stated before, an occasional failure is almost a fact of life. The system must be designed to withstand the disruptions caused by insert burnouts.

Fault Scenario

The structural support for Hybrid III, an overview of which is shown in Fig. 3.5, was designed according to the following postulated fault scenario:

1. At $t = 0$, with the superconducting magnet fully energized and generating a field of 12.2 T and the insert operating at the full current of 40 kA, an arc occurs at the midplane between the two Bitter stacks. (Normally, the two stacks are connected in series at the top, and the current flows helically upward through the inner one and downward through the outer one.) Now current is suddenly flowing through the arc instead of through the top halves of the insert.

2. At $t \sim 0.1$ s, the arc which has been travelling downward because of the axial voltage gradient between the coils reaches the bottom. Since the motor-generator power supply used in FBNML behaves as a voltage source, the decreasing resistance allows the current to increase manyfold.

3. At $t \sim 0.2$ s, the "bad news" has worked its way through the plant protection logic to initiate an emergency stop. The generator excitation is inverted to force the current down before the breakers open 0.3 s later. In that time, the current through the shorted insert can surge from 40 to 250 kA.

4. At $t \sim 0.5$ s, the generator voltage returns to zero.

The arc-induced short thus effectively halves the insert length, shifting the insert's magnetic center downward by one quarter the length of the insert.

a) Show that between two concentric nested solenoidal coils, an interaction force in the axial direction is restoring (stable). Assume the central axial field from each coil points upwards, in the $+z$-direction.

b) Show that between two concentric nested solenoidal coils, an interaction force in the radial direction is unstable.

Solution to Problem 3.14

a) Figure 3.20 shows an arrangement of two nested solenoidal coils, Coil 1 (inner) displaced axially upward by Δz with respect to Coil 2 (outer). Each coil generates an axial field that points upward.

Coil 2 generates radial field, $B_{r2}\,\vec{i}_r$, that points radially outward above its midplane and inward below its midplane. Note that $B_{r2}\,\vec{i}_r = 0$ along its axis ($r = 0$).

The net axial force acting on Coil 1 due to the $J_{\theta1} \times B_{r2}$ interaction, where $J_{\theta1}$, the $-\theta$-directed current density in Coil 1, is zero when the two coils are concentric and the midplane of each coil coincides.

When the midplane of Coil 1 is placed Δz above that of Coil 2, F_{z1} is given, algebraically, by:

$$F_{z1} =< \text{downward force on winding between 0 and } b + \Delta z >$$
$$- < \text{upward force on winding between } -b + \Delta z \text{ and } 0 > \quad (S14.1)$$

However, the net force generated in the upper winding between 0 and $b - \Delta z$ and that generated in the lower winding between $-b + \Delta z$ and 0 cancel each other. This leaves an unbalanced downward force from the upper winding between $b - \Delta z$ and $b + \Delta z$, given roughly by:

$$F_{z1} = -2\pi \int_{b-\Delta z}^{b+\Delta z} \int_{a_1}^{a_2} J_{\theta1}(r) B_{r2}(r, z) r^2 \, dr \, dz$$
$$\simeq -4\pi \Delta z \int_{a_1}^{a_2} J_{\theta1}(r) B_{r2}(r, b) r^2 \, dr \quad (S14.2)$$

Because F_{z1} is $-z$-directed, it is restoring (stable).

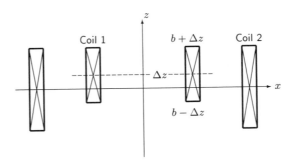

Fig. 3.20   Two nested solenoidal coils, one displaced axially from the other.

Solution to Problem 3.14

b)   When Coils 1 and 2 are concentric, the $J_\theta \times B_z$ interaction force is r-directed and the net radial force is zero. If Coil 1 is displaced by Δr in the x-direction ($\theta = 0$), then its winding section within an arc between $\theta \simeq -90°$ and $\theta \simeq +90°$ is exposed to B_{z2} that is slightly higher than that to which the winding section within an arc between $\simeq +90°$ and $\simeq 270°$ is exposed. The resulting net unbalanced force, F_{x1}, may be given approximately by:

$$F_{x1} \simeq 4\pi\Delta r \int_0^b \int_{a_1}^{a_2} J_{\theta1} \frac{\partial B_{z2}(r,z)}{\partial r} r^2 \, dr \, dz \qquad (S14.3)$$

Because F_{x1} is positive, it is unstable.

Another way of looking at this and the previous question is to recognize that an energized coil is always attracted to the highest field region. Thus, if Coil 1 is displaced axially by Δz, it seeks to align its maximum field region with that of Coil 2, resulting in a stable condition. If Coil 1 is displaced radially, however, it continues to move radially towards the innermost winding of Coil 2 because that is where the field generated by Coil 2 is highest.

Vertical Magnetic Force during Hybrid III Insert Burnout

A large magnetic interaction force, F_{fc}, appears in the vertical direction in Hybrid III during an insert burnout which follows the scenario outlined above. Figure 3.21 shows an analytical $F_{fc}(t)$ plot for an insert burnout event in Hybrid III when both are fully energized—the insert generating 22.7 T and superconducting magnet 12.3 T [3.30]. Note that a peak force of 1.1 MN is reached at 0.1 s after the start of the burnout, at which point the traveling arc reaches the bottom of the Bitter stack.

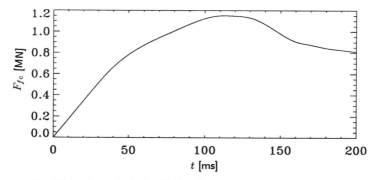

Fig. 3.21   Analytical $F_{fc}(T)$ plot during an insert burnout when Hybrid III is fully energized [3.30].

Problem 3.15: Fault condition in hybrid magnets

2. Mechanical support requirements

Figure 3.22 depicts the dynamic model of Hybrid III used to analyze responses of major components during an insert burnout fault [3.7, 3.30]. It shows the insert and the cryostat connected by a stretched spring representing the magnetic force between them during a burnout event. Motion in the system can be retarded by drag forces. The elements in the figure are as follows:

M_1: mass of the insert, 1000 kg;
M_2: mass of the superconducting magnet and cryostat, 5500 kg;
M_3: mass of the superconducting magnet (Nb_3Sn coil and Nb-Ti coil), 3200 kg;
M_4: mass of the cryostat mass, 2300 kg;
k_1: "magnetic" spring constant between the insert and the SCM, 9 MN/m;
k_2: spring constant of the cryostat support (stiff), 45 MN/m;
k_3: spring constant of the insert support (stiff), 15 MN/m;
D_1: friction drag force provided to retard the cryostat motion, 130 kN;
D_2: friction drag force (friction hinge in Fig. 3.5) for the insert motion, 45 kN.

a) Comment on the consequences of cryostat movement from an overall system consideration.

b) The insert and the cryostat are supported so as to allow them to move. Why?

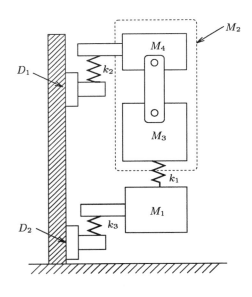

Fig. 3.22   Dynamic model of Hybrid III for fault analysis [3.7, 3.30]

Solution to Problem 3.15

The model is essentially a two-mass system connected by a spring. By allowing the masses (M_1 and M_2) to move under the action of the "magnetic" spring, the designer sought to greatly reduce the forces within the cryostat and the external support forces. Otherwise, the magnitude of the force would be several MN and incompatible with cryostat construction, where small cross sections are needed to minimize thermal conduction. It is necessary to extract energy from the oscillatory system in order to limit the motion excursion; it is also necessary to examine floor reactions to make sure the cryostat support structure remain in contact with the floor at all times. The only parameters within the designer's control are k_2, k_3, D_1, and D_2. The Hybrid III suspension system is designed to limit insert excursions to ~10 cm and those of the cryostat to ~2 cm and to dissipate energy through frictional drag devices provided for the purpose.

a) Obviously motion of the cryostat is undesirable, as it complicates the connection of various services and it interferes with the experiment occupying the magnet. Not only must motion be accommodated but the accommodation must anticipate the potential violence from large accelerations—close to 10 gee on the insert and 5 gee on the cryostat.

b) By allowing the two systems to move under the action of the magnetic spring, the designer reduced the forces within the cryostat and on the external support.

Figure 3.23 shows analytical plots of the insert (solid) and cryostat (dotted) displacements *vs* time during the first 0.2 s of an insert burnout based on the model shown in Fig. 3.22 [3.30]. As before, Hybrid III is fully energized. The plots show that the insert moves upward as much as ~10 cm, while the cryostat moves downward by ~1 cm. As remarked above, the system is designed for cryostat displacements up to about ±2 cm.

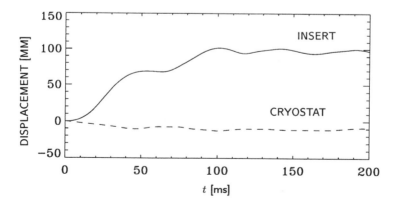

Fig. 3.23 Analytical plots of insert and cryostat locations during the first 0.2 s of a Hybrid III insert burnout event [3.30].

Problem 3.16: Fault condition in hybrid magnets
3. Fault force transmission

This problem considers a structural requirement that is unique to hybrid magnets; it is not present in either a resistive magnet or a superconducting magnet alone. Structure is needed to restrain the large axial interaction force that could arise between the resistive (insert) and the superconducting magnet in the event of an insert burnout. This interaction force is a severe complication, inasmuch as it appears in the cold mass (consisting of the superconducting magnet and cryostat), and therefore must also be transmitted over a large temperature difference with minimum heat transmission.

In the most severe burnout scenario, as described in Problem 3.14, an arc develops in the insert at its midplane, essentially shorting out half of the insert length. Suddenly, only one half is still carrying current and continues to generate field. This shift in current distribution displaces the magnetic center of the insert relative to that of the superconducting magnet, thereby creating a large restoring force between them. When Hybrid III is operating at its maximum field of 35 T, as shown in Fig. 3.21, a peak axial force of 1.1 N would result from such an insert burnout.

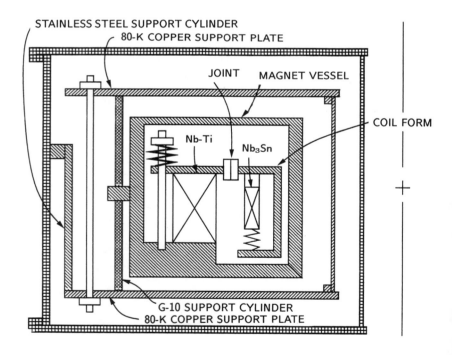

Fig. 3.24   Schematic structural diagram of the Hybrid III cryostat.

Problem 3.16: 3. Fault force transmission

Figure 3.24 is a schematic of the structure inside the Hybrid III cryostat. The center line of the cryostat bore is at the right; the figure thus represents only half of the cryostat cross section. As remarked earlier and illustrated in the figure, the superconducting magnet consists of two components, an inner Nb$_3$Sn coil placed in the bore of an outer coil wound with Nb-Ti conductor. The total interaction force is shared between them according to the system's overall field distribution. The basic structural problem is to firmly fix both coil windings to the magnet vessel which is ultimately anchored to a room temperature support, all with enough strength to withstand the maximum fault load.

As with the other hybrid magnets built at FBNML, the Hybrid III Nb-Ti coil is assembled from double pancakes, and the assembly is clamped by an array of tie rods around the outside. The tie rods clamp the stack of 32 double-pancakes between a stiff plate at the top and the floor of the magnet vessel. The purpose of the clamping is to provide a preload that will immobilize the winding as it is subjected both to its own Lorentz forces and to fault forces. To maintain the preload, a spring follow-up of the differing thermal contractions of the structural elements is necessary. Otherwise the clamping force can vary between zero and infinite as something "out-shrinks" something else. Elasticity is provided by stacks of cone-spring (Belleville) washers on the tie rods.

The Nb-Ti coil is thus fixed to the magnet vessel bottom and forces exerted upon the coil are transmitted directly to the magnet vessel without relative motion. The Nb$_3$Sn coil is wound on a form that is firmly attached to the magnet vessel, hence the winding has to remain one with the form throughout the same temperature excursions and forces. The coil form consists of a stainless steel tube with end flanges. To keep it from becoming either too tight or too loose between the flanges, the winding pack is spring loaded against the top flange. The spring load is set to have a force equal to its share of the total fault force. Mechanically this is accomplished with jacking screws and springs, preloading the winding against the top flange. The layer-wound coil is thus a self-contained unit. It is fixed to the magnet vessel indirectly in that its top flange is attached to the outer coil's upper clamping plate. In this arrangement, field misalignment from thermal changes are reduced as the coil contractions tend to compensate each other.

a) Examining Fig. 3.24 carefully, explain how fault forces experienced by each coil are transmitted to the room-temperature shell of the cryostat. Also, speculate on the structural design of the conductor joint between the two coils, Nb$_3$Sn and Nb-Ti.

b) Although radiation shielding is not shown in the diagram, the approach to minimizing thermal conduction to the magnet vessel is evident. Describe the design concept, tracing the principal path of conductive heat to the magnet vessel.

c) When the system is cooled from room temperature to liquid helium temperatures, thermal contractions arise. How do they affect the alignment between the insert and the superconducting magnet?

Solution to Problem 3.16

a) The Nb-Ti coil is firmly anchored to the bottom of the magnet vessel; a (fault) force appearing on it is transmitted directly to the magnet vessel's outer shell. The force is then transmitted to the G-10 cylinder, which is attached to the midpoint of the vessel's outer wall. The G-10 cylinder rests on a support plate which is anchored to an 80-K radiation shield. The force is finally transmitted to the room-temperature shell through the stainless steel support cylinder.

The Nb_3Sn coil and Nb-Ti coil are joined at their top ends, as shown in Fig. 3.24. Because the Nb-Ti coil is immobilized, both by anchoring and pre-clamping, its top end remains motionless, immobilizing the splice joining the LF Nb_3Sn conductor and HF Nb-Ti conductor.

b) The design concept adopted for Hybrid III is to use a thin-walled shell made of a material with high strength but low thermal conductivity. The thin wall takes up a small radial space. To minimize conductive heat input, the use of low thermal conductivity alone is not sufficient; it must be supplemented by making the conductive distance between the temperature points long and the conductive cross section small. Force transmission usually dictates the conductive cross section to be large because it is through this cross section that the fault force is transmitted; this leaves only the conductive distance at the designer's option. For Hybrid III, a 5-mm thick, 100-cm dia. 80-cm long G-10 cylinder was selected to transmit force between the magnet vessel and the 80-K radiation panels.

The conduction heat input to the magnet vessel thus comes through the G-10 cylinder, whose warm ends (top and bottom) are anchored to 80-K radiation panels. This means the "80-K" heat is conducted through the G-10 cylinder over a distance of $\sim$40 cm and enters the magnet vessel at its midpoint. The inside diameter of the G-10 cylinder is "weakly" anchored to a 25-K panel. The total computed heat input to the magnet vessel is $\sim$3 W.

c) A shrinkage of the outer stainless steel tube tends to lift the bottom copper support plate anchored at 80 K. The G-10 support tube also shrinks, out-shrinking the stainless steel tube about twice on a unit length basis. However, because the length of the stainless steel support cylinder is about twice as long as the distance of the G-10 support tube from the bottom copper plate to its anchor position at the magnet vessel's midpoint, the net result is hardly any vertical displacement of the bottom plate of the magnet vessel to which the two coils are anchored. Alignment is thus minimally affected.

"Quality is remembered long after price is forgotten." —Anonymous

Problem 3.17: Stresses in an epoxy-impregnated solenoid

This problem deals with simple but approximate stress computations applicable for epoxy-impregnated magnets. We will use a 500-MHz (12 T) NMR superconducting magnet built in the late 1970s at FBNML as an example [3.31]. The magnet consists of a high-field insert, a main coil, and several correction and shim coils. The main coil's winding inner radius (a_1) is 72.6 mm, winding outer radius (a_2) is 102 mm, and the winding length ($2b$) is 488 mm. The main coil is wound with a Nb-Ti multifilamentary conductor; the volumetric ratio of copper to Nb-Ti is 2.1. The composite wire has diameters of 0.63 mm bare (D_{cd}) and 0.71 mm insulated (D_{ov}). The winding has a close-packed hexagonal configuration. The space between the wires is filled with epoxy resin. Figure 3.25 shows three neighboring wires in a close-packed hexagonal winding configuration.

When all the coils are energized, the axial (z) component of the magnetic induction, B_z, decreases linearly with radial distance (r) through the build of the main coil. At the midplane ($z = 0$) of the main coil, B_z varies from 8.22 T at $r = a_1$ to -0.21 T at $r = a_2$. The overall operating current density (J_{ov}) in the main coil is 248 MA/m^2.

The analytical solution for an anisotropic cylinder loaded with body forces was used to calculate the stresses at the midplane of the main coil. The hoop stresses at the inner and outer radii were found to be, respectively, 105 MPa and 65 MPa; the hoop stress decreases more or less linearly from the inner radius to the outer radius.

a) Using simple force equilibrium considerations, show that these stress values are consistent with the loading situation.

b) Assuming the winding pattern is a close-packed (wires touching) hexagonal configuration, compute the area fraction for each of the three constituents, Nb-Ti, copper, and organic materials (epoxy plus insulation).

c) Based on these area fractions and approximate values of Young's moduli for these materials at 4.2 K ($E_{sc} = 85$ GPa, $E_{cu} = 100$ GPa, and $E_{in} = 30$ GPa), find the hoop stresses in the Nb-Ti and copper at the innermost layer of the winding.

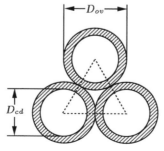

Fig. 3.25 Three neighboring conductors in a close-packed hexagonal winding configuration.

Solution to Problem 3.17

a) The average hoop stress in the winding is given by:

$$\tilde{\sigma} = \frac{\sigma_i + \sigma_o}{2} = \frac{105\,\text{MPa} + 65\,\text{MPa}}{2} \qquad (S17.1)$$

$$= 87.5\,\text{MPa}$$

The mean winding radius, $\tilde{R} = (a_1 + a_2)/2$ is 87.3 mm; the mean magnetic induction, $\tilde{B}_z = (B_i + B_o)/2$, is 4.0 T. The mean hoop stress in the winding may thus be given by:

$$\tilde{\sigma} = \tilde{R}J_{ov}\tilde{B}_z = (87.3\times10^{-3}\,\text{m})(248\times10^6\,\text{A/m}^2)(4.0\,\text{T}) \qquad (S17.2)$$

$$= 86.7\,\text{MPa}$$

which is pretty close to $\tilde{\sigma}$ computed (Eq. S17.1) by taking an average of σ_i and σ_o.

b) See Fig. 3.25 for the close-packed hexagonal configuration. The triangular area, A_{tr}, defined by the dotted lines is given in terms of the overall conductor diameter, D_{ov}, by: $A_{tr} = \sqrt{3}D_{ov}^2/4$. The conductor area, A_{cd}, within the triangle is given by: $A_{cd} = \pi D_{cd}^2/8$, of which 2.1/3.1 is copper area, A_{cu}, and 1/3.1 is Nb-Ti area, A_{sc}. The epoxy and insulation area, A_{in}, within the triangle is given by: $A_{in} = A_{tr} - A_{cd}$. Thus:

$$f_{cu} = \frac{A_{cu}}{A_{tr}} = \frac{\frac{2.1}{3.1}\frac{\pi D_{cd}^2}{8}}{\frac{\sqrt{3}}{4}D_{ov}^2} = \frac{(2.1)(\pi)(4)D_{cd}^2}{(3.1)(\sqrt{3})(8)D_{ov}^2} = 0.48 \qquad (S17.3a)$$

$$f_{sc} = \frac{A_{sc}}{A_{tr}} = \frac{1}{2.1}\frac{A_{cu}}{A_{tr}} = 0.23 \qquad (S17.3b)$$

$$f_{in} = \frac{A_{in}}{A_{tr}} = 1 - \frac{A_{cu} + A_{sc}}{A_{tr}} = 1 - 0.48 - 0.23 = 0.29 \qquad (S17.3c)$$

c) The Young's modulus for the composite, $\tilde{E}$, may be given from the parallel mixture rule:

$$\tilde{E} = f_{cu}E_{cu} + f_{sc}E_{sc} + f_{in}E_{in} \qquad (S17.4)$$

$$= (0.48)(100\,\text{GPa}) + (0.23)(85\,\text{GPa}) + (0.29)(30\,\text{GPa})$$

$$\simeq 76\,\text{GPa}$$

We may calculate the stress of each component at the innermost winding radius:

$$\sigma_{cu} = \sigma_i\frac{E_{cu}}{\tilde{E}} = (105\,\text{MPa})\frac{100\,\text{GPa}}{76\,\text{GPa}} \qquad (S17.5a)$$

$$\simeq 138\,\text{MPa}$$

$$\sigma_{sc} = \sigma_i\frac{E_{sc}}{\tilde{E}} = (105\,\text{MPa})\frac{85\,\text{GPa}}{76\,\text{GPa}} \qquad (S17.5b)$$

$$\simeq 117\,\text{MPa}$$

Problem 3.18: Stresses in a composite Nb₃Sn conductor

In this problem we shall study the stress state in a composite Nb_3Sn conductor, focusing on the stress relationships among bronze, copper, and Nb_3Sn—the three major constituents of the composite [3.32]. Because the superconducting properties of Nb_3Sn filaments are highly sensitive to strain, a thorough understanding of the stress state in composite Nb_3Sn conductors is very important in designing successful magnets wound with this material. Ekin discusses in detail the effects of strain on conductor performance [3.33, 3.34].

When a Nb_3Sn composite is cooled to 4.2 K, each constituent experiences a temperature reduction of ~1000 K, from the reaction temperature of ~1000 K to the operating temperature of 4.2 K. Because each constituent has a different coefficient of thermal contraction, residual stress arises in each constituent.

Figure 3.26 shows, schematically, strain states for three cases of interest: a) the composite at a reaction temperature ~1000 K, b) at 4.2 K if the three constituents can contract *individually*, c) the composite at 4.2 K. Though exaggerated, the figure indicates the relative sizes of the individual thermal strains of bronze, copper, and Nb_3Sn, given respectively by ϵ_{br_o}, ϵ_{cu_o}, ϵ_{s_o}, after cooldown from ~1000 K to 4.2 K. Correspondingly, their residual strains, ϵ_{br_r}, ϵ_{cu_r}, ϵ_{s_r}, in the composite at 4.2 K are as shown in the figure. That is, both bronze and copper will be in *tension*, while Nb_3Sn will be in *compression*.

In the figure E and A refer, respectively, to Young's modulus and cross section, with subscripts indicating constituents.

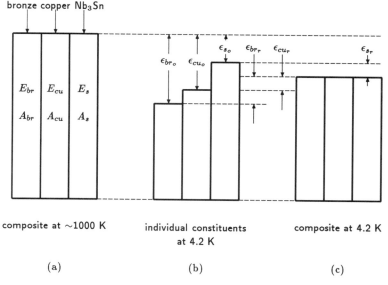

Fig. 3.26 Schematic strain states in the Nb₃Sn composite after cooldown.

Problem 3.18: Stresses in a composite Nb₃Sn conductor

a) Explain the following equations:

$$\epsilon_{br_r} A_{br} E_{br} + \epsilon_{cu_r} A_{cu} E_{cu} + \epsilon_{s_r} A_s E_s = 0 \qquad (3.58a)$$

$$\epsilon_{br_r} + \epsilon_{s_r} = \epsilon_{br_0} - \epsilon_{s_0} \qquad (3.58b)$$

$$\epsilon_{cu_r} + \epsilon_{s_r} = \epsilon_{cu_0} - \epsilon_{s_0} \qquad (3.58c)$$

$$\epsilon_{cu_r} - \epsilon_{br_r} = \epsilon_{cu_0} - \epsilon_{br_0} \qquad (3.58d)$$

Equation 3.58a implicitly assumes that each constituent is in its elastic range. This assumption, however, is usually not valid.

b) By combining Eqs. 3.58a, 3.58b, and 3.58c, show that the following expressions are valid for residual strains for the three constituents.

$$\epsilon_{br_r} = \frac{(\epsilon_{br_0} - \epsilon_{cu_0}) A_{cu} E_{cu} - (\epsilon_{br_0} - \epsilon_{s_0}) A_s E_s}{A_{cu} E_{cu} + A_{br} E_{br} - A_s E_s} \qquad (3.59a)$$

$$\epsilon_{cu_r} = \frac{(\epsilon_{cu_0} - \epsilon_{br_0}) A_{br} E_{br} - (\epsilon_{cu_0} - \epsilon_{s_0}) A_s E_s}{A_{cu} E_{cu} + A_{br} E_{br} - A_s E_s} \qquad (3.59b)$$

$$\epsilon_{s_r} = \frac{(\epsilon_{cu_0} - \epsilon_{s_0}) A_{cu} E_{cu} + (\epsilon_{br_0} - \epsilon_{s_0}) A_{br} E_{br}}{A_{cu} E_{cu} + A_{br} E_{br} - A_s E_s} \qquad (3.59c)$$

c) By using Eqs. 3.59a, 3.59b, and 3.59c and the values given in Table 3.2 below for each constituent, compute the residual strains, ϵ_{br_r}, ϵ_{cu_r}, ϵ_{s_r}.

d) From ϵ_{br_r} and ϵ_{cu_r} computed above, compute the corresponding stresses in the bronze and copper. You should find that these stresses put the metals into the plastic range.

Table 3.2: Properties of Bronze, Copper, and Nb₃Sn at 4.2 K
—Approximate Values—

Constituent	ϵ_0^* [%]	E [GPa]	A†
bronze	−1.66	100	0.24
copper	−1.62	100	0.62
Nb₃Sn	−0.72	165	0.14

* Thermal contraction strain from ~1000 to 4.2 K.
† Fraction of the total composite cross section.

Solution to Problem 3.18

a) Equation 3.58*a* states that the net internal force within the composite is zero.
Equation 3.58*b* gives the strain compatibility for the bronze and Nb₃Sn.
Equation 3.58*c* gives the strain compatibility for the copper and Nb₃Sn.
Equation 3.58*d* gives strain compatibility for copper and bronze.

b) Referring to Fig. 3.26, we have:

$$\epsilon_{s_r} = \epsilon_{br_0} - \epsilon_{s_0} - \epsilon_{br_r} \tag{$S18.1a$}$$

$$\epsilon_{cu_r} = \epsilon_{cu_0} - \epsilon_{s_0} - \epsilon_{s_r}$$

$$= \epsilon_{cu_0} - \epsilon_{br_0} + \epsilon_{br_r} \tag{$S18.1b$}$$

Thus,

$$\epsilon_{br_r} A_{br} E_{br} + (\epsilon_{cu_0} - \epsilon_{br_0} + \epsilon_{br_r}) A_{cu} E_{cu}$$
$$+ (\epsilon_{br_0} - \epsilon_{s_0} - \epsilon_{br_r}) A_s E_s = 0 \tag{$S18.2a$}$$

$$\epsilon_{br_r}(A_{br} E_{br} + A_{cu} E_{cu} - A_s E_s)$$
$$= (\epsilon_{br_0} - \epsilon_{cu_0}) A_{cu} E_{cu} - (\epsilon_{br_0} - \epsilon_{s_0}) A_s E_s \tag{$S18.2b$}$$

$$\epsilon_{br_r} = \frac{(\epsilon_{br_0} - \epsilon_{cu_0}) A_{cu} E_{cu} - (\epsilon_{br_0} - \epsilon_{s_0}) A_s E_s}{A_{br} E_{br} + A_{cu} E_{cu} - A_s E_s} \tag{3.59a}$$

Similarly,

$$\epsilon_{s_r} = \epsilon_{cu_0} - \epsilon_{s_0} - \epsilon_{cu_r} \tag{$S18.3a$}$$

$$\epsilon_{br_r} = \epsilon_{br_0} - \epsilon_{s_0} - \epsilon_{cu_0} + \epsilon_{s_0} + \epsilon_{cu_r} \tag{$S18.3b$}$$

$$\epsilon_{br_r} = \epsilon_{br_0} - \epsilon_{cu_0} + \epsilon_{cu_r} \tag{$S18.3c$}$$

Thus,

$$(\epsilon_{br_0} - \epsilon_{cu_0} + \epsilon_{cu_r}) A_{br} E_{br} + \epsilon_{cu_r} A_{cu} E_{cu}$$
$$+ (\epsilon_{cu_0} - \epsilon_{s_0} - \epsilon_{cu_r}) A_s E_s = 0 \tag{$S18.4a$}$$

$$\epsilon_{cu_r}(A_{cu} E_{cu} + A_{br} E_{br} - A_s E_s)$$
$$= (\epsilon_{cu_0} - \epsilon_{br_0}) A_{br} E_{br} - (\epsilon_{cu_0} - \epsilon_{s_0}) A_s E_s \tag{$S18.4b$}$$

$$\epsilon_{cu_r} = \frac{(\epsilon_{cu_0} - \epsilon_{br_0}) A_{br} E_{br} - (\epsilon_{cu_0} - \epsilon_{s_0}) A_s E_s}{A_{cu} E_{cu} + A_{br} E_{br} - A_s E_s} \tag{3.59b}$$

Also,

$$\epsilon_{br_r} = \epsilon_{br_0} - \epsilon_{s_0} - \epsilon_{s_r} \tag{$S18.5a$}$$

$$\epsilon_{cu_r} = \epsilon_{cu_0} - \epsilon_{s_0} - \epsilon_{s_r} \tag{$S18.5b$}$$

Solution to Problem 3.18

Thus,

$$
(\epsilon_{br_0} - \epsilon_{s_0} - \epsilon_{s_r})A_{br}E_{br}
$$
$$
+ (\epsilon_{cu_0} - \epsilon_{s_0} - \epsilon_{s_r})A_{cu}E_{cu} + \epsilon_{s_r}A_sE_s = 0 \qquad (S18.6a)
$$

$$
\epsilon_{s_r}(A_{cu}E_{cu} + A_{br}E_{br} - A_sE_s)
$$
$$
= (\epsilon_{cu_0} - \epsilon_{s_0})A_{cu}E_{cu} + (\epsilon_{br_0} - \epsilon_{s_0})A_{br}E_{br} \qquad (S18.6b)
$$

$$
\epsilon_{s_r} = \frac{(\epsilon_{cu_0} - \epsilon_{s_0})A_{cu}E_{cu} + (\epsilon_{br_0} - \epsilon_{s_0})A_{br}E_{br}}{A_{cu}E_{cu} + A_{br}E_{br} - A_sE_s} \qquad (3.59c)
$$

c) Inserting appropriate values from Table 3.2 into Eqs. 3.59a, 3.59b, and 3.59c, we have:

$$
\epsilon_{br_r} = \frac{(-1.66 + 1.62)(0.62)(100\,\text{GPa}) - (-1.66 + 0.72)(0.14)(165\,\text{GPa})}{(0.24)(100\,\text{GPa}) + (0.62)(100\,\text{GPa}) - (0.14)(165\,\text{GPa})}
$$
$$
= \frac{(-2.48\,\text{GPa}) - (-21.71\,\text{GPa})}{(24.0\,\text{GPa}) + (62.0\,\text{GPa}) - (23.1\,\text{GPa})} = \frac{19.23\,\text{GPa}}{62.9\,\text{GPa}} \simeq 0.31\%
$$

$$
\epsilon_{cu_r} = \frac{(-1.62 + 1.66)(0.24)(100\,\text{GPa}) - (-1.62 + 0.72)(0.14)(165\,\text{GPa})}{62.9\,\text{GPa}}
$$
$$
= \frac{21.75\,\text{GPa}}{62.9\,\text{GPa}} \simeq 0.35\%
$$

$$
\epsilon_{s_r} = \frac{(-1.62 + 0.72)(0.62)(100\,\text{GPa}) + (-1.66 + 0.72)(0.24)(100\,\text{GPa})}{62.9\,\text{GPa}}
$$
$$
= \frac{-78.36\,\text{GPa}}{62.9\,\text{GPa}} \simeq -1.25\%
$$

Note that both matrix materials are in tension, while Nb$_3$Sn is in compression; ϵ_{s_r} of -1.25% is of course too severe and would certainly damage the conductor [3.33, 3.34]. Note also that when the magnet is energized, the conductor is subjected mostly to a tensile stress, which tends to place ϵ_{s_r} towards zero strain; usually, the Lorentz tensile stresses are sufficiently great to make Nb$_3$Sn stretched when the magnet is the energized.

d) We have:

$$
\sigma_{br_r} = \epsilon_{br_r}E_{br} \simeq (3.1 \times 10^{-4})(100 \times 10^9\,\text{Pa}) \simeq 310\,\text{MPa} \qquad (S18.7a)
$$

$$
\sigma_{cu_r} = \epsilon_{cu_r}E_{cu} \simeq (3.5 \times 10^{-4})(100 \times 10^9\,\text{Pa}) \simeq 350\,\text{MPa} \qquad (S18.7b)
$$

The yield stress of annealed bronze, σ_{br_y}, and that of annealed copper, σ_{cu_y}, are about the same and equal to $\sim$100 MPa. Both bronze and copper are thus yielded after cooldown.

References

[3.1] For example, *Soldesign*, created by and available from R.D. Pillsbury, Jr. of the Plasma Fusion Center, MIT, Cambridge MA.

[3.2] D. Bruce Montgomery, *Solenoid Magnet Design* (Robert E. Krieger Publishing, New York, 1980).

[3.3] R.J. Weggel (personal communication, 1993).

[3.4] D.B. Montgomery, J.E.C. Williams, N.T. Pierce, R. Weggel, and M.J. Leupold, "A high field magnet combining superconductors with water-cooled conductors," *Adv. Cryogenic Eng.* **14**, 88 (1969).

[3.5] M.J. Leupold, J.R. Hale, Y. Iwasa, L.G. Rubin, and R.J. Weggel, "30 tesla hybrid magnet facility at the Francis Bitter National Magnet Laboratory," *IEEE Trans. Magn.* **MAG-17**, 1779 (1981).

[3.6] M.L. Leupold, Y. Iwasa, J.R. Hale, R.J. Weggel, and K. van Hulst, "Testing a 1.8 K hybrid magnet system," *Proc. 9th Intl. Conf. Magnet Tech. (MT-9)* (Swiss Institute for Nuclear Research, Villegen, 1986), 215.

[3.7] M.J. Leupold, Y. Iwasa, and R.J. Weggel, "Hybrid III system," *IEEE Trans. Mag.* **MAG-24**, 1070 (1988).

[3.8] Y. Iwasa, M.J. Leupold, R.J. Weggel, J.E.C. Williams, and Susumu Itoh, "Hybrid III: the system, test results, the next step," *IEEE Trans. Appl. Superconduc.* **3**, 58 (1993).

[3.9] Y. Nakagawa, K. Noto, A. Hoshi, S. Miura, K. Watanabe, and Y. Muto, "Hybrid magnet project at Tohoku University," *Proc. 8th Intl. Conf. Magnet Tech. (MT-8)* (Journal de Physique, Colloque C1 supplément au no° 1, 1984), C1-23.

[3.10] K. van Hulst and J.A.A.J. Perenboom, "Status and development at the High Field Magnet Laboratory of the University of Nijmegen," *IEEE Trans. Magn.* **24**, 1397 (1988).

[3.11] H.-J. Schneider-Muntau and J.C. Vallier, "The Grenoble hybrid magnet," *ibid.*, 1067.

[3.12] K. Inoue, T. Takeuchi, T. Kiyoshi, K. Itoh, H. Wada, H. Maeda, K. Nii, T. Fujioka, Y. Sumiyoshi, S. Hanai, T. Hamajima, and H. Maeda, "Primary design of 40 tesla class hybrid magnet system," *Proc. 11th Int'l Conf. Magnet Tech. (MT11)* (Elsevier Applied Science, London, 1990), 651.

[3.13] J.R. Miller, M.D. Bird, S. Bole, A. Bonito-Oliva, Y. Eyssa, W.J. Kenney, T.A. Painter, H.-J. Schneider-Muntau, L.T. Summers, S.W. Van Sciver, S. Welton, R.J. Wood, J.E.C. Williams, E. Bobrov, Y. Iwasa, M. Leupold, V. Stejskal, and R. Weggel, "An overview of the 45-T hybrid magnet system for the New National High Magnetic Field Laboratory," *IEEE Trans. Magn.* **30**, 1563 (1994).

[3.14] M. Suenaga, "Metallurgy of continuous filamentary A15 superconductors," in *Superconductor Materials Science – Metallurgy, Fabrication, and Applications*, Eds. S. Foner and B.B. Schwartz (Plenum Press, New York, 1981), 201.

[3.15] D.C. Larbalestier, "Niobium-titanium superconducting materials," *ibid.*, 133.

[3.16] E.W. Collings, *A Sourcebook of Titanium Alloy Superconductivity* (Plenum Press, New York, 1983).

[3.17] M.W. Garret, "Thick cylindrical coil systems for strong magnetic fields or gradient homogeneities of the 6th to 20th order," *J. Appl. Phys.* **38**, 2563 (1967).

[3.18] W.B. Sampson, "Superconducting magnets for beam handling and accelerators," *Proc. Int'l Conf. Magnet Tech.* (Rutherford Laboratory, Chilton, 1967), 574.

[3.19] See, for example, Hiromi Hirabayashi, "Dipole magnet development in Japan," *IEEE Trans. Magn.* **MAG-19**, 198 (1983).

[3.20] Z.J.J. Stekly, A.M. Hatch, J.L. Zar, W.N. Latham, C. Borchert, A. El Bindari, R.E. Bernert, T.A. deWinter, "A large experimental superconducting magnet for MHD power generation," *Int'l Inst. Refrig. Commission I Meeting*, (1966).

[3.21] R. Nieman, S. Wang, W. Pelczarski, J. Gonczy, K. Mataya, H. Ludwig, D. Hillis, H. Phillips, L. Turner, J. Purcell, D. Montgomery, J. Williams, A. Hatch, P. Marston, P. Smelser, V. Zenkevitch, L. Kirjenin, W. Young, "Superconducting magnet system U-25 MHD facility," *IEEE Trans. Magn.* **MAG-13**, 632 (1977).

[3.22] P. Thullen, J.C. Dudley, D.L. Green, J.L. Smith, Jr., and H.H. Woodson, "An experimental alternator with a superconducting rotating field winding," *IEEE Trans. Power Apparatus and Systems* **PAS-90**, 611 (1971).

[3.23] C.E. Oberly, "Air Force applications of lightweight superconducting machinery," *IEEE Trans. Magn.* **MAG-13**, 260 (1977).

[3.24] See, for example, J.R. Powell and G.T. Danby, "Magnetic suspension for levitated tracked vehicles," *Cryogenics* **21**, 192 (1971).

[3.25] See, for example, H. Ogiwara, N. Takano, and H. Yonemitsu, "Experimental studies on magnetically suspended high speed trains using large superconducting magnets," *Proc. 4th Int'l Conf. Magnet Tech. (MT-4)* (Brookhaven National Laboratory, Upton, NY, 1972), 70.

[3.26] See, for example, G. Bogner, "Preliminary results on a[n] electrodynamically levitated superconducting coil," *Proc. 2nd Sympo. on Electro-Magnetic Suspension* (Southampton, England, 1971), M.1.

[3.27] See, for example, T. Ohtsuka and Y. Kyotani, "Superconducting maglev test," *IEEE Trans. Magn.* **MAG-15**, 1416 (1979).

[3.28] Y. Aiyama, K. Fushimi, K. Yasukochi, T. Kasahara, R. Saito, N. Tada, H. Kimura, and S. Sato, "A large superconducting MHD magnet," *Proc. 5th Int'l Cryogenic Engr. Conf.* (IPC Science and Technology Press, London, 1974), 300.

[3.29] See, for example, D. Dell'Orco, S. Caspi, J. O'Neill, A. Lietzke, R. Scanlan, C.E. Taylor, A. Wandesforde, "A 50 mm bore superconducting dipole with a unique iron yoke structure," *IEEE Trans. Appl. Superconduc.* **3**, 637 (1993).

[3.30] R. Weggel and M. Leupold (an internal memo, FBNML, unpublished 1988.)

[3.31] J.E.C. Williams, L.J. Neuringer, E.S. Bobrov, R. Weggel, and W.G. Harrison, "Magnet system of the 500 MHz spectrometer at the FBNML: 1. Design and development of the magnet," *Rev. Sci. Instrum.* **52**, 649 (1981).

[3.32] D.S. Easton, D.M. Kroeger, W. Specking, and C.C. Koch, "A prediction of the stress state in Nb_3Sn superconducting composites," *J. Appl. Phys.* **51**, 2748 (1980).

[3.33] J.W. Ekin, "Strain scaling law for flux pinning in practical superconductors. Part 1: Basic relationship and application to Nb_3Sn conductors," *Cryogenics* **20**, 611 (1980).

[3.34] J.W. Ekin, "Mechanical properties and strain effects in superconductors," in *Superconductor Materials Science – Metallurgy, Fabrication, and Applications*, Eds. S. Foner and B.B. Schwartz (Plenum Press, New York, 1981), 455.

CHAPTER 4
CRYOGENICS

4.1 Introduction

Cryogenics deals with temperatures below 150~200 K. Two principal areas of cryogenics are: 1) processes and equipment and 2) experimental determination of physical properties of materials at these temperatures. For any equipment that must be operated in a cryogenic environment the ultimate goal of cryogenic engineering is to achieve with a high degree of reliability, the most efficient means of producing and maintaining that environment

A superconducting magnet is cryogenic equipment. Cryogenics thus plays an important role in superconducting magnet technology. However, it is also important to put cryogenics in perspective and not *overemphasize* its role. For example, from a purely cryogenic standpoint, a superconducting magnet would operate more efficiently at liquid nitrogen temperature (77 K) than at the temperature of liquid helium (4.2 K); but if this superconducting magnet is to be part of a complex system, the impact of the higher operating temperature on the system as a whole must be evaluated. In this case, the magnet at 77 K may require considerably more superconductor than at 4.2 K. As a result, the cost reduction achieved in cryogenics may be insufficient to offset the increased superconductor cost.

This chapter presents problems dealing with some important issues of cryogenics as applied to the operation of superconducting magnets. In this introductory section, we shall briefly review important properties of the cryogens most relevant to superconducting magnets. Purely cryogenic issues, such as heat inputs to a cryostat by conduction, radiation, and convection, are not discussed in this section; they are studied in the Problem Section.

4.2 Cryogens

Because liquid helium, as any liquid, boils at a constant temperature as long as pressure is kept constant, the most widely used mode of operation has been simply to immerse a superconducting magnet in a bath of boiling helium under atmospheric pressure. Fortunately for superconducting magnets, liquid helium boils at 4.2 K, a temperature sufficiently cold for Nb-Ti and Nb_3Sn conductors. Although we study design and operational issues of magnets using mostly LTS systems as examples, we shall cover here four other cryogens whose normal boiling temperatures are all below 100 K; some of them may replace liquid helium as the primary cryogen for HTS magnets.

Liquefaction of gases closely followed the development of thermodynamics. By the mid 1850s most gases had been liquefied, exceptions being the so-called "permanent gases," *i.e.* oxygen, nitrogen, and hydrogen. At that time noble gases had not been discovered. Helium ("sun") was discovered by Janssen in 1868; argon ("inert") was discovered by Ramsey in 1894, who also discovered neon ("new"), krypton ("hidden"), and xenon ("stranger") in 1898.

4.2.1 Boiling Temperatures

Boiling temperatures of five cryogens having normal boiling temperatures below 100 K are presented in Table 4.1. T_s in the table is saturation (boiling) temperature under atmospheric pressure. T_{mn} is the "practical" lower limit achievable under the "pumped" condition; T_{mx} is the practical upper limit considering that the cryostat must be pressurized to achieve it. P_{mn} and P_{mx} are saturation pressures corresponding, respectively, to T_{mn} and T_{mx}. Because of its explosive nature, hydrogen, though listed here, would unlikely be used in most applications. Neon is at the right temperature range for the first generation of HTS magnets, expected to be operated below $\sim$30 K. Its drawbacks are a narrow practical temperature range, as seen from Table 4.1, and high price; in large quantity it may be purchased at $\sim$\$100 per liquid liter, which is, at least in the United States, more than 20 times more expensive than liquid helium. Because of its wide availability and virtual inertness, nitrogen would be the best cryogen for superconducting magnets operable in the temperature range 64$\sim$80 K. Because of its reactivity, oxygen is less desirable; however, it offers a wide practical boiling temperature range (55$\sim$94 K) and may be suitable for "laboratory" use. Argon (87.3 K) is not included in the list, because, though quite widely available, it has, like neon, a narrow practical boiling temperature range (85$\sim$90 K). It should be noted that, except in a few special situations where liquid nitrogen is used to maintain the operating temperature of HTS magnets, HTS magnets will most likely rely on cryocoolers to maintain their temperatures.

4.2.2 Latent Heat of Vaporization

The volumetric latent heat of vaporization (h_L) is one of the most important thermodynamic properties in magnet operation. It is the energy required to vaporize a unit volume of cryogen; the smaller it is, the more quickly the cryogen boils away for a given heat load to the magnet vessel containing the liquid. The last column in Table 4.1 presents values of h_L (per unit liquid volume) for the five cryogens; h_L increases with boiling temperature. Helium thus has the lowest value; nitrogen's is $\sim$60 times helium's. (Oxygen is nearly 100 times helium's.) Because nitrogen boils at 77 K, heat input to a magnet vessel containing liquid nitrogen would naturally be much less than that to a vessel containing liquid helium, making it even easier to maintain a liquid level with nitrogen than with helium.

Table 4.1: "Practical" Boiling Temperatures and Heat of Vaporization

Cryogen	T_s [K]	$T_{mn} \sim T_{mx}$ [K]	$P_{mn} \sim P_{mx}$ [torr]	h_L [J/cm^3]
Helium	4.22	1.6$\sim$4.5	6$\sim$984	2.6
Hydrogen	20.4	14$\sim$21	59$\sim$937	31.4
Neon	27.1	25$\sim$28	383$\sim$992	104
Nitrogen	77.4	64$\sim$80	109$\sim$1026	161
Oxygen	90.2	55$\sim$94	1.4$\sim$950	243

4.2.3 Thermodynamic Properties

Appendix II presents selected thermodynamic properties of helium, hydrogen, neon, nitrogen, and oxygen useful for magnet design.

4.2.4 Nucleate Boiling Heat Transfer

Nucleate boiling heat transfer curves are key input data for "cryostable" superconducting magnets which are usually operated in a bath of boiling liquid. (Cryostable magnets will be discussed and studied in Chapter 6.) Figure 4.1 shows steady-state heat transfer flux data for bare copper surface immersed in a bath of liquid helium boiling at 4.2 K under atmospheric pressure [4.1, 4.2]; the data span both the nucleate boiling and film boiling regimes. It should be emphasized here that transfer data scatter—easily by a factor of 2 both ways—influenced by important factors of magnet design such as surface orientation and finish, and channel gap and height (in most windings, channels are long and narrow). Transient heat flux data, applicable in the nucleate boiling regime, are generally an order of magnitude greater than steady-state values [4.3].

Because the liquid, or more precisely vaporization of liquid provides the cooling in the nucleate boiling regime, vapor must be replaced with liquid continuously at the surface for the curve shown in Fig. 4.1 to be valid. This means the winding of a cryostable magnet must provide not only a sufficient cooling area but also passages (ventilation) for the vapor to leave the winding. This point is always a dilemma for the magnet designer: more space allocated for cooling means less space available for the conductor and structural materials, and a structurally less rigid winding.

Another important point to note is that the shape of the heat transfer curve, shown in Fig. 4.1, is essentially the same for other liquids, including water. Obviously key parameters on both axes, usually in log scale, need to be adjusted for a specific liquid. The key parameters are T_s (saturation temperature), q_{pk}, ΔT_{qp}, and q_{fm}. q_{pk} is the peak nucleate boiling heat transfer flux, which for liquid helium, from the figure, is ~1 W/cm^2. ΔT_{qp} is the ΔT at which q_{pk} occurs. q_{fm} is the minimum film boiling heat flux. As noted above, q_{pk} and q_{fm} are particularly sensitive to cooling channel configuration and surface finish of the metal. Table 4.2 presents these key parameters for the five liquids; values, all typical, are for each liquid boiling under atmospheric pressure.

Table 4.2: Boiling Heat Transfer Parameters

Cryogen	T_s [K]	q_{pk} [W/cm^2]	ΔT_{qp} [K]	q_{fm} [W/cm^2]
Helium	4.22	~1	~1	~0.3
Hydrogen	20.4	~10	~5	~0.5
Neon	27.1	~15	~5	~1
Nitrogen	77.4	~15	~10	~1
Oxygen	90.2	~25	~30	~2

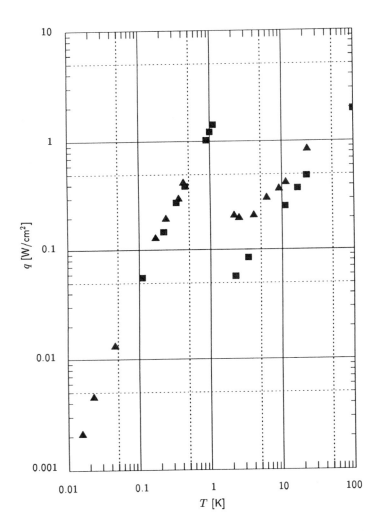

Fig. 4.1   Steady-state nucleate and film boiling heat transfer
data for bare copper surface in liquid helium at 4.2 K [4.1, 4.2].

4.3 Superfluidity

Figure 4.2 shows the phase diagram of ordinary helium (He⁴), in which two forms of liquid are present, He I and He II [1.1]. Because of its unique properties of extremely high thermal conductivity (k) and low viscosity (ν), He II is known as superfluid helium; superfluidity has been compared to superconductivity. As may be inferred from the phase diagram, ordinary liquid helium (He I) boiling at 4.2 K can readily be transformed to He II simply by pumping on the liquid. When the saturation pressure of 38 torr (0.050 atm) is reached, the liquid is at 2.18 K and "enters" into the superfluid phase. According to the "two-fluid" model, the fraction of superfluid is zero at 2.18 K and increases monotonically as temperature is lowered. Figure 4.3 shows the liquid specific heat, c_p, as a function of temperature [1.1]. c_p remains nearly constant as the temperature is lowered from 4.2 K towards 2.18 K; however, just as the liquid enters into the superfluid phase, c_p dramatically increases. It has been speculated that under an ideal experiment c_p may indeed be infinite. Below 2.18 K, c_p decreases sharply. Because the shape of $c_p(T)$ resembles the Greek letter lambda (λ), the temperature 2.18 K is known as the λ-point and designated by T_λ. The extraordinary thermal conductivity and viscosity of this phase can best be appreciated by comparing these properties with those of common materials, as summarized in Table 4.3.

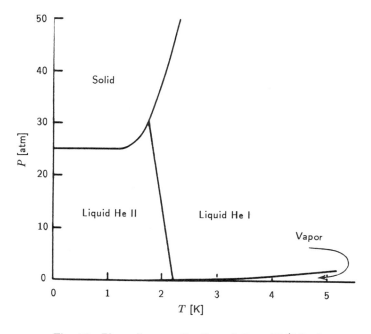

Fig. 4.2 Phase diagram of ordinary helium (He⁴) [1.1].

Table 4.3: Thermal Conductivity and Viscosity of He II
– Approximate Values and Comparison –

Material	k [W/m K]	ν [μPa s*]
He II	~100,000	0.01~0.1
He I (4.2 K, liq.)	0.02	~3
Copper (4.2 K)	~400	—
Water (RT)	~1	~1,000
Air (RT)	~0.05	~20

* 1 Pa s = 10 poise.

4.3.1 Transport Properties

Because of its extremely high thermal conductivity, superfluid helium is sometimes used for the operation of superconducting magnets. In magnet applications,
a temperature of 1.8 K ($< T_\lambda$) is considered the normal operating temperature.
As mentioned earlier, the liquid space within the magnet winding is quite limited
and for He I boiling above T_λ it is necessary to provide ventilation passages to
allow the vapor to flow out of the winding. The high thermal conductivity of
He II does not allow a temperature gradient in the liquid sufficient for creation of
vapor; ventilation passages for vapor are thus not needed in the winding. However,

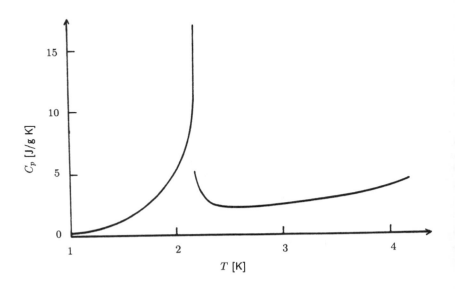

Fig. 4.3   Specific heat of liquid helium (He⁴) in contact with its vapor [1.1].

this does not mean that He II can transport unlimited heat fluxes through narrow channels. Analogous to the critical current density in superconductors, He II has a critical heat flux.

Bon Mardion, Claudet, and Seyfert studied heat flux in He II through narrow channels [4.4]. Figure 4.4 presents their results in the form of a parameter $X(T)$ designated by them; it is is given by:

$$X(T_{cl}) - X(T_{wm}) = q^{3.4}L \tag{4.1}$$

where T_{cl} [K] is the cold-end temperature and T_{wm} [K] is the warm-end temperature. q [W/cm^2] is the heat flux through a channel L [cm] long connecting the two ends. Note that Eq. 4.1 is applicable for the case when there is no additional heating introduced to the liquid from the channel itself between the two ends. Under normal operating conditions, $T_{cl} = T_b$, where T_b is the bath temperature; T_{wm} would be the liquid temperature adjacent to a heated region within the winding and T_{wm} cannot exceed T_λ; $T_{wm} = T_\lambda$ or from Fig. 4.4, $X(T_{wm}) = 0$. Considering these operating conditions, we can simplify Eq. 4.1 to:

$$X(T_b) = q_c^{3.4}L \tag{4.2}$$

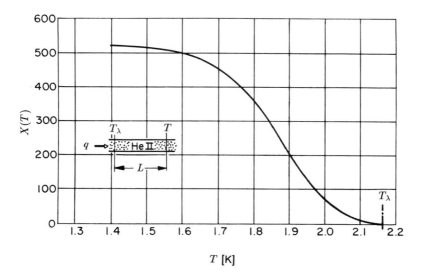

Fig. 4.4 Bon Mardion-Claudet-Seyfert plot of $X(T)$ for a channel of length L [cm] filled with 1-atm, superfluid helium. The fluid temperature is T_{cl} [K] at the cold end and T_{wm} [K] at the warm end [4.4].

In designing channel configuration and dimensions, we must make sure that the operating heat flux, q_{op}, corresponding to a particular channel design, does not exceed q_c given by Eq. 4.2.

Heated Channel

When heating is uniformly introduced over the entire length of channel, L, rather than at the hot end as discussed above, Eq. 4.2 is modified [4.5]:

$$X(T_b) = \frac{q_c^{3.4}}{4.4}L \tag{4.3}$$

4.3.2 Heat Transfer—Kapitza Resistance

Heat transfer between normal metal and He II is controlled by Kapitza resistance. Heat transfer flux, q_k [W/cm^2], between a metal whose surface is at T_{cd} [K] and liquid at T_b [K] is given by:

$$q_k = a_k(T_{cd}^{n_k} - T_b^{n_k}) \tag{4.4}$$

Table 4.4 gives typical values of a_k and n_k.

Table 4.4: Approximate Values for Kapitza Resistance*

Metal (surface)	a_k [W/cm^2]	n_k
Aluminum (polished)	0.05	3.4
Copper (polished)	0.02	4.0
Copper[†] (polished)	0.02	3.8
Copper (as-received)	0.05	2.8
Copper (solder-coated)	0.08	3.4
Copper (varnish-coated)	0.07	2.1
Silver (polished)	0.06	3.0

* Based on values given in references [4.6, 4.7].
† Annealed.

"The perpetual motion? Nonsense! It can never be discovered. It is a dream that may delude men whose brains are mystified with matter, but not me." —Owen Warland

Problem 4.1: Carnot refrigerator

This problem deals with a Carnot refrigerator. The Carnot cycle is composed of two reversible adiabatic (isentropic) and two reversible isothermal processes, in which a working fluid operates between two thermal reservoirs to produce work or refrigeration at the most efficient level.

While Carnot efficiency can never be attained in practice, it sets an upper bound on what is possible. Since superconductivity occurs at very low temperatures, refrigeration is required to achieve and maintain the cryogenic environment for magnet operation. Currently, most superconducting magnets operate at liquid helium temperatures ($\sim 4\,\mathrm{K}$).

a) Draw a Carnot cycle on a T vs S plot. Use the following notation: T_1 is the temperature of the cold reservoir; S_1 is the entropy leaving the cold-temperature reservoir; T_2 is the temperature of the warm-temperature reservoir; and S_2 is the entropy entering the warm-temperature reservoir.

b) Show that for an ideal Carnot refrigerator, the work input W_{ca} required to extract heat Q_1 from the reservoir at T_1 and release it to the reservoir at $T_2 > T_1$ (Fig. 4.5) is given by:

$$W_{ca} = Q_1 \left(\frac{T_2}{T_1} - 1 \right) \qquad (4.5)$$

c) Show that $W_{ca}/Q_1 \simeq 74$ and $W_{ca}/Q_1 \simeq 3$ for a Carnot refrigerator operating in the temperature ranges, respectively, 4 to $300\,\mathrm{K}$ and 77 to $300\,\mathrm{K}$.

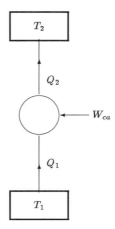

Fig. 4.5   Carnot refrigerator operating between two heat reservoirs.

Solution to Problem 4.1

a) The Carnot refrigerator operates between two reservoirs, the lower one at T_1 and the higher one at T_2, extracting heat Q_1 at T_1 and rejecting heat Q_2 at T_2. As illustrated schematically in Fig. 4.5, work W_{ca} is needed to run the refrigerator.

The Carnot refrigeration cycle consists of 4 reversible processes performed on a working fluid, as shown in the T *vs* S plot of Fig. 4.6.

- an isentropic compression of working fluid, starting at state 1 (S_2, T_1);
- an isothermal compression, starting at state 2 (S_2, T_2);
- an isentropic expansion, starting at state 3 (S_1, T_2);
- an isothermal expansion, starting at state 4 (S_1, T_1) and ending at state 1.

b) From the 1st law of thermodynamics, we have:

$$Q_1 + W_{ca} = Q_2 \qquad (S1.1)$$

Q_1 for the reversible process is given by $T_1(S_2 - S_1)$. W_{ca} is the work input to the refrigerator, graphically equal to the area enclosed on the T-S diagram. Q_2 for the reversible process is given by $T_2(S_2 - S_1)$.

Because the entire cycle consists of 4 reversible processes, the net entropy generated is zero and there is a constant entropy flow from the cold-temperature reservoir at T_1, given by Q_1/T_1, to the warm-temperature reservoir at T_2, given by Q_2/T_2. Thus:

$$\frac{Q_1}{T_1} = \frac{Q_2}{T_2} \qquad (S1.2)$$

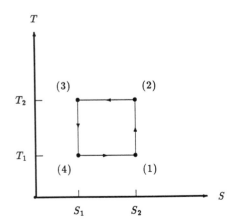

Fig. 4.6   Temperature *vs* entropy plot for the Carnot refrigerator.

Solution to Problem 4.1

Combining Eqs. $S1.1$ and $S1.2$, we have:

$$\frac{Q_1}{T_1} = \frac{Q_1 + W_{ca}}{T_2} \qquad (S1.3)$$

By solving Eq. $S1.3$ for W_{ca}, we obtain:

$$W_{ca} = Q_1 \left(\frac{T_2}{T_1} - 1 \right) \qquad (4.5)$$

c) *Temperature range 4~300 K*: With $T_1 = 4\,K$ and $T_2 = 300\,K$, Eq. 4.5 gives $W_{ca}/Q_1 \simeq 74$. That is, for each 1 W of refrigeration at 4 K, the refrigerator requires 74 W of work input. In a real refrigerator, the ratio varies from about 300 for large systems to about 2000 for small systems.

Temperature range 77~300 K: With $T_1 = 77\,K$ and $T_2 = 300\,K$ inserted into Eq. 4.5, we find $W_{ca}/Q_1 \simeq 3$. In a real refrigerator, the ratio is close to ~10.

Clearly, it is much more expensive to provide refrigeration at 4.2 K than at 77 K. Even with the reduced refrigeration cost to motivate an increase in the magnet operating temperature, however, we must ask what percentage of the *total* system cost, both capital and operating, is directly related to the operating temperature of the magnet. The answer to such a difficult but important question invariably depends on numerous parameters.

Joule-Thomson Process

In the Joule-Thomson (J-T) process, a working gas expands adiabatically and isenthalpically through a restricted passage, *e.g.* a needle valve, changing its temperature. Whether the change is positive, negative, or zero depends on the gas properties, the starting temperature, and initial and final pressures. For helium at initial and final pressures of 10 atm and 1 atm, respectively, liquefaction results if the initial temperature is below ~7.5 K. Because the process is irreversible, liquefaction by the J-T process always results in a smaller portion of liquid produced than if the gas was expanded isentropically. For example, if the helium gas is expanded at a starting temperature of 6 K and a pressure of 10 atm—typical values for liquefiers—to liquefaction at 4.2 K and 1 atm, the following isenthalpic relationship may be used to compute a fraction of helium liquefied (x_ℓ):

$$h_g(6\,K, 10\,atm) = x_\ell h_\ell(4.2\,K, 1\,atm) + (1 - x_\ell)h_g(4.2\,K, 1\,atm) \qquad (4.6)$$

From Eq. 4.6 we find $x_\ell = 0.47$. If the same gas is expanded isentropically, an entropy relationship similar to the enthalpy relationship of Eq. 4.6 gives $x_\ell = 0.53$. Despite this reduction in liquid production efficiency, the J-T expansion, because of its mechanical simplicity, is used in the final stage of many helium liquefiers.

Problem 4.2: Cooling modes of a magnet

This problem deals with the cooldown of a magnet using liquid helium. A super-conducting magnet is often operated while immersed in a bath of liquid helium. Because liquid nitrogen is less expensive than liquid helium, a common magnet cooldown procedure consists of the following sequence: 1) fill the cryostat with liquid nitrogen at 77 K, immersing the magnet in a bath of liquid nitrogen; 2) flush out the liquid nitrogen from the cryostat; 3) while the magnet is still at ~77 K, transfer liquid helium into the cryostat until the magnet is completely submersed. In some cases involving large magnets (>1 ton), there can be an additional step between steps 1 and 2 in which liquid nitrogen is pumped down to a pressure of 109 torr to bring the boiling temperature to 64 K thereby further reducing the magnet enthalpy. For each of the following questions, you may use the thermodynamic properties of copper for those of the magnet.

a) **"Perfect" Cooldown Mode**: Under a "perfect" cooldown mode, the magnet is cooled by a series of infinitesimal energy exchanges with cold helium. At the nth step, for example, the magnet at temperature T_n is cooled an infinitesimal amount ΔT by exchanging heat with an infinitesimal amount of cold helium mass ΔM_{he}, which is heated from 4.2 K (liquid) to T_n. All of the "coldness" in the helium between 4.2 K (liquid) and T_n is assumed to be used to cool the magnet by ΔT. Note that the "coldness" in the helium between T_n and room temperature is not usable in the cooling.

If M_{he} is the mass of helium required to cool a magnet of mass M_{mg} from T_i to 4.2 K, show that an expression for M_{he}/M_{mg} is given by:

$$\frac{M_{he}}{M_{mg}} = \int_{4.2\,\mathrm{K}}^{T_i} \frac{c_{cu}(T)\, dT}{h_{he}(T) - h_{he}(4.2\,\mathrm{K, liq.})} \tag{4.7}$$

where $c_{cu}(T)$ is the specific heat of copper (representing all the materials in the winding) and $h_{he}(T)$ is the specific enthalpy of helium

b) **"Dunk" Mode**: An extreme mode of cooldown is to "dunk" the whole magnet initially at T_i into a bath of liquid helium boiling at 4.2 K. Show that an expression for M_{he}/M_{mg} under this "dunk" mode to cooldown the magnet initially at T_i to 4.2 K is given by:

$$\frac{M_{he}}{M_{mg}} = \frac{1}{h_L}[h_{cu}(T_i) - h_{cu}(4.2\,\mathrm{K})] \simeq \frac{h_{cu}(T_i)}{h_L} \tag{4.8}$$

where h_L is the heat of vaporization of liquid helium at 4.2 K. $h_{cu}(T_i)$ is the specific enthalpy of copper at T_i and $h_{cu}(T_i) \gg h_{cu}(4.2\,\mathrm{K})$ when $T_i > 10\,\mathrm{K}$.

c) A 1000-kg copper block is to be cooled to 4.2 K from an initial temperature of T_i. Use Eqs. 4.7 and 4.8 to construct a table of liquid helium required (in liters) vs T_i for the two cooling modes in the temperature range 10 K to 300 K. The values obtained are good estimates for a 1000-kg superconducting magnet.

Solution to Problem 4.2

a) In the perfect cooling mode, we have:

dq_{mg} = infinitesimal heat removed from magnet = $M_{mg}c_{cu}(T)\,dT$

dq_{he} = infinitesimal heat added to helium = $dM_{he}[h_{he}(T) - h_{he}(4.2\,\text{K}, \text{liq.})]$

For perfect heat exchange between magnet and helium, we have $dq_{mg} = dq_{he}$, and hence:

$$M_{mg}c_{cu}(T)\,dT = dM_{he}[h_{he}(T) - h_{he}(4.2\,\text{K}, \text{liq.})] \qquad (S2.1)$$

From Eq. $S2.1$, we obtain:

$$dM_{he} = M_{mg}\frac{c_{cu}(T)\,dT}{h_{he}(T) - h_{he}(4.2\,\text{K}, \text{liq.})} \qquad (S2.2)$$

Integrating Eq. $S2.2$ and dividing both sides by M_{mg}, we obtain:

$$\frac{M_{he}}{M_{mg}} = \int_{4.2\,\text{K}}^{T_i} \frac{c_{cu}(T)\,dT}{h_{he}(T) - h_{he}(4.2\,\text{K}, \text{liq.})} \qquad (4.7)$$

This cooling mode may be approached, but never realized in practice, by having liquid helium introduced in the cryostat space *underneath* the magnet at a very *slow* rate. However, the cooling rate cannot be too slow because of two heat sources: 1) the cryostat containing the magnet, which itself is not completely adiabatic; and 2) the transfer line that brings in the liquid from the storage dewar which also introduces heat into the liquid.

b) In the dunk cooling mode, only the latent heat of vaporization of liquid helium (h_L) is used to cool down the magnet. Thus:

$$M_{he}h_L = M_{mg}\int_{T_f}^{T_i} c_{cu}(T)\,dT \qquad (S2.3)$$

From Eq. $S2.3$ we obtain:

$$\frac{M_{he}}{M_{mg}} = \frac{1}{h_L}[h_{cu}(T_i) - h_{cu}(4.2\,\text{K})] \simeq \frac{h_{cu}(T_i)}{h_L} \qquad (4.8)$$

c) Table 4.5 presents liquid helium (LHe) requirements (in liters) to cool a 1000-kg copper block from T_i to 4.2 K for both cooling modes. The values in Table 4.5 clearly suggest that it is important to precool a magnet with liquid nitrogen first. For certain experiments, *e.g.* cryotribology, it is not permissible to contaminate or "wet" the system with liquid nitrogen and the liquid helium must be used from the beginning. Under such a condition it is extremely important to remember the tremendous difference—a factor of almost 40—in liquid helium requirements between the two cooling modes and to transfer the liquid at the slowest rate practical.

Solution to Problem 4.2

Table 4.5: Liquid Helium Required to Cool
a 1000-kg Copper Block from T_i to 4.2 K

T_i	LHe Requirement [liter]	
[K]	"Perfect"	"Dunk"
300	790	31200
280	750	28210
260	710	25230
240	660	22290
220	610	19430
200	560	16610
180	500	13830
160	440	11165
140	375	8660
120	305	6310
100	230	4150
90	190	2030
80	150	1640
70	113	1620
60	77	1010
50	44	550
40	37	240
30	19	76
25	6	35
20	3.5	13
15	1.4	4
10	0.5	1

Problem 4.3: Optimum gas-cooled leads—Part 1

This problem deals with a quantitative design approach for "optimum" gas-cooled leads. Because the power supply to a superconducting magnet is usually located outside the magnet's cryogenic environment, current leads that span over a large temperature range between power supply and magnet are an essential component in any superconducting system. Unfortunately, these leads are also a heat load on the cryogenic environment through both conduction from the room-temperature environment and Joule heating generated by the transport current they carry. The basic design concept for optimum leads is to make the Joule heating and conduction heating about equal and to remove them by funneling cold vapor rising from the boiling helium through the lead.

Because this problem consists of many (17) questions, it is divided into Part 1 (7 questions) and Part 2 (10 questions).

a) Using a one-dimensional model of a differential element dz, show that the steady-state power equation [W/m] within a gas-cooled copper current lead carrying transport current I_t is given by:

$$\frac{d}{dz}\left[Ak(T)\frac{dT}{dz}\right] - \dot{m}_I c_p(T)\frac{dT}{dz} + \frac{\rho(T)I_t^2}{A} = 0 \qquad (4.9)$$

where z is the axial distance along the lead, $z = 0$ at the lead cold end. $k(T)$, A, and $\rho(T)$ are, respectively, the lead (usually copper) thermal conductivity, cross section, and electrical resistivity. $\dot{m}_I$ and $c_p(T)$ are, respectively, the helium mass flow rate and specific heat. Note that heat transfer between the helium and the lead is assumed perfect, and that T is the temperature of both lead and helium at z.

b) In the "high" current limit, it is possible to neglect the first term in Eq. 4.9 which represents axial conduction along the lead. Under this assumption, show that the total heat Q_I transferred into the cold reservoir (liquid helium bath) by conduction is given by:

$$Q_I = \frac{k_0 \rho_0 I_t^2}{\dot{m}_I c_{p0}} \qquad (4.10)$$

k_0, ρ_0, and c_{p0} are evaluated at the lead bottom-end temperature, T_0 ($\sim 6\,\text{K}$).

In real systems, the current lead's bottom end ($z = 0$) is *above* the liquid helium level and connected by a bus bar to a terminal of the magnet, immersed in the liquid. Often the bus bar consists of a Nb-Ti composite superconductor to which an additional copper bar is soldered. The composite carries I_t with zero Joule dissipation; the copper bar carries Q_I, given by Eq. 4.10, from the lead's bottom end to the liquid. To keep the composite superconducting over the entire length of the bus bar, the temperature at the bus bar's top end ($z = 0$) must be below $\sim 9\,\text{K}$ (Nb-Ti's T_c); at the same time, the copper bar must conduct Q_I with a small temperature difference (at most $\sim 5\,\text{K}$). These two conditions usually translate to a copper bar having a considerable cross section and a temperature of $\sim 6\,\text{K}$ at the current lead's bottom end.

Problem 4.3: Optimum gas-cooled leads—Part 1

c) By equating Q_I given by Eq. 4.10 with the power necessary to generate a liquid boil-off rate of $\dot{m}_I$, show that $\dot{m}_I$ is given by:

$$\dot{m}_I = I_t \sqrt{\frac{k_0 \rho_0}{c_{p0} h_L}} \tag{4.11}$$

where h_L is the liquid helium's volumetric heat of vaporization.

d) By combining Eqs. 4.10 and 4.11, derive the following equation:

$$Q_I = I_t \sqrt{\frac{h_L k_0 \rho_0}{c_{p0}}} \tag{4.12}$$

e) By using $\rho_0 = 2.5 \times 10^{-10}\,\Omega\,\mathrm{m}$ and values for c_{p0} and k_0 corresponding to the bottom-end temperature ($T_0 = 6\,\mathrm{K}$), show:

$$\frac{Q_I}{I_t} \sim 1\,\mathrm{mW/A} \tag{4.13}$$

f) To derive an appropriate design value of $(I_t \ell / A)_{ot}$ for an "optimum" lead, where ℓ is the lead length and I_t is its designed transport current, one begins with the high-current approximation of Eq. 4.9 and combines it with Eq. 4.11. Derive the following integral expression:

$$\int_{T_0}^{T_\ell} \frac{dT}{\rho(T)} \simeq \left(\frac{I_t \ell}{A} \right)_{ot} \sqrt{\frac{h_L}{c_{p0} k_0 \rho_0}} \tag{4.14a}$$

T_ℓ (at $z = \ell$) is the lead top-end temperature. Assume $c_p(T) = c_{p0}$.

g) For copper, the integral of the left-hand side of Eq. 4.14a is given approximately by:

$$\int_{T_0}^{T_\ell} \frac{dT}{\rho(T)} \simeq 1.2 \times 10^{11}\,\mathrm{K/\Omega\,m} \tag{4.14b}$$

for $T_0 = 6\,\mathrm{K}$ and $T_\ell = 273\,\mathrm{K}$. By combining Eqs. 4.14a and 4.14b, show that for copper:

$$\left(\frac{I_t \ell}{A} \right)_{ot} \equiv \zeta_\circ = 1.2 \times 10^{11} \sqrt{\frac{c_{p0} k_0 \rho_0}{h_L}} \tag{4.15a}$$

$$\left(\frac{I_t \ell}{A} \right)_{ot} \simeq 2.5 \times 10^7\,\mathrm{A/m} \tag{4.15b}$$

Equation 4.15b is our criterion for an optimum current lead for given I_t. In practice, I_t is usually set by the magnet design; the appropriate (ℓ/A) ratio may then be selected for the optimum current lead.

Solution to Problem 4.3—Part 1

a) The total power Q_{in} flowing into and generated within a differential volume $A\Delta z$ of the lead, shown in Fig. 4.7, is given by:

$$Q_{in} = \left[Ak(T)\frac{dT}{dz}\right]_{z+\Delta z} + \dot{m}_I c_p(T)T + \frac{\rho(T)I_t^2}{A}\Delta z \qquad (S3.1)$$

The total power Q_{out} flowing out of the differential volume of $A\Delta z$ is given by:

$$Q_{out} = \left[Ak(T)\frac{dT}{dz}\right]_z + \dot{m}_I c_p(T)(T + \Delta T) \qquad (S3.2)$$

In the steady-state condition, we have $Q_{in} = Q_{out}$. Combining Eqs. $S3.1$ and $S3.2$, we have:

$$\left[Ak(T)\frac{dT}{dz}\right]_{z+\Delta z} - \left[Ak(T)\frac{dT}{dz}\right]_z - \dot{m}_I c_p(T)\Delta T + \frac{\rho(T)I_t^2}{A}\Delta z = 0 \qquad (S3.3)$$

By dividing Eq. $S3.3$ by Δz and letting Δz go to zero, we obtain Eq. 4.9:

$$\frac{d}{dz}\left[Ak(T)\frac{dT}{dz}\right] - \dot{m}_I c_p(T)\frac{dT}{dz} + \frac{\rho(T)I_t^2}{A} = 0 \qquad (4.9)$$

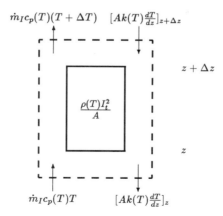

Fig. 4.7   Heat balance for the differential volume of a gas-cooled lead.

Solution to Problem 4.3—Part 1

b) In the "high" current limit, Eq. 4.9, evaluated at $z = 0$, where $T(0) = T_0$, is simplified to:

$$-\dot{m}_I c_{p0} \frac{dT}{dz}\bigg|_{z=0} + \frac{\rho_0}{A} I_t^2 \simeq 0 \qquad (S3.4)$$

where $c_{p0} = c_p(T_0)$ and $\rho_0 = \rho(T_0)$. From the above equation, we can solve for $(dT/dz)_{z=0}$:

$$\frac{dT}{dz}\bigg|_{z=0} \simeq \frac{\rho_0 I_t^2}{A\dot{m}_I c_{p0}} \qquad (S3.5)$$

Because Q_I at $z = 0$ is purely by conduction, we have:

$$Q_I = Ak(T)\frac{dT}{dz}\bigg|_{z=0}$$

$$= \frac{k_0 \rho_0 I_t^2}{\dot{m}_I c_{p0}} \qquad (4.10)$$

c) The power input Q_I into the liquid boils off the liquid at a rate $\dot{m}_I$ given by:

$$\dot{m}_I = \frac{Q_I}{h_L} \qquad (S3.6)$$

Combining Eqs. 4.10 and S3.6 and solving for $\dot{m}_I$, we obtain:

$$\dot{m}_I = I_t \sqrt{\frac{k_0 \rho_0}{c_{p0} h_L}} \qquad (4.11)$$

d) Inserting $\dot{m}_I$ given by Eq. 4.11 into Eq. 4.10, we have:

$$Q_I = \left(\frac{k_0 \rho_0 I_t^2}{c_{p0}}\right) \frac{1}{I_t} \sqrt{\frac{c_{p0} h_L}{k_0 \rho_0}}$$

$$= I_t \sqrt{\frac{h_L k_0 \rho_0}{c_{p0}}} \qquad (4.12)$$

Note that Q_I depends neither on ℓ, the lead active length between the bottom end and top end, nor on A, the lead conductor cross section. It is, however, directly proportional to transport current.

e) From Eq. 4.12, we have:

$$\frac{Q_I}{I_t} = \sqrt{\frac{h_L k_0 \rho_0}{c_{p0}}} \qquad (S3.7a)$$

Solution to Problem 4.3—Part 1

With $h_L = 20.4 \times 10^3$ J/kg and $c_{p0} \simeq 6.0 \times 10^3$ J/kg K for helium; $k_0 = 600$ W/m K and $\rho_0 = 2.5 \times 10^{-10}$ Ω m for copper, we obtain:

$$\frac{Q_I}{I_t} = \sqrt{\frac{(20.4 \times 10^3 \text{ J/kg})(600 \text{ W/m})(2.5 \times 10^{-10} \text{ } \Omega \text{ m})}{6.0 \times 10^3 \text{ J/kg K}}} \quad (S3.7b)$$

$$= 7.1 \times 10^{-4} \text{ W/A}$$

$$\sim 1 \text{ mW/A} \quad (4.13)$$

For an optimum gas-cooled lead, 1 mW/A is a rule-of-thumb value useful for estimating the heat input into liquid helium. For a 10-kA lead operating at 10 kA, the heat load will be $\sim$10 W or $\sim$20 W for a pair of such leads.

f) The high-current approximation of Eq. 4.9 is given by:

$$-\dot{m}_I c_p(T)\frac{dT}{dz} + \frac{\rho(T)I_t^2}{A} = 0 \quad (S3.8)$$

Solving for $dT/\rho(T)$ from Eq. $S3.8$ with $c_p(T) \simeq c_{p0}$, and integrating both sides over appropriate limits (T_0 at $z = 0$ and T_ℓ at $z = \ell$) we obtain:

$$\int_{T_0}^{T_\ell} \frac{dT}{\rho(T)} = \int_0^\ell \frac{I_t^2 \, dz}{A\dot{m}_I c_{p0}} = \frac{I_t^2 \ell}{A\dot{m}_I c_{p0}} \quad (S3.9)$$

Inserting $\dot{m}_I$ (Eq. 4.11) in Eq. $S3.9$ and rearranging the right-hand side, we obtain:

$$\int_{T_0}^{T_\ell} \frac{dT}{\rho(T)} = \left(\frac{I_t^2 \ell}{A c_{p0}}\right)\frac{1}{I_t}\sqrt{\frac{c_{p0}h_L}{k_0\rho_0}}$$

$$\int_{T_0}^{T_\ell} \frac{dT}{\rho(T)} = \left(\frac{I_t\ell}{A}\right)_{ot}\sqrt{\frac{h_L}{c_{p0}k_0\rho_0}} \quad (4.14a)$$

g) Combining Eqs. 4.14a and 4.14b, we obtain:

$$\left(\frac{I_t\ell}{A}\right)_{ot} \equiv \zeta_\circ = 1.2 \times 10^{11}\sqrt{\frac{c_{p0}k_0\rho_0}{h_L}} \quad (4.15a)$$

Inserting appropriate values for the parameters in the above equation, we can numerically solve for $\zeta_\circ$:

$$\zeta_\circ = (1.2 \times 10^{11} \text{ K}/\Omega \text{ m})$$

$$\times \sqrt{\frac{(6.0 \times 10^3 \text{ J/kg K})(600 \text{ W/m})(2.5 \times 10^{-10} \text{ } \Omega \text{ m})}{20.4 \times 10^3 \text{ J/kg}}} \quad (S3.10)$$

$$\zeta_\circ \simeq 2.5 \times 10^7 \text{ A/m} \quad (4.15b)$$

Problem 4.3: Optimum gas-cooled leads—Part 2

h) Now let us estimate the boil-off rate of liquid helium when the optimized lead carries no current. This boil-off, called the "standing" boil-off rate, $\dot{m}_0$, is due to heat input into the helium by conduction only. Solve Eq. 4.9 for $T(z)$ when $I_t = 0$, using the following boundary conditions:

$$T(z = 0) = T_0 \tag{4.16a}$$

$$Ak_0 \frac{dT}{dz}\bigg|_{z=0} = \dot{m}_0 h_L \tag{4.16b}$$

Replace $k(T)$ appearing in Eq. 4.9 with an average value, $\tilde{k}$, given by:

$$\tilde{k} = \frac{1}{T_\ell - T_0} \int_{T_0}^{T_\ell} k(T)\, dT \tag{4.17}$$

$\tilde{k}$ is a thermal conductivity value useful for computing thermal conduction rates over a given temperature range. Table 4.6 (p. 140) presents values of $\tilde{k}$ for three materials (G-10, stainless steel 304, and copper) over three common temperature ranges in cryogenic applications: 4~80 K; 4~300 K; and 80~300 K.

With $T_\ell = 273\,\text{K}$ and $T_0 = 6\,\text{K}$, $\tilde{k} = 660\,\text{W/m\,K}$ for copper. Also assume $c_p(T) = c_{p0}$. Under these conditions, show that $T(z)$ is given by:

$$T(z) = T_0 + \frac{h_L \tilde{k}}{c_{p0} k_0} \left[\exp\left(\frac{\dot{m}_0 c_{p0} z}{\tilde{k} A} \right) - 1 \right] \tag{4.18}$$

Note that $T(z)$ increases exponentially with z.

i) Using another boundary condition, $T(\ell) = T_\ell$, show that the standing heat input at $z = 0$, Q_0, for the optimum lead is given by:

$$Q_0 = \frac{\tilde{k} h_L}{c_{p0}} \left(\frac{A}{\ell} \right) \ln \left[\frac{c_{p0} k_0 (T_\ell - T_0)}{h_L \tilde{k}} + 1 \right] \tag{4.19}$$

j) Combining Eqs. 4.12 and 4.19, show that an expression for the ratio of Q_0 (standing boil-off rate) to Q_I (boil-off rate with I_t) is given by:

$$\frac{Q_0}{Q_I} = \frac{\tilde{k} h_L}{k_0 c_{p0} \rho_0 1.2 \times 10^{11}} \ln \left[\frac{c_{p0} k_0 (T_\ell - T_0)}{h_L \tilde{k}} + 1 \right] \tag{4.20}$$

Problem 4.3: Optimum gas-cooled leads—Part 2

k) Solve Q_0/Q_I numerically for $T_0 = 6\,\text{K}$ and $T_\ell = 273\,\text{K}$ and show that:

$$\left(\frac{Q_0}{Q_I}\right) \sim 0.5 \tag{4.21}$$

l) The catalogue of American Magnetics Inc., a manufacturer of gas-cooled leads, indicates that the boil-off rate of their pair of 500-A leads at 500 A is 1.4 liters (liquid)/h. Is their claim consistent with an estimate based on Eq. 4.11? (AMI's leads are manufactured according to the design principle developed by Efferson [4.8].)

m) The catalogue also claims the standing boil-off rate of the leads is 0.9 liter (liquid)/h. Is this claim consistent according to the analysis studied here?

n) Show that irrespective of current rating or size, the voltage drop across an optimum lead, V_{ot}, at its rated current, I_{ot}, is independent of current rating or size. That is, the voltage drop across an optimum 100-A lead at a current of 100 A and that across an optimum 10-kA lead at a current of 10 kA are essentially identical. Specifically, show that for each optimum lead at the rated current:

$$V_{ot} \sim 250\,\text{mV} \tag{4.22}$$

o) For the same designed operating current, optimum gas-cooled leads of different lengths (ℓ) give rise to the same liquid helium boil-off rate under equilibrium conditions: $Q_I/I_t \sim 1\,\text{mW/A}$ (Eq. 4.13), which as given by $S3.7a$ is independent of ℓ.

When the cooling gas flow is abruptly discontinued, the steady-state solution on which the design of optimum leads is based is no longer valid. If the lead continues to carry its operating current without cooling, "flow stoppage meltdown" of the lead may occur, most often near the top end. By deriving and solving the *time-dependent* heat equation, show that the thermal time constant, τ_ℓ, of the lead is given by:

$$\tau_\ell = \frac{C_o \ell^2}{b\zeta_o^2} \tag{4.23}$$

where C_o is the heat capacity of copper, assumed constant. b is a coefficient appearing in the resistivity function: $\rho(T) = a + bT$. Equation 4.23 indicates that τ_ℓ is proportional to the square of the lead length, making a longer optimum lead safer than a shorter optimum lead in the event of cooling gas flow stoppage.

p) Compute an approximate value of τ_ℓ for an optimum 10-kA lead 1-m long. Use $C_o = 3.4\,\text{MJ/m}^3\,\text{K}$ and $b = 68\,\text{p}\Omega\,\text{m/K}$.

q) Compute the mass of the *conductive part* of the above lead.

Solution to Problem 4.3—Part 2

h)   With $I_t = 0$ inserted into Eq. 4.9, we have:

$$A\tilde{k}\frac{d^2T}{dz^2} - \dot{m}_0 c_{p0}\frac{dT}{dz} = 0 \qquad (S3.11)$$

By letting $dT/dz = \theta(z)$, we rewrite Eq. $S3.11$ as:

$$A\tilde{k}\frac{d\theta}{dz} - \dot{m}_0 c_{p0}\theta = 0 \qquad (S3.12)$$

from which we obtain:

$$\theta = \theta_0 \exp\left(\frac{\dot{m}_0 c_{p0} z}{A\tilde{k}}\right)$$

or

$$\frac{dT}{dz} = \theta_0 \exp\left(\frac{\dot{m}_0 c_{p0} z}{A\tilde{k}}\right) \qquad (S3.13)$$

Equation $S3.13$ may be solved to obtain $T(z)$:

$$T(z) = C_0 \exp\left(\frac{\dot{m}_0 c_{p0} z}{A\tilde{k}}\right) + C_1 \qquad (S3.14)$$

where C_0 and C_1 are constants to be determined from boundary conditions given by Eqs. 4.16a and 4.16b.

Combining the boundary condition given by Eq. 4.16a and $T(z = 0)$ given by Eq. $S3.14$, we have:

$$T(z = 0) = T_0 = C_0 + C_1 \qquad (S3.15a)$$

Equating the boundary condition given by Eq. 4.16b with $(dT/dz)_{z=0}$ given by Eq. $S3.14$, we have:

$$\left.\frac{dT}{dz}\right|_{z=0} = \frac{\dot{m}_0 h_L}{A k_0} = C_0\frac{\dot{m}_0 c_{p0}}{A\tilde{k}} \qquad (S3.15b)$$

From Eq. $S3.15b$, we obtain $C_0 = h_L\tilde{k}/(c_{p0}k_0)$. Inserting it and $C_1 = T_0 - h_L\tilde{k}/c_{p0}k_0$ into $T(z)$ (Eq. $S3.14$), we obtain:

$$T(z) = T_0 + \frac{h_L\tilde{k}}{c_{p0}k_0}\left[\exp\left(\frac{\dot{m}_0 c_{p0} z}{A\tilde{k}}\right) - 1\right] \qquad (4.18)$$

Solution to Problem 4.3—Part 2

i) From Eq. 4.18, we have $T(\ell)$:

$$T(\ell) = T_\ell = T_0 + \frac{h_L \tilde{k}}{c_{p0} k_0}\left[\exp\left(\frac{\dot{m}_0 c_{p0}\ell}{A\tilde{k}}\right) - 1\right] \qquad (S3.16)$$

From Eq. $S3.16$, we can obtain the expression for $\dot{m}c_{p0}\ell/A\tilde{k}$:

$$\frac{c_{p0} k_0 (T_\ell - T_0)}{h_L \tilde{k}} + 1 = \exp\left(\frac{\dot{m}_0 c_{p0}\ell}{A\tilde{k}}\right) \qquad (S3.17)$$

Solving Eq. $S3.17$ for $\dot{m}_0 c_{p0}\ell/A\tilde{k}$, we have:

$$\frac{\dot{m}_0 c_{p0}\ell}{A\tilde{k}} = \ln\left[\frac{c_{p0} k_0 (T_\ell - T_0)}{h_L \tilde{k}} + 1\right] \qquad (S3.18)$$

Solving Eq. $S3.18$ for $\dot{m}_0$, we obtain:

$$\dot{m}_0 = \frac{\tilde{k}}{c_{p0}}\left(\frac{A}{\ell}\right)\ln\left[\frac{c_{p0} k_0 (T_\ell - T_0)}{h_L \tilde{k}} + 1\right] \qquad (S3.19)$$

With $Q_0 = \dot{m}_0 h_L$ and inserting an expression for $\dot{m}_0$ given by the above equation, we obtain:

$$Q_0 = \frac{\tilde{k} h_L}{c_{p0}}\left(\frac{A}{\ell}\right)\ln\left[\frac{c_{p0} k_0 (T_\ell - T_0)}{h_L \tilde{k}} + 1\right] \qquad (4.19)$$

j) By inserting $A/\ell = I_t/\zeta_\circ$ into Eq. 4.19, we obtain:

$$Q_0 = \frac{\tilde{k} h_L}{c_{p0}}\left(\frac{I_t}{\zeta_\circ}\right)\ln\left[\frac{c_{p0} k_0 (T_\ell - T_0)}{h_L \tilde{k}} + 1\right] \qquad (S3.20)$$

Combining Eqs. $S3.20$ and 4.12, we can also obtain an expression for Q_0/Q_I:

$$\frac{Q_0}{Q_I} = \frac{\tilde{k} h_L}{c_{p0}}\left(\frac{I_t}{\zeta_\circ}\right)\ln\left[\frac{c_{p0} k_0 (T_\ell - T_0)}{h_L \tilde{k}} + 1\right] \times \frac{1}{I_t}\sqrt{\frac{c_{p0}}{h_L k_0 \rho_0}} \qquad (S3.21)$$

Inserting $\zeta_\circ$ given by Eq. 4.15a (p. 126) into Eqs. $S3.21$, we obtain:

$$\frac{Q_0}{Q_I} = \frac{\tilde{k} h_L}{k_0 c_{p0}\rho_0 1.2\times 10^{11}}\ln\left[\frac{c_{p0} k_0 (T_\ell - T_0)}{h_L \tilde{k}} + 1\right] \qquad (4.20)$$

Solution to Problem 4.3—Part 2

k) With $\tilde{k} = 660\,\text{W/m K}$, $h_L = 20.4 \times 10^3\,\text{J/kg}$, $k_0 = 600\,\text{W/m K}$, $c_{p0} = 6.0 \times 10^3\,\text{J/kg K}$, $\rho_0 = 2.5 \times 10^{-10}\,\Omega\,\text{m}$, $T_\ell = 270\,\text{K}$, and $T_0 = 6\,\text{K}$ inserted into 4.20, the ratio becomes:

$$\frac{Q_0}{Q_I} = \frac{(660\,\text{W/m K})(20.4 \times 10^3\,\text{J/kg})}{(600\,\text{W/m K})(6.0 \times 10^3\,\text{J/kg K})(2.5 \times 10^{-10}\,\Omega\,\text{m})(1.2 \times 10^{11})}$$

$$\times \ln \left[\frac{(6 \times 10^3\,\text{J/kg K})(600\,\text{W/m K})(264\,\text{K})}{(20.4 \times 10^3\,\text{J/kg})(660\,\text{W/m K})} + 1 \right] \qquad (S3.22)$$

$$= 0.125 \times \ln(70.6) = 0.125 \times 4.25$$

$$\frac{Q_0}{Q_I} = 0.53 \qquad (4.21)$$

That is, a standing boil-off rate is roughly half that corresponding to the full current state; this is reasonable because an optimum lead is designed to make heat input at the bottom end equally divided between conduction and Joule heating.

l) By inserting proper values into Eq. 4.11, we obtain:

$$\dot{m}_I = I_t \sqrt{\frac{k_0 \rho_0}{c_{p0} h_L}} \qquad (4.11)$$

$$= (500\,\text{A}) \sqrt{\frac{(600\,\text{W/m K})(2.5 \times 10^{-10}\,\Omega\,\text{m})}{(6.0 \times 10^3\,\text{J/kg K})(20.4 \times 10^3\,\text{J/kg})}} \qquad (S3.23)$$

$$\simeq 1.8 \times 10^{-5}\,\text{kg/s} = 18\,\text{mg/s}$$

Because liquid helium density at $4.2\,\text{K}$ is $0.125\,\text{g/cm}^3$, a mass flow rate of $18\,\text{mg/s}$ is equivalent to a boil-off rate of $\sim 0.5\,\text{liter/h}$. With two leads, a boil-off rate at $500\,\text{A}$ becomes $\sim 1\,\text{liter/h}$, which is $\sim 70\%$ of the boil-off rate quoted by AMI for their 500-A leads. This discrepancy results in part by our high-current approximation which neglects conduction through the lead.

m) Combining Eq. 4.21 and the above result, $Q_0 = 9\,\text{mg/s}$, which is equivalent to a boil-off rate of $0.27\,\text{liter/h}$ for a 500-A lead or a boil-off rate of $0.54\,\text{liter/h}$ for a pair. These rates are $\sim 60\%$ of corresponding rates quoted by AMI.

n) Equation 4.15b relates I_t, ℓ, and A for an optimum lead:

$$(I_t \ell / A)_{ot} \equiv \zeta_\circ = 2.5 \times 10^7\,\text{A/m} \qquad (4.15b)$$

The voltage drop V_{ot} over the entire length of the lead at I_{ot} is given by:

$$V_{ot} = \frac{I_{ot}}{A} \int_0^\ell \rho_{cu}\,dz \qquad (S3.24)$$

Solution to Problem 4.3—Part 2

Integration over lead length ℓ is necessary because ρ_{cu}, the electrical resistivity of copper, is temperature-dependent and thus varies with z. Equation $S3.24$ may be expressed as:

$$V_{ot} = \tilde{\rho}_{cu} \left(\frac{I_t \ell}{A} \right)_{ot} = \tilde{\rho}_{cu} \zeta_o \qquad (S3.25)$$

where

$$\tilde{\rho}_{cu} = \frac{1}{\ell} \int_0^\ell \rho_{cu} \, dz$$

$$\simeq \frac{1}{T_\ell - T_0} \int_{T_0}^{T_\ell} \rho_{cu}(T) \, dT \qquad (S3.26)$$

Note that Eq. $S3.26$ assumes linear temperature gradient along current lead. Combining Eqs. $S3.25$ and $S3.26$, we have:

$$V_{ot} = \tilde{\rho}_{cu} 2.5 \times 10^7 \, \text{V} \qquad (S3.27)$$

That is, V_{ot} is the same among leads optimized for a given value of I_t.

For copper, ρ_{cu} varies from a low value of $\sim 2.5 \times 10^{-10} \, \Omega \, \text{m}$ at 6 K to a high value of $\sim 2 \times 10^{-8} \, \Omega \, \text{m}$ at 273 K; it is essentially linear with T above ~ 50 K. $\tilde{\rho}_{cu}$ is thus $\sim 1 \times 10^{-8} \, \Omega \, \text{m}$. For either an optimum 100-A lead at 100 A or an optimum 10-kA lead at 10 kA, we have:

$$V_{ot} \sim 250 \, \text{mV} \qquad (4.22)$$

o) A time-dependent power equation [W/m] for a differential element of an optimum lead is given by:

$$AC_{cu}(T) \frac{dT}{dt} = \frac{d}{dz} \left[Ak(T) \frac{dT}{dz} \right] - \dot{m}_I c_p(T) \frac{dT}{dz} + \frac{\rho_{cu}(T)}{A} I_t^2 \qquad (S3.28)$$

where $C_{cu}(T)$ is the heat capacity of the lead metal (copper). With no cooling ($\dot{m}_I = 0$) and the conduction term set equal to zero, Eq. $S3.28$ becomes:

$$AC_{cu}(T) \frac{dT}{dt} = \frac{\rho_{cu}(T)}{A} I_t^2 \qquad (S3.29)$$

We also know that an optimum lead satisfies $(I_t \ell / A)_{ot} = \zeta_o = 2.5 \times 10^7 \, \text{A/m}$. We thus have:

$$C_{cu}(T) \frac{dT}{dt} = \frac{\rho_{cu}(T) \zeta_o^2}{\ell^2} \qquad (S3.30)$$

Solution to Problem 4.3—Part 2

Because flow stoppage meltdown usually happens near the top end, we replace $C_{cu}(T)$ with C_o (a constant) and $\rho_{cu}(T) = \rho_o + bT$ (b another constant):

$$\frac{dT}{dt} = \frac{\rho_o \zeta_o^2}{C_o \ell^2} + \frac{b \zeta_o^2}{C_o \ell^2} T$$

$$= \frac{\rho_o}{b \tau_\ell} + \frac{T}{\tau_\ell} \qquad (S3.31)$$

The thermal time constant of the lead, τ_ℓ, is given by:

$$\tau_\ell = \frac{C_o \ell^2}{b \zeta_o^2} \qquad (4.23)$$

The solution to Eq. $S3.31$ is given by:

$$T(t) = A_0 e^{t/\tau_\ell} - \frac{\rho_o}{b} \qquad (S3.32)$$

Equation $S3.32$ states that upon gas flow stoppage $T(t)$ rises exponentially with a time constant of τ_ℓ. Because τ_ℓ is proportional to ℓ^2, a longer optimum lead takes a longer time to reach the metal's meltdown temperature than does a shorter optimum lead. That is, a longer optimum lead is safer than a shorter optimum lead in the event of cooling gas flow stoppage.

p) Inserting $C_0 = 3.5 \times 10^6 \, \mathrm{J/m^3\,K}$, $\ell = 1\,\mathrm{m}$, $b = 68 \times 10^{-12}\,\Omega\,\mathrm{m/K}$, and $\zeta_o = 2.5 \times 10^7\,\mathrm{A/m}$ into Eq. 4.23, we obtain:

$$\tau_\ell = \frac{(3.5 \times 10^6 \, \mathrm{J/m^3\,K})(1\,\mathrm{m})^2}{(68 \times 10^{-12}\,\Omega\,\mathrm{m/K})(2.5 \times 10^7\,\mathrm{A/m})} \qquad (S3.33)$$

$$= 80\,\mathrm{s}$$

Note that we can double the time constant by selecting an optimum 10-kA lead 1.4-m long.

q) Solving Eq. 4.15b for lead cross sectional area A, we obtain:

$$A = \frac{I_t \ell}{\zeta_o} \qquad (S3.34)$$

With $I_t = 10^4\,\mathrm{A}$, $\ell = 1\,\mathrm{m}$, and $\zeta_o = 2.5 \times 10^7\,\mathrm{A/m}$ inserted into Eq. $S3.34$:

$$A = \frac{(10^4\,\mathrm{A})(1\,\mathrm{m})}{(2.5 \times 10^7\,\mathrm{A/m})} = 4 \times 10^{-4}\,\mathrm{m^2} \qquad (S3.35)$$

The conductive part of the lead mass, M_ℓ, is given by:

$$M_\ell = \varrho_{cu} A \ell = (8900\,\mathrm{kg/m^3})(4 \times 10^{-4}\,\mathrm{m^2})(1\,\mathrm{m}) = 3.6\,\mathrm{kg} \qquad (S3.36)$$

The conductive element of an optimum 10-kA lead has a mass of 3.6 kg.

Problem 4.4: Optimum leads for a vacuum environment— Normal conductive metal vs HTS

This problem deals with the design of optimum leads applicable for cryocooler-cooled superconducting magnets operated at $T_{op} > 10\,\text{K}$ in a vacuum environment. We shall approach the problem by examining first normal conductive metal leads and then HTS leads.

a) With $\dot{m}_I = 0$ in Eq. 4.9 of Problem 4.3, show that an expression for Q_I is given by:

$$Q_I = \tilde{k}(T_\ell - T_0)\left(\frac{A}{\ell}\right) + \frac{\tilde{\rho}I_t^2}{2}\left(\frac{\ell}{A}\right) \tag{4.24}$$

Assume $k(T)$ and $\rho(T)$ to be constant and represented by $\tilde{k}$ and $\tilde{\rho}$ respectively. Define $T(z = 0) = T_0$ and $T(z = \ell) = T_\ell$. Note that $T_0 < T_\ell$.

b) Optimize Eq. 4.24 in terms of ℓ/A to minimize Q_I. Show that the value of $(I_t\ell/A)_{ov} \equiv \zeta_{ov}$ that minimizes Q_I is given by:

$$\left(\frac{I_t\ell}{A}\right)_{ov} \equiv \zeta_{ov} = \sqrt{\frac{2\tilde{k}(T_\ell - T_0)}{\tilde{\rho}}} \tag{4.25}$$

c) Show an expression for heat input to the cold end by a normal conductive metal lead, $[Q_{Iov}]_{ncm}$, is given by:

$$[Q_{Iov}]_{ncm} = I_t\sqrt{2\tilde{k}\tilde{\rho}(T_\ell - T_0)} \tag{4.26}$$

d) For temperatures up to $\sim$80 K, it is now possible to consider using HTS as a lead material; effective HTS leads have been successfully demonstrated [4.9, 4.10]. As is evident from Table A5.4 (Appendix V), thermal conductivities of HTS materials, like those of LTS materials, are quite low—comparable with those of alloys, e.g. brass. Also like the normal-state resistivities of LTS materials, the normal-state resistivities of HTS materials are several orders of magnitude greater than those of normal conductive metals. Therefore, HTS materials cannot be used alone for leads; they must be "stabilized" in case they are driven to the normal state—during this time Eq. 4.24 is valid. Because this stability requirement makes Eq. 4.24 valid even for HTS leads, they must also satisfy the same $(I_t\ell/A)_{ov}$ given by Eq. 4.25.

Show that an expression for heat input by a stabilized HTS lead at the cold end, $[Q_{Iov}]_{HTS}$, is given by:

$$[Q_{Iov}]_{HTS} = I_t\sqrt{\frac{\tilde{k}\tilde{\rho}(T_\ell - T_0)}{2}} \tag{4.27}$$

Solution to Problem 4.4

a) With $\dot{m}_I = 0$ and $k(T) \simeq \tilde{k}$, Eq. 4.9 of Problem 4.3 reduces to:

$$A\tilde{k}\frac{d^2T}{dz^2} + \frac{\rho_{cu}(T)I_t^2}{A} = 0 \qquad (S4.1)$$

Integrating Eq. $S4.1$ twice with respect to z and dividing the resulting equation by $A\tilde{k}$, we obtain:

$$T(z) = -\frac{\tilde{\rho}I_t^2}{2A^2\tilde{k}}z^2 + az + b \qquad (S4.2)$$

where a and b are constant. Using appropriate boundary conditions, $T(z = 0) = T_0$ and $T(z = \ell) = T_\ell$, we can reduce Eq. $S4.2$ to:

$$T(z) = -\frac{\tilde{\rho}I_t^2}{2A^2\tilde{k}}z^2 + \left[\frac{(T_\ell - T_0)}{\ell} + \frac{\tilde{\rho}I_t^2\ell}{2A^2\tilde{k}}\right]z + T_0 \qquad (S4.3)$$

Q_I is given by:

$$Q_I = A\tilde{k}\frac{dT}{dz}\bigg|_{z=0} = \tilde{k}(T_\ell - T_0)\left(\frac{A}{\ell}\right) + \frac{\tilde{\rho}I_t^2}{2}\left(\frac{\ell}{A}\right) \qquad (4.24)$$

b) Differentiating Eq. 4.24 with respect to ℓ/A and setting it to 0 for $(I_t\ell/A)_{ov} = \zeta_{ov}$, we can solve for ζ_{ov} that minimizes Q_I:

$$-\tilde{k}(T_\ell - T_0)\left(\frac{A}{\ell}\right)^2 + \frac{\tilde{\rho}I_t^2}{2} = 0 \qquad (S4.4)$$

Solving for $(I_t\ell/A)_{ov}$ from Eq. $S4.4$, we obtain:

$$\left(\frac{I_t\ell}{A}\right)_{ov} \equiv \zeta_{ov} = \sqrt{\frac{2\tilde{k}(T_\ell - T_0)}{\tilde{\rho}}} \qquad (4.25)$$

c) From Eqs. 4.24 and 4.25, we obtain an expression for $[Q_{Iov}]_{ncm}$:

$$[Q_{Iov}]_{ncm} = I_t\sqrt{2\tilde{k}\tilde{\rho}(T_\ell - T_0)} \qquad (4.26)$$

d) Because current in an HTS lead is carried superconductively, heat input at the cold end, $[Q_{Iov}]_{HTS}$, is by conduction only, which, for optimum leads, is half that given by Eq. 4.26. Thus:

$$[Q_{Iov}]_{HTS} = \tfrac{1}{2}[Q_{Iov}]_{ncm} = \frac{I_t\sqrt{2\tilde{k}\tilde{\rho}(T_\ell - T_0)}}{2} \qquad (S4.5)$$

$$= I_t\sqrt{\frac{\tilde{k}\tilde{\rho}(T_\ell - T_0)}{2}} \qquad (4.27)$$

In addition to the HTS leads considered here, which depend only on conduction for cooling, there also are HTS versions of gas-cooled leads [4.9, 4.10].

Wiedemann-Franz-Lorenz Law and Lorenz number

The product of electrical resistivity and thermal conductivity, $\rho(T)k(T)$, in most conductive metals is proportional to temperature. This relationship is known as the Wiedemann-Franz-Lorenz (WFL) law:

$$k(T)\rho(T) = L_\ell T \qquad (4.28)$$

where L_ℓ is the Lorenz number. According to the free electron model of metals, L_ℓ is the same for all metals:

$$L_\ell = (\pi k_B/\sqrt{3}e)^2 = 24.5\,\mathrm{nW}\,\Omega/\mathrm{K}^2 \qquad (4.29)$$

where k_B is the Boltzmann's constant and e is the electronic charge.

Because of this universality of L_ℓ, the variation in $\rho(T)k(T)$ among conductive metals is generally minimal. Figure 4.8 shows $L_\ell(T)$ plots for copper (RRR=100), silver (99.99%), and aluminum (99.99%). The plots are based on property values presented in Appendices III and IV. The dotted line is the theoretical value of the Lorenz number based on the free electron model.

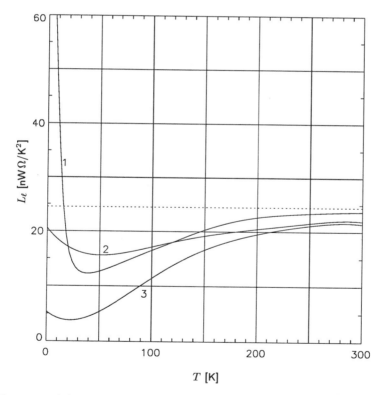

Fig. 4.8 $L_\ell(T)$ plots. 1: copper (RRR=100); 2: silver (99.99%); and 3: aluminum (99.99%). The dotted line is the theoretical value based on the free electron model.

Problem 4.5: Gas-cooled support rods

This problem deals with conduction loss associated with support rods. Structural supports inside the cryostat represent a conductive heat load on the cryogenic environment. Like the current leads of Problem 4.3, the use of a stream of helium gas to cool a support rod spanning two temperatures (T_0 at the bottom end and T_ℓ at the top end) can greatly reduce the conductive heat loss.

a) Equation 4.19 may be used to show that it is possible to reduce significantly the conduction heat input through a support rod by means of the cold helium gas that removes the conduction heat. The rod has a cross-sectional area of A and the distance between the bottom and top ends is ℓ. Assume its thermal conductivity is temperature-independent and given by $\tilde{k}$. Assume also that heat transfer between the helium stream and the rod is perfect. Specifically, show that the ratio of conduction heat input through the rod from T_ℓ to T_0 without helium cooling, $Q_{\bar{g}}$, to that with helium cooling, Q_g, is given by:

$$\frac{Q_{\bar{g}}}{Q_g} = \frac{c_{p0}(T_\ell - T_0)}{h_L \ln\left[\dfrac{c_{p0}k_0(T_\ell - T_0)}{h_L\tilde{k}} + 1\right]} \tag{4.30}$$

b) Compute Eq. 4.30 for stainless steel 304 with $T_0 = 4\,\mathrm{K}$ and $T_\ell = 300\,\mathrm{K}$. Use $k_0 = 0.25\ \mathrm{W/m\,K}$ and a value of $\tilde{k}$ given in Table 4.6.

c) Repeat b) for copper. Use $k_0 = 560\,\mathrm{W/m\,K}$ for copper.

Structural Materials for Cryogenic Applications

Structural materials for cryogenic applications must withstand a large tensile or compressive stress, while conducting as little heat flux as possible over a temperature range. A material property that can be used to quantify the suitability for cryogenic applications is $\tilde{k}/\sigma_U$, the ratio of the material's average thermal conductivity (from T_{cd} to T_{wm}) to its tensile strength.

Table 4.6 presents values of appropriate properties for G-10, stainless steel 304, and copper to gauge their suitability as structural materials for cryogenic applications; copper is included to demonstrate its unsuitability. Based on $\tilde{k}/\sigma_U$ values, G-10 appears more suitable than stainless steel.

Table 4.6: Values of Thermal Conductivity-to-Tensile Strength Ratio For G-10, Stainless Steel 304, and Copper

	$\tilde{k}$ [W/m K]			σ_U	$\tilde{k}/\sigma_U$
	Temperature Range [K]			[MPa]	[m²/K s]
	4~80	4~300	80~300	295 K	(80~300 K)
G-10	0.25	0.50	0.56	280	2×10^{-9}
SS 304	4.5	11	13	1300	10×10^{-9}
Copper	1300	660	460	250	2×10^{-6}

Solution to Problem 4.5

a) The conduction heat input through a gas-cooled support rod, Q_g, is identical to Q_0 (Eq. 4.19, Problem 4.3), the standing loss of the optimum gas-cooled lead. Thus:

$$Q_g = Q_0 = \frac{\tilde{k}h_L}{c_{p0}} \left(\frac{A}{\ell}\right) \ln\left[\frac{c_{p0}k_0(T_\ell - T_0)}{h_L\tilde{k}} + 1\right] \qquad (4.19)$$

For a noncooled support rod of cross section A and length ℓ with a linear temperature variation (T_0 at $z = 0$ and T_ℓ at $z = \ell$) and constant thermal conductivity $\tilde{k}$, the conduction heat input, $Q_{\bar{g}}$, is given by:

$$Q_{\bar{g}} = A\tilde{k}\frac{T_\ell - T_0}{\ell} \qquad (S5.1)$$

Taking the ratio of $Q_{\bar{g}}$ and Q_g, we have:

$$\frac{Q_{\bar{g}}}{Q_g} = \frac{c_{p0}(T_\ell - T_0)}{h_L \ln\left[\frac{c_{p0}k_0(T_\ell - T_0)}{h_L\tilde{k}} + 1\right]} \qquad (4.30)$$

Note that the ratio is independent of A and ℓ.

b) With $c_{p0} = 6.0 \times 10^3\,\text{J/kg K}$, $h_L = 20.4 \times 10^3\,\text{J/kg}$, $k_0 = 0.25\,\text{W/m K}$, $\tilde{k} = 11\,\text{W/m K}$ for stainless steel 304 and with $T_0 = 4\,\text{K}$ and $T_\ell = 300\,\text{K}$, we have:

$$\frac{Q_{\bar{g}}}{Q_g} = \frac{(6.0 \times 10^3\,\text{J/g K})(296\,\text{K})}{(20.4 \times 10^3\,\text{J/kg})\ln(3.0)} \qquad (S5.2)$$

$$= 80$$

$$\sim 100$$

That is, it is possible to greatly reduce the conduction heat input through structural elements by an effective use of cold helium gas.

c) With $c_{p0} = 6.0 \times 10^3\,\text{J/kg K}$, $h_L = 20.4 \times 10^3\,\text{J/kg}$, $k_0 = 560\,\text{W/m K}$, $\tilde{k} = 660\,\text{W/m K}$ for copper and with $T_0 = 4\,\text{K}$ and $T_\ell = 300\,\text{K}$, we have:

$$\frac{Q_{\bar{g}}}{Q_g} = \frac{(6.0 \times 10^3\,\text{J/g K})(296\,\text{K})}{(20.4 \times 10^3\,\text{J/kg})\ln(74.9)} \qquad (S5.3)$$

$$= 20.2$$

$$\sim 20$$

The impact of gas cooling is greater for stainless steel than for copper; however, because copper conducts much more so than stainless steel in absolute term, gas cooling is particularly important with copper—of course the best example is gas-cooled leads studied in Problem 4.3

Problem 4.6: Subcooled 1.8-K cryostat

This problem deals with a cryostat for the operation of superconducting magnets in a bath of subcooled, 1.8-K superfluid helium at 1 atm. Magnet performance at 1.8 K is improved significantly over that of operation at 4.2 K, particularly in bath-cooled Nb-Ti magnets [4.11], because of significant improvement in: 1) critical current density; and 2) heat transfer between conductor and coolant.

Figure 4.9 presents a schematic of the subcooled 1.8-K cryostat for the Hybrid III system [4.12]; it is a modified version of the design invented by Claudet, Roubeau, and Verdier [4.13]. The 1-atm, 1.8-K magnet vessel is connected hydraulically to the 1-atm, 4.2-K reservoir located above through a narrow channel sufficient both to keep the magnet vessel essentially at 1 atm (subcooled) and to minimize heat input from the reservoir to the magnet vessel.

Refrigeration is provided by an evaporator located inside the magnet vessel. The helium to be pumped in the evaporator is fed from the 4.2-K reservoir. Helium at point 1 in the figure is thus 1-atm liquid at 4.2 K; it is cooled by the J-T heat exchanger and flows through the J-T valves, isenthalpically dropping its pressure to 12.6 torr as it enters the evaporator. The helium leaves the evaporator as vapor at 1.8 K. After passing through the J-T heat exchanger, the vapor leaves the cryostat and goes into a pumping station located outside the cryostat system. The helium is purified and stored in a pressure tank. The exhaust helium gas from the 4.2-K reservoir is funneled through the current leads and it too is stored in the pressure tank. Room-temperature helium from the tank is liquefied and transferred into a 500-liter storage dewar, from which it is continuously transferred to the 4.2-K reservoir to maintain the reservoir's liquid level. The 1.8-K cryostat system, therefore, is a closed system.

Under normal operating conditions, the superfluid helium inside the evaporator is at a saturation temperature of 1.8 K and pressure of 12.6 torr and the liquid level is maintained inside the evaporator. The total combined heat load on the magnet vessel, $Q_{1.8}$, is thus matched by refrigeration produced by the evaporator. $Q_{1.8}$ enters into the evaporator from the vessel through the evaporator wall.

The current leads pass through the 4.2-K reservoir and then enter the magnet vessel through current links connecting the two liquid vessels. Note that important components not directly related to the refrigeration cycle, e.g. structural elements (Fig. 3.24), are not included in Fig. 4.9; that is, the cryogenic system is more complex than indicated in the figure.

a) Using the first law of thermodynamics, explain the following basic power equation for the 1.8-K circuit:

$$Q_{1.8} = \dot{m}_{he}(h_3 - h_2) \tag{4.31}$$

where $\dot{m}_{he}$ is the mass flow rate through the evaporator and the hs designate helium enthalpies at designated points in the figure.

b) What role does the J-T heat exchanger play in the evaporator circuit?

Problem 4.6: Subcooled 1.8-K cryostat

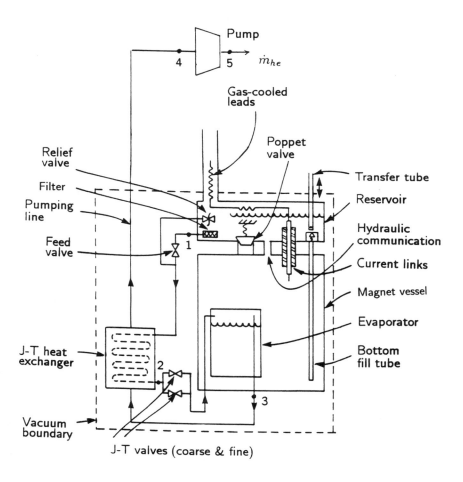

Fig. 4.9   Schematic of a subcooled 1.8-K cryostat [4.12].

Problem 4.6: Subcooled 1.8-K cryostat

c) Using helium data given in Appendix II, determine $\dot{m}_{he}$ (in g/s) to make $Q_{1.8} = 20\,\text{W}$. Assume $P_2 = 1\,\text{atm}$, $T_2 = 3.0\,\text{K}$, P_3 and T_3 are saturation values corresponding to 1.8 K, and helium at point 3 is 100% vapor.

d) What is the minimum input power required to the pump when $\dot{m}_{he} = 1\,\text{g/s}$, $P_4 = 12.6\,\text{torr}$, $T_4 = 300\,\text{K}$, and $P_5 = 1\,\text{atm}$ (760 torr)? Assume the pumping process to be isentropic.

e) What precaution(s) must the magnet designer take in designing the plumbing system connecting points 3 and 4?

f) For the Hybrid III cryostat, the hydraulic communication has an effective area of $2.6\,\text{mm}^2$ to keep the magnet vessel at 1 atm. Its effective length, L, connecting the helium in the reservoir and that in the magnet vessel is 10 cm. Using a Bon Mardoin-Claudet-Seyfert plot of $X(T)$ given in Fig. 4.4, compute the heat input into the magnet vessel, Q_{hh}, conducted by the superfluid helium in the channel. Assume the helium temperature at the bottom of the reservoir to be T_λ and that at the magnet vessel to be 1.8 K.

g) One of final steps in the "cool-down mode" of Hybrid III SCM involves cooling the liquid helium in the magnet vessel from 4.2 K to 1.8 K. The necessary cooling is provided by the evaporator, which is continuously fed 4.2-K liquid from the reservoir and at the same time pumped. Assuming that the liquid volume in the magnet vessel is 250 liters, estimate the total volume of "replenishment" liquid for the magnet vessel as the liquid is cooled from 4.2 K to 1.8 K.

h) The current leads must reach from the bottom of the reservoir to the terminals of the magnet located in the magnet vessel. It is customary to use Nb-Ti composite superconductors for the leads. Clearly each composite superconductor must have a small cross sectional area occupied by copper to minimize conduction heat input to the vessel from the reservoir but still large enough to keep the superconducting filaments stable during magnet operation. The lead criterion developed in Problem 4.4 is applicable for this case because between the two ends, the lead is essentially insulated—in the Hybrid III cryostat, the space separating the two vessels is vacuum. Show that the steady-state peak temperature in the lead when it is in the normal state and carrying I_t occurs at the lead's reservoir end and because its temperature is close to T_λ, well below Nb-Ti's critical temperature, the entire lead will recover.

"He was right, he is right; Murphy will always be right."
—An exasperated cryogenic engineer

Solution to Problem 4.6

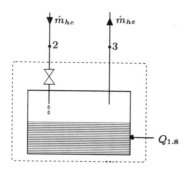

Fig. 4.10   Heat balance for an evaporator.

a)   Equation 4.31 is derived from the first law of thermodynamics applied to the control volume (c.v.) enclosing the evaporator (Fig. 4.10). Under the steady-state condition, the heat input to the evaporator, Q_{in}, is given by:

$$Q_{in} = Q_{1.8} + \dot{m}_{he}h_2 \qquad (S6.1)$$

where $Q_{1.8}$ is the refrigeration load to the evaporator. $Q_{1.8}$ consists principally of:

- dissipation within the magnet—AC losses during field change, splice losses— these losses are discussed in Chapter 7.

- heat input to the magnet vessel. For example, conduction through structural supports, radiation, conduction through leads between the 4.2-K reservoir and the magnet vessel, superfluid conduction through the pressure communication channel (Q_{hh} in **f**)).

Heat output, Q_{out}, from the control volume is given by $\dot{m}_{he}h_3$. By equating Q_{in} with Q_{out} and solving for $Q_{1.8}$, we obtain:

$$Q_{1.8} = \dot{m}_{he}(h_3 - h_2) \qquad (4.31)$$

b)   Helium leaving the reservoir (at point 1) is at 4.2 K; helium leaving the evaporator (at point 3) is nominally at 1.8 K. To maximize $Q_{1.8}$ at a given helium flow rate, as seen from Eq. 4.31, $(h_3 - h_2)$ must be maximized, or the helium temperature at point 2 must be as close to 1.8 K as possible. The heat exchanger is used to cool down this incoming helium.

c)   Appropriate values for h_3 (1.8 K, 1 atm) and h_2 (3.0 K, 1 atm), both from Table A2.2 (Appendix II) are:

$$h_3 = 24.0\,\text{J/g}$$
$$h_2 = 5.23\,\text{J/g}$$

Solution to Problem 4.6

With these values inserted into Eq. 4.31, we obtain:

$$Q_{1.8} = \dot{m}_{he}(24.0\,\text{J/g} - 5.23\,\text{J/g}) = 20\,\text{W}$$

Solving for $\dot{m}_{he}$, we obtain $\dot{m}_{he} = 1.07\,\text{g/s}$, which corresponds to a liquid helium (at 4.2 K, 1 atm) supply rate of 31 liter/h.

Note that in addition to this 31 liter/h liquid replenishment rate, it is necessary to transfer liquid into the reservoir to replenish that boiled off by heat inputs caused by current leads, *etc.*

d) For an isentropic pump, the pump power requirement, $\mathbb{P}_s$, is given by:

$$\mathbb{P}_s = \dot{m}_{he}\left(\frac{\gamma}{\gamma-1}\right)(P_4 v_4)\left[\left(\frac{P_5}{P_4}\right)^{\frac{\gamma-1}{\gamma}} - 1\right] \qquad (S6.2)$$

$\gamma = C_p/C_v$ for helium is 1.67; v_4 is the specific volume at point 4, which for helium at 300 K and 12.6 torr is 371 m^3/kg. With $P_4 = 12.6\,\text{torr} = 1.68 \times 10^3\,\text{Pa}$, $P_5/P_4 = 60.3$, and $\dot{m}_{he} = 0.001\,\text{kg/s}$ (1 g/s), we obtain: $\mathbb{P}_s = 6487\,\text{W}$ or $\sim$9 hp (horsepower). Note that this is for an ideal case; the power requirement for a real pump would be roughly $\sim$30 hp.

e) The pressure drop between points 3 and 4 for a necessary helium flow rate must be much less than 12.6 torr, the operating pressure at point 3. Note that the lower P_4 gets (below 12.6 torr), the more $\mathbb{P}_s$ is needed.

For the Hybrid III system, a piping system connecting the top of the "chimney" to the pump consists basically of a 15-cm i.d. pipe, 13 m long. It has a total of five 90° bends and one shut-off valve and can handle mass flow rates up to $\sim$2 g/s with a total pressure drop less than 1 torr.

Also, precautions should be taken not to introduce contaminants to the evaporator through the piping system. Such contaminants usually get solidified at the narrowest passage areas, *e.g.* J-T valves, and block the line. In Hybrid III each J-T valve has a heater attached to melt away solidified contaminants.

f) For narrow channels filled with 1-atm superfluid helium, as discussed in **4.3.1**, a function $X(T)$ given by Eq. 4.1 and plotted in Fig. 4.4 relates conduction heat flux q [W/cm^2], channel length L [cm], and two end temperatures, T_{wm} (warm) and T_{cl} (cold). With $T_{wm} = T_\lambda$, $T_{cl} = 1.8\,\text{K}$, and $L = 10\,\text{cm}$, we have: $X(T_\lambda) = 0$, and $X(1.8) = 360$ (in appropriate units). Thus, we obtain:

$$360 = q^{3.4}10 \qquad (S6.3)$$

Solving Eq. $S6.3$ for q, we have: $q = 1.46\,\text{W/cm}^2$. With the channel cross sectional area of 2.6 mm^2, total conduction heat input to the magnet vessel through the hydraulic communication channel becomes $\sim$40 mW.

Solution to Problem 4.6

Three carbon resistors are placed in the bottom area of the reservoir to measure liquid temperature in that area. (Problem 4.10 studies carbon resistors as thermometers.) Measurements indicate that the liquid temperature at the reservoir bottom is $\sim$3 K, clearly above T_λ. Often with subcooled 1.8-K cryostats of this basic design configuration, the liquid at the reservoir bottom is very close to T_λ.

Because a cross section of 2.6 mm^2 is insufficient to maintain atmospheric pressure in the magnet vessel in the event of a magnet quench, the Hybrid III cryostat has a "poppet" valve with a cross section of 40 mm^2. Under normal conditions the poppet valve is kept shut by means of a spring; it opens only when a pressure is built up in the magnet vessel.

g) From Table A2.2 (Appendix II), the liquid densities at 1 atm are 125 kg/m^3 at 4.2 K and 145 kg/m^3 at 1.8 K. Thus for Hybrid III, the 250-liter vessel starts with about 31 kg of liquid at 4.2 K and ends up with about 36 kg of liquid at 1.8 K. That is, about 5 kg of liquid must be supplied to the vessel. In terms of volume at 4.2 K, this translates to 40 liters.

Although a cross section of 2.6 mm^2 provided by the hydraulic communication channel is adequate to transport this additional mass of liquid over a cooldown period of $\sim$2 hr, the poppet valve with a total flow passage area of 40 mm^2 is kept open until the liquid in the vessel reaches T_λ.

h) An expression for the steady-state temperature profile along the lead that is in the normal state and generating Joule heating is given by a modified expression of Eq. S4.3 (Problem 4.4, p. 138):

$$T(z) = -\frac{\rho I_t^2}{2A^2 k}z^2 + \left[\frac{(T_\ell - T_0)}{\ell} + \frac{\rho I_t^2 \ell}{2A^2 k}\right]z + T_0 \qquad (S6.4)$$

where ρ and k are the electrical and thermal conductivity of copper used in the lead. A and ℓ are the lead's cross sectional area and length separating the cold and warm ends, respectively, at T_0 (1.8 K) and T_ℓ ($\sim T_\lambda$). According to Problem 4.4, an optimal lead satisfies the condition given by Eq. 4.25 (p. 137):

$$\left(\frac{I_t \ell}{A}\right)_{ov} \equiv \zeta_{ov} = \sqrt{\frac{2\tilde{k}(T_\ell - T_0)}{\tilde{\rho}}} \qquad (4.25)$$

Combining Eqs. S6.4 and 4.25 and defining a new variable $y \equiv z/\ell$, we have:

$$T(y) = -(T_\ell - T_0)y^2 + 2(T_\ell - T_0)y + T_0 \qquad (S6.5)$$

Either by actually plotting Eq. S6.5 or finding the location at which $dT/dy = 0$ occurs, we can show that a peak in the temperature distribution given by Eq. S6.5 occurs at $y = 1$ and thus the peak temperature is T_λ. That is, even if the lead is driven normal, if the lead satisfies the criterion given by Eq. 4.25, the conduction cooling is sufficient to recover the lead.

Problem 4.7: Residual gas heat transfer into a cryostat

This problem deals with heat input into a cryostat by residual helium gas present within the space in the cryostat's multi-walled structure. The Hybrid III cryostat is used as an example.

Heat Input by Residual Gas: "High" Pressure Limit

When the pressure of a gas is sufficiently high, the mean free path (λ_g) of the gas is generally much shorter than a typical distance (d) separating the two surfaces at different temperatures in a cryostat. Under this condition of $\lambda_g \ll d$, the thermal conductivity of a gas (k_g), according to the kinetic theory, is proportional only to the mean velocity of the molecule ($\bar{v}$), which in turn varies as $\sqrt{T}$. The important point here is that when $\lambda_g \ll d$, k_g is independent of gas pressure, P_g. For a cryostat this condition is of secondary interest because the space in the cryostat's multi-walled structure is usually kept at "vacuum," which means P_g of $\sim 10^{-4}$ torr ($\sim 10^{-2}$ Pa) or less.

The kinetic theory also shows that λ_g is proportional to T and inversely proportional to P_g:

$$\lambda_g \propto \frac{T}{P_g} \tag{4.32}$$

For helium gas at $T = 300$ K and $P_g = 10^5$ Pa (1 atm), λ_g is $\sim 0.1\,\mu$m, or the condition $\lambda_g \ll d$ is clearly satisfied in this "high" pressure limit. At a "vacuum" pressure of $P_g \sim 10^{-4}$ torr, however, λ_g becomes ~ 1 m and the condition $\lambda_g \ll d$ is violated. That is, when $\lambda_g \gg d$, we have a "low" pressure limit, discussed below.

Heat Input by Residual Gas: "Low" Pressure Limit

Under a vacuum pressure P_g of $\sim 10^{-4}$ torr or less, k_g is no longer independent of pressure: k_g becomes directly proportional to P_g. For a parallel-plate configuration with one plate at a cold temperature T_{cl} [K] and the other at a warm temperature T_{wm} [K], heat flux (q_g [W/m^2]) from the warm plate to the cold plate by a "residual" helium gas at pressure P_g [Pa] may be given by [4.14]:

$$q_g = \eta_g P_g (T_{wm} - T_{cl}) \tag{4.33}$$

η_g [W/m^2 Pa K] depends not only on T_{wm} and T_{cl} but also on the so-called accommodation coefficient, which for helium varies from 0.3 at room temperature to 1 at 4.2 K. Table 4.7 presents η_g and q_g at a pressure P_g of 10^{-5} torr (1.33 mPa) across two parallel plates, respectively, at T_{cl} and T_{wm}.

Compute total heat input into the Hybrid III magnet vessel at 4.2 K by residual helium gas of pressure 10^{-5} torr present in the vacuum space. Use the "parallel-plates" approximation, *i.e.* the surface area of one plate at a cold temperature T_{cl} and that of the other plate at a warm temperature T_{wm} are the same. The magnet vessel surface areas are: 1) 7.3 m^2 facing the 20-K radiation shields; 2) 2.8 m^2 facing the 80-K radiation shields.

Problem 4.7: Residual gas heat transfer into a cryostat

Table 4.7: Heat Conduction by
Residual He Gas at P_g of 10^{-5} Torr [4.14]

$T_{cl} \sim T_{wm}$ [K]	η_g [W/m^2 Pa K]	q_g [W/m^2]
4~20	0.35	8
4~80	0.21	21
4~300	0.12	47
80~300	0.04	11

Vacuum Pumping System

Figure 4.11 shows a schematic diagram of a typical vacuum system used in the operation of a superconducting magnet. The cryostat vacuum space outlet is connected to a diffusion pump via a cold trap and the diffusion pump is in turn connected to a mechanical pump. The cold trap—a 77-K surface one side of which is in direct contact with boiling liquid nitrogen—serves to condense oil, water, and other contaminant vapors that would otherwise be condensed on the cold surfaces of the cryostat. A procedure customarily used to evacuate the cryostat vacuum space is to reach a vacuum of $\sim 10^{-2}$ torr by the mechanical pump alone and then use the diffusion pump, eventually reaching a vacuum of $10^{-5} \sim 10^{-6}$ torr.

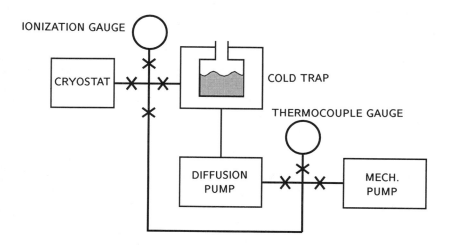

Fig. 4.11   Schematic diagram of a typical pumping vacuum system used in the operation of a superconducting magnet.

Solution to Problem 4.7

For a magnet vessel surface at 4.2 K exposed to 20-K radiation shields, q_g at $P_{rg} = 10^{-5}$ torr is 8 mW/m². Thus, for a surface area of 7.3 m² under the parallel-plates approximation, the total heat input becomes 58 mW. Similarly, for a magnet vessel surface at 4.2 K exposed to 80-K radiation shields, q_g is 21 mW/m², or a heat input of ~60 mW for a surface area of 2.8 m². Thus the combined heat input to the 4.2-K magnet vessel becomes ~120 mW, which is quite acceptable.

At a poorer vacuum, e.g. $P_{rg} = 10^{-3}$ torr ($\lambda_g \sim 0.1$ m, still greater than "d" in most cryostats), the heat input becomes ~12 W, which is intolerably high. It is therefore important to keep the cryostat vacuum pressure to less than ~10^{-4} torr.

Vacuum Gauges

Two types of vacuum gauges are commonly used for cryostats: 1) thermocouple; and 2) ionization. Below is a brief description of each type.

Thermocouple: The thermocouple gauge relies on the dependance on pressure of the thermal conductivity of a gas, valid in the "low" pressure limit discussed above. The thermocouple junction is situated in a tube connected to the vacuum space to be measured and its temperature is set by a heater. The gas provides cooling, which varies with the gas pressure; the change in the induced current through the junction circuit is a measure of the vacuum pressure. The gauge's range of applicability is 10^{-3}~1 torr or the range covered by mechanical pumps.

Ionization: For the vacuum range between ~10^{-6} and ~10^{-3} torr, the operating range in most cryostats, the ionization gauge is used most widely. There are two versions: 1) hot-cathode; and 2) cold-cathode.

Hot-Cathode: This gauge consists of a heated filament (hot cathode), an anode, and a negatively biased ion collector plate, all housed in a tube that connects to the vacuum space to be measured. The electrons flowing from the filament to the anode collide with gas molecules, creating ionized molecules that are drawn to the collector plate and measured as a current through the measurement circuit. Because the molecules are ionized by electrons, the ion current depends also on the number of electrons bombarding the molecules: accurate pressure measurement thus requires careful control of filament current. The hot-cathode gauge used in the Hybrid III cryostat is turned off during magnet operation to minimize "filament fatigue" caused by the filament's oscillating motion that results from the Lorentz interaction of the filament supply current (60 Hz) and the magnet's fringing field.

Cold-Cathode: Known as the Philips gauge, it uses a cold cathode and two parallel anode plates with a magnetic field applied in the direction normal to the anode plates, which are turned on (~2 kV) one at a time. The small number of electrons produced by the cold cathode are thus made to travel in a helical path alternatively toward one of the two plates. This configuration effectively increases the collision chances between the small number of electrons and gas molecules. Unlike hot filaments, cold cathodes do not contaminate the gas nor are they destroyed in the event of a loss of vacuum.

Problem 4.8: Radiation heat transfer into a cryostat

This problem deals with heat input to a cryostat by radiation. The Hybrid III cryostat is used as an example.

Radiation Heat Transfer: Applications to a Cryostat

The theory of radiative heat transfer begins with the Stefan-Boltzmann equation:

$$q_r = e_r \sigma_{SB} T^4 \tag{4.34}$$

q_r is the radiative heat flux [W/m^2] from a surface at temperature T [K]. e_r is the total emissivity at T. σ_{SB} is the Stefan-Boltzmann constant, $5.67{\times}10^{-8}$ W/m^2 K^4. What makes computation of radiation heat input to a cryostat usually less straight-forward than suggested by Eq. 4.34 is the task of having to choose a correct value of e_r for each of the two surfaces that are radiating heat according to Eq. 4.34. For the parallel-plates configuration with emissivities of $[e_r]_{cl}$ and $[e_r]_{wm}$, respectively, for one surface at cold temperature T_{cl} and the other at warm temperature T_{wm}, the effective total emissivity, $[e_r]_{cw}$ is given by:

$$[e_r]_{cw} = \frac{[e_r]_{cl}[e_r]_{wm}}{[e_r]_{cl} + [e_r]_{wm} - [e_r]_{cl}[e_r]_{wm}} \tag{4.35}$$

Although theory distinguishes among "parallel-plate," "cylindrical," and "spherical" configurations, in most cryostat applications, the parallel-plate configuration suffices even for nonparallel-plate configurations. (Nonparallel-plate configurations usually mean that the two surfaces have different areas.) This is because: 1) in most cryostats, the distance separating the two surfaces is generally much less than a characteristic surface length; and 2) an error that would invariably be introduced in an estimate of radiative heat input through this geometric approximation is still likely to be considerably less than that introduced by uncertainties associated with emissivities of the surfaces in question. Equation 4.34 is thus modified to:

$$q_r = [e_r]_{cw} \sigma_{SB} (T_{wm}^4 - T_{cl}^4) \tag{4.36}$$

Effect of Superinsulation Layers

Examination of Table 4.8 reveals that the largest radiative heat load in a cryostat is usually on an 80-K shield, which receives heat from the 300-K surface. Thus, it is quite customary to place a number of aluminized Mylar sheets, known as superinsulation, in the vacuum space between the 80-K and 300-K surfaces. Assuming $[e_r]_{cw}$ of all surfaces to be nearly identical, the presence of N_i layers of superinsulation modifies Eq. 4.36 to:

$$q_r = \frac{[e_r]_{cw}}{N_i + 1} \sigma_{SB} (T_{wm}^4 - T_{cl}^4) \tag{4.37}$$

Equation 4.37 indicates that the presence of even one layer of superinsulation between the 300-K and 80-K surfaces reduces q_r by a factor of 2. A rule of thumb is to use about 10 layers for each 1-cm spacing.

Problem 4.8: Radiation heat transfer into a cryostat

Table 4.8: "Typical" Values of Radiation Heat Flux [4.14]

Material	$T_{cl} \sim T_{wm}$ [K]	$[e_r]_{cw}$	q_r [mW/m^2]
Copper			
as received	4~20	0.03*	0.3
	4~80	0.06	140
	80~300	0.12	55,000
mechanically	4~20	0.01*	0.1
polished	4~80	0.02	46
	80~300	0.06	27,000
Stainless steel			
as received	4~20	0.06*	0.6
	4~80	0.12	280
	80~300	0.34	155,000
mechanically	4~20	0.04*	0.4
polished	4~80	0.07	160
	80~300	0.12	55,000
electropolished	4~20	0.03*	0.3
	4~80	0.06	140
	80~300	0.10	46,000
Aluminum			
as received	4~20	0.04*	0.4
	4~80	0.07	160
	80~300	0.49	224,000
mechanically	4~20	0.03*	0.3
polished	4~80	0.06	140
	80~300	0.10	46,000
electropolished	4~20	0.02*	0.2
	4~80	0.03	70
	80~300	0.08	37,000
foil	4~80	0.01	20
	80~300	0.06	28,000

* One half of the 4.2~80 K value.

Problem 4.8: Radiation heat transfer into a cryostat

Table 4.8 presents "typical" values of $[e_r]_{cw}$ over three different temperature spans for three common materials used in cryostats [4.14] and values of corresponding q_r [mW/m²]. As indicated in the footnote of Table 4.8, values of $[e_r]_{cw}$ over a temperature span of 4.2 to 20 K, not given by Noguchi [4.14], are taken to be half of those corresponding to a temperature span of 4.2 to 80 K. It should be stressed that values given in Table 4.8 are typical; actual values to be used for the surfaces of a specific cryostat can vary at least by a factor of 2 in either directions.

a) Compute the total heat input to the Hybrid III magnet vessel at 4.2 K by radiation from the 20-K and 80-K radiation shields. The magnet vessel surface areas are: 1) 7.3 m² facing the 20-K radiation shields; 2) 2.8 m² facing the 80-K radiation shields.

Assume that the total surface area of the 20-K panels facing the magnet vessel is 7.3 m² and the total surface area of the 80-K panels facing the magnet vessel is 2.8 m². Also assume all three surfaces to be of mechanically polished stainless steel.

b) Compute the net heat input to the 80-K radiation shield facing the 300-K surface. The total area of the 80-K panels facing the 300-K surface is 11.7 m².

Assume that the total 300-K surface facing the 80-K panels is also 11.7 m². The 80-K panels also face the 20-K panels and the magnet vessel. The 80-K panels receive cooling from these cold surfaces; it may be neglected for computing the total heat load to the 80-K panels.

c) The Hybrid III cryostat does *not* contain "superinsulation" layers. Discuss the effect of superinsulation in reducing radiation heat input.

Practical Considerations of Emissivity

It is important to note that radiation is an electromagnetic phenomenon: emissivity e_r increases with "surface" electrical resistivity of the material. The material's emissivity is thus affected in the same way the material's surface electrical resistivity is affected. Based on this principle, we may list the following rules of thumb on emissivity:

- For the same temperature range, e_r values of copper are smaller than those of aluminum, which in turn are smaller than those of stainless steel.

- For the same surface, e_r decreases with temperature. Thus, copper's e_r decreases more markedly than does stainless steel's.

- e_r of conductive metal is more sensitive to surface contamination than that of nonconductive metal. Contamination includes oxidation and alloying.

- Mechanical polishing sometimes improves (decreases) e_r and sometimes degrades (increases) e_r. If an oxide layer from conductive metal's surface is removed by mechanical polishing, the result is an improvement. If the metal's resistivity is increased by work-hardening, the result is degradation.

Solution to Problem 4.8

a)   Using appropriate values of q_r given in Table 4.8 for the mechanically polished stainless steel surface, we have:

20-K panels to vessel

$$Q_r = (0.4 \times 10^{-3}\,\mathrm{W/m^2})(7.3\,\mathrm{m^2}) \qquad (S8.1)$$

$$\simeq 3\,\mathrm{mW}$$

80-K panels to vessel

$$Q_r = (160 \times 10^{-3}\,\mathrm{W})(2.8\,\mathrm{m^2}) \qquad (S8.2)$$

$$= 448\,\mathrm{mW}$$

Thus the total radiative heat input into the vessel is $\sim$0.5 W or unlikely to be more than $\sim$1 W.

b)   Again, using appropriate values of q_r given in Table 4.8 for the mechanically polished stainless steel surface, we have:

300-K to 80-K panels

$$Q_r = (55\,\mathrm{W/m^2})(11.7\,\mathrm{m^2}) \qquad (S8.3)$$

$$= 644\,\mathrm{W}$$

Thus the net radiative heat input into the 80-K panels is $\sim$650 W.

c)   Unfortunately—by oversight—the superinsulation wrapping around the 80-K panels was not installed by the vendor contracted by MIT to supply the Hybrid III cryostat. The absence of superinsulation causes a much higher heat load than the designed heat load on the 80-K panels. Even the presence of one layer of superinsulation would have reduced the heat load to $\sim$150 W. A factor of 4 reduction is possible because there are two factors of 2 reductions—one from 0.12 to 0.06 in $[e_r]_{cw}$ (Table 4.8) and the other $1/(N_i + 1)$ where $N_i = 1$ (Eq. 4.37). Because of this oversight, these panels operate at higher temperatures, ranging from 95 K to as high as 120 K, in turn causing the 20-K panels to operate at close to 25 K.

"We may not arrive at our port within a calculable period,
but we would preserve the true course." —Henry D. Thoreau

Problem 4.9: Laboratory-scale hydrogen (neon) condenser

This problem deals with design principles for a laboratory-scale condenser for liquefaction of hydrogen; the principles are also applicable for liquefaction of neon. A small quantity of liquid hydrogen or neon is useful for measuring the current-carrying capacities of HTS, because immersing a test sample in boiling liquid at a known pressure is the easiest way to maintain (and be sure of) the superconductor's temperature, particularly for currents above 10 A.

Because of the explosive nature of hydrogen in both liquid and vapor phases, the procedure enforced at FBNML for liquid-hydrogen experiments consists of five stages: 1) liquefaction in a cryostat; 2) sample holder insertion into the cryostat after the completion of liquefaction; 3) magnet run; 4) sample holder extraction from the cryostat; and 5) discharge from the cryostat of liquid unused in the magnet run. Safety rules stipulate that Stages 1, 2, 4, and 5 be performed outdoors, that in Stage 3 the cryostat be completely sealed to keep air from condensing, and that the hydrogen vapor be vented outdoors through a pumped line to keep it from mixing with air inside the building.

Figure 4.12 illustrates the liquefaction setup in Stage 1, with the condenser placed in a cryostat [4.15]. A regulated high-pressure tank supplies the hydrogen gas, which is cooled and liquefied through the condenser by a stream of liquid helium supplied from a standard 25-liter storage dewar.

Note that this hydrogen liquefier can also be used to condense about the same amount of liquid neon, again using liquid helium as a refrigerant [4.16].

a) Using an enthalpy balance argument, show that an expression for the ratio of 4.2-K liquid helium mass required to the mass of hydrogen liquefied at 20 K, $M_{he}(T)/M_{h2}$, is given by:

$$\frac{M_{he}(T)}{M_{h2}} = \frac{(h_L)_{h2}}{[h_{he}(T) - h_{he}(4.2\,\text{K}, \text{liq.})]} \qquad (4.38)$$

where $(h_L)_{h2}$ is the volumetric heat of vaporization of hydrogen. Assume that the helium entering the condenser is 100% liquid. Note that the ratio is a function of temperature, T, because the upper temperature limit of the helium can be as low as 4.2 K to as high as 20 K, depending on condenser design.

b) Using Eq. 4.38, compute the liquid helium transfer rates (liter/h) required to liquefy hydrogen at a rate of 1 liter/h with: 1) an ideal condenser; and 2) a condenser with T at 10 K.

c) The condenser, in addition to condensing hydrogen, must be able to cooldown the hydrogen gas from room-temperature to its condensation temperature (20 K) at a specified rate. For a condensation rate of 1 liter/h, what would the required length be if the condenser is to be made of 5-mm o.d. copper tubing? The question really is to find the condenser surface area sufficient to cooldown the gas from room-temperature to 20 K at this condensation rate.

d) The hydrogen condenser built at FBNML was used to condense neon to make J_c measurements for HTS samples [4.16]. Show that the condenser is almost as effective for neon as it is for hydrogen.

Problem 4.9: Laboratory-scale hydrogen (neon) condenser

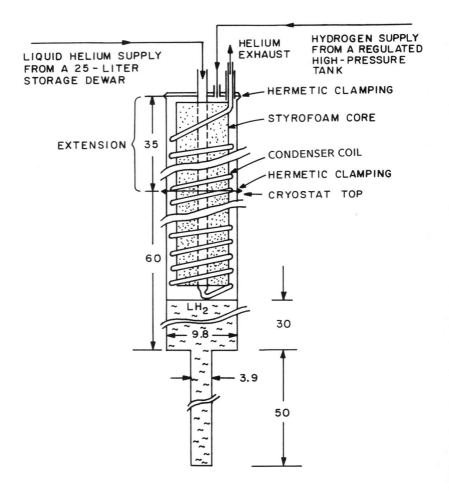

Fig. 4.12   Hydrogen liquefaction setup in stage 1. Dimensions are in mm.

Solution to Problem 4.9

a) The hydrogen gas entering the cryostat at the top end is cooled gradually as it moves downwards. Condensation takes place on the cold part of the condenser where the surface temperature of the coil is at a temperature below 20 K, the condensation temperature of hydrogen. (Most of the condenser surface is needed to cool the room-temperature hydrogen to 20 K. Condenser design is thus dictated really by the cooldown process as discussed in **c)** of this problem.)

The only refrigeration available to condense hydrogen at 20 K is thus the total enthalpy increase of helium from 4.2 K liquid to a vapor temperature T, which cannot exceed 20 K: $M_{he}[h_{he}(T) - h_{he}(4.2\,\mathrm{K, liq.})]$, where M_{he} [g] is the total mass of helium needed for hydrogen condensation and $h_{he}(T)$ [J/g] and $h_{he}(4.2\,\mathrm{K, liq.})$ [J/g] are the specific enthalpies of helium, respectively, at T and 4.2 K (liquid phase). This enthalpy increase must match the latent heat of vaporization of hydrogen: $M_{h2}(h_L)_{h2}$, where M_{h2} [g] is the total mass of hydrogen condensed and $(h_L)_{h2}$ [J/g] is the latent heat of vaporization of hydrogen. Thus:

$$M_{he}[h_{he}(T) - h_{he}(4.2\,\mathrm{K, liq.})] = M_{h2}(h_L)_{h2} \qquad (S9.1)$$

Solving for $M_{he}(T)/M_{h2}$ from the above equation, we obtain:

$$\frac{M_{he}(T)}{M_{h2}} = \frac{(h_L)_{h2}}{[h_{he}(T) - h_{he}(4.2\,\mathrm{K, liq.})]} \qquad (4.38)$$

The mass ratio is minimum at $T = 20$ K, achievable only with an "ideal" condenser. Typically $T \sim 10$ K, or a temperature difference of $\sim$10 K, is required for real condensers.

b) With $(h_L)_{h2} = 443$ J/g, $h_{he}(20\,\mathrm{K}) = 117.9$ J/g, and $h_{he}(4.2\,\mathrm{K, liq}) \simeq 9.7$ J/g, we have, from Eq. 4.38, the minimum mass ratio of 4.1. For $T = 10$ K, which corresponds to a temperature difference of 10 K between the helium vapor and condensing hydrogen, the mass ratio becomes 8.0, or almost twice that for the ideal condenser.

Because the density of "normal" hydrogen at 20 K is 0.071 g/cm³, 1 liter of liquid hydrogen has a mass of 71 g. The M_{he} required to liquefy 1 liter of hydrogen is 291 g for the ideal condenser; it is 568 g for the 10-K condenser. Corresponding liquid helium transfer rates are 2.3 liter/h and 4.5 liter/h.

c) The total heat transferred, Q_{cn}, from hydrogen to helium in the condenser, cooling hydrogen from 300 to 20 K, is given by:

$$Q_{cn} = \tilde{h}_{cn} A_{cn} \Delta T_{cn} \qquad (S9.2)$$

where $\tilde{h}_{cn}$, A_{cn}, and ΔT_{cn}, all applicable at the hydrogen side of the condenser surface, are, respectively, the convective heat transfer coefficient, total area, and temperature difference. $\tilde{h}_{cn}$ is related to the Nusselt number, Nu, of the hydrogen flow and the hydrogen thermal conductivity, k, as:

Solution to Problem 4.9

$$\tilde{h}_{cn} = \frac{\text{Nu}\, k}{D_{cn}} \tag{S9.3}$$

D_{cn} is the outside tube diameter of the condenser coil. Nu is related to the Prandtl number Pr and the Reynolds number Re of hydrogen flow by:

$$\text{Nu} = 0.8\text{Pr}^{0.36}\text{Re}^{0.4} \tag{S9.4}$$

Reynolds number Re is given by:

$$\text{Re} = \frac{\rho V D_{cn}}{\mu} \tag{S9.5}$$

To produce 1 liter (71 g) of liquid hydrogen from room-temperature hydrogen gas in 1 hour, Q_{cn} is given by:

$$Q_{cn} = \frac{(0.071\,\text{g})[h_{h2}(293\,\text{K}) - h_{h2}(20\,\text{K}, \text{liq.})]}{3600\,\text{s}} = 76\,\text{W}$$

Although an appropriate value of Re can vary with condenser design, typically it is in the range of 1~100. With $D_{cn} \simeq 5\,\text{mm}$, we find $\tilde{h}_{cn}$ to be ~500 W/m² K for $\Delta T_{cn} \sim 1\,\text{K}$. Thus the required A_{cn} becomes ~0.15 m² which, for a 5-mm o.d. copper tubing, translates to a length of ~10 m.

d) Because $\tilde{h}_{cn}$ is proportional to Nu k (Eq. $S9.3$), we can compare this for hydrogen and neon. Thus:

$$\frac{\tilde{h}_{cn}|_{h2}}{\tilde{h}_{cn}|_{ne}} = \left(\frac{\text{Nu}|_{h2}}{\text{Nu}|_{ne}}\right)\left(\frac{k|_{h2}}{k|_{ne}}\right) \tag{S9.6}$$

Combining Eqs. $S9.4$~$S9.6$, we have:

$$\frac{\tilde{h}_{cn}|_{h2}}{\tilde{h}_{cn}|_{ne}} = \left(\frac{\text{Pr}|_{h2}}{\text{Pr}|_{ne}}\right)^{0.36}\left(\frac{\rho|_{h2}}{\rho|_{ne}}\right)^{0.4}\left(\frac{\mu|_{ne}}{\mu|_{h2}}\right)^{0.4}\left(\frac{k|_{h2}}{k|_{ne}}\right) \tag{S9.7}$$

Inserting values of each gas' parameters near, for example, its saturation temperature (Table A2.4, Appendix II), we have:

$$\frac{\tilde{h}_{cn}|_{h2}}{\tilde{h}_{cn}|_{ne}} = \left(\frac{0.765}{0.557}\right)^{0.36}\left(\frac{1.33\,\text{kg/m}^3}{9.37\,\text{kg/m}^3}\right)^{0.4}\left(\frac{4.77\,\mu\text{Pa}\,\text{s}}{1.11\,\mu\text{Pa}\,\text{s}}\right)^{0.4}\left(\frac{15.8\,\text{mW/m}\,\text{K}}{9.9\,\text{mW/m}\,\text{K}}\right)$$

$$= (1.12)(0.46)(1.79)(1.6) = 1.47$$

Thus, the condenser is less effective for neon than for hydrogen; nevertheless it should be adequate for condensing neon, as it turned out in actual use at FBNML.

Problem 4.10: Carbon resistor thermometers

Temperature measurement is one of the key requirements in the operation of superconducting magnets; it is not a straightforward matter in the cryogenic environment, particularly in the presence of magnetic field. This problem deals with low-temperature thermometry based on carbon resistors.

Despite their magnetoresistance effect, carbon resistors are often used in the operation of superconducting magnets, because they: 1) are sensitive (below ~20 K); 2) occupy small space; 3) are easy to calibrate and use; and 4) are inexpensive. The absolute uncertainty in temperature measurement, although dependent on temperature, is ~1 mK at 1.8 K.

The following logarithmic polynomial form relating T and R has been found to fit well with experimental points:

$$\frac{1}{T} = \sum_{j}^{n} m_j (\ln R)^j \tag{4.39}$$

where T is in K and R is in Ω. The m_js are constants (with appropriate units) derived from calibrated points. Typically, j runs from 0 to 2 or from -1 to 1 (meaning at least three calibration points are required):

$$\frac{1}{T} = m_{0a} + m_{1a} \ln R + m_{2a} (\ln R)^2 \tag{4.40a}$$

$$\frac{1}{T} = \frac{m_{-1b}}{\ln R} + m_{0b} + m_{1b} \ln R \tag{4.40b}$$

For this problem, a nominal 47-Ω (1/8 W) resistor with the following sets of coefficients are used: 1) $m_{0a} = -0.647403121$; $m_{1a} = 0.115776702$; $m_{2a} = 0.00650089$ and 2) $m_{-1b} = 1.749782525$; $m_{0b} = -1.464991212$; $m_{1b} = 0.242426753$ [4.17]. These m_js were determined from 7 calibration points taken between 4.23 K and 1.58 K [4.17]. Equation 4.40 in either form with these m_js is thus accurate only over this temperature range.

a) Using Eq. 4.40a, determine the value of R [Ω] at $T = 4.23$ K to within $\pm 0.1\,\Omega$. You may use the following approximate values for the coefficients: $m_{0a} = -0.6474$; $m_{1a} = 0.1158$; $m_{2a} = 0.0065$.

b) Now use Eq. 4.40b and determine R [Ω] at $T = 4.23$ K to within $\pm 0.1\,\Omega$. You may use the following approximate values for the coefficients: $m_{-1b} = 1.7498$; $m_{0b} = -1.4650$; $m_{1b} = 0.2424$.

c) What is the sensitivity ($S \equiv |dR/dT|$, in Ω/K) of this carbon resistor at 3.35 K ($R \simeq 439\,\Omega$) and at 1.80 K ($R \simeq 1577\,\Omega$) ?

d) The resistance R is determined by measuring the voltage V across the resistor at a given current I supplied by a stable current source. Estimate the experimental uncertainty in temperature, ΔT, for this resistor when the uncertainties in voltage $\Delta V/V$ and supply current $\Delta I/I$ are, respectively, $\pm 0.3\%$ and $\pm 0.1\%$. Compute values of ΔT for 4.23 K and 1.80 K.

Solution to Problem 4.10

a) With $T = 4.23\,\mathrm{K}$ inserted into Eq. 4.40a, we have:

$$0.2364 = -0.6474 + 0.1158 \ln R + 0.0065(\ln R)^2 \qquad (S10.1)$$

Equation $S10.1$ can be written as:

$$(\ln R)^2 + 17.8154 \ln R - 135.9692 = 0 \qquad (S10.2)$$

Solving Eq. $S10.2$ for R, we find: $R = 319.8\,\Omega$. The actual measured value at $4.23\,\mathrm{K}$ (one of the calibrated points) is $319.8\,\Omega$.

b) Similarly, Eq. 4.40b becomes:

$$0.2364 = \frac{1.7498}{\ln R} - 1.4650 + 0.2424 \ln R \qquad (S10.3)$$

which can be written as:

$$(\ln R)^2 - 7.0190 \ln R + 7.2186 = 0 \qquad (S10.4)$$

Solving Eq. $S10.4$ for R, we find: $R = 319.7\,\Omega$.

c) We differentiate Eq. 4.40a with respect to R and obtain dR/dT:

$$-\frac{1}{T^2}\frac{dT}{dR} = \frac{m_{1a}}{R} + \frac{2m_{2a}\ln R}{R}$$

$$\frac{dT}{dR} = -T^2 \left(\frac{m_{1a} + 2m_{2a}\ln R}{R} \right) \qquad (S10.5a)$$

$$S \equiv \left| \frac{dR}{dT} \right| = \frac{R}{T^2(m_{1a} + 2m_{2a}\ln R)} \qquad (S10.5b)$$

At $T = 3.35\,\mathrm{K}$ ($R \simeq 439\,\Omega$), we have:

$$S_{3.35} = \frac{439}{(3.35)^2[0.1158 + 2(0.0065)\ln(439)]} \simeq 201\,\Omega/\mathrm{K}$$

At $T = 1.80$ K ($R = 1535\,\Omega$), we have:

$$S_{1.80} = \frac{1577}{(1.80)^2[0.1158 + 2(0.0065)\ln(1577)]} = 2301\,\Omega/\mathrm{K}$$

The sensitivity clearly *decreases* with *increasing* temperature.

Solution to Problem 4.10

Because R increases rapidly as temperature is decreased, one important point to remember is to use a low value of supply current I to keep the resistor from being "overheated." A rule of thumb for use in the liquid helium temperature range of 1.8~4.2 K is to select a supply current so that V across the resistor is in the range 1~10 mV. With $R \sim 2000\,\Omega$, a voltage of 10 mV implies a supply current of no greater than $\sim 5\,\mu$A or a total dissipation within the resistor of ~ 50 nW.

c) Once S is defined, we have:

$$\Delta T = \frac{\Delta R}{S} \qquad (S10.6)$$

Because $R = V/I$, uncertainties in R, ΔR, may be related to ΔV and ΔI as:

$$\frac{\Delta R}{R} = \sqrt{\left|\frac{\Delta V}{V}\right|^2 + \left|\frac{\Delta I}{I}\right|^2} \qquad (S10.7)$$

Combining Eqs. $S10.6$ and $S10.7$, we obtain:

$$\Delta T = \frac{R(T)}{S(T)}\sqrt{\left|\frac{\Delta V}{V}\right|^2 + \left|\frac{\Delta I}{I}\right|^2} \qquad (S10.8)$$

As indicated by Eq. $S10.8$, ΔT is clearly dependent on T through $R(T)$ and $S(T)$. For $T = 4.23$ K, we have $R \simeq 320\,\Omega$ and $S = 201\,\Omega/$K, and:

$$\Delta T_{4.23} = \frac{320}{201}\sqrt{(0.003)^2 + (0.001)^2} = \pm 5.0\,\text{mK}$$

For $T = 1.80$ K, we have $R = 1577\,\Omega$ and $S = 2301\,\Omega/$K, and:

$$\Delta T_{1.8} = \frac{1577}{2301}\sqrt{(0.003)^2 + (0.001)^2} = \pm 2.2\,\text{mK}$$

These values indicate that measurement error in terms of absolute temperature *increases* with *increasing* temperature. In fact carbon resistors are not usable as accurate thermometers above ~ 20 K.

Effects of a Magnetic Field on Thermometers

A magnetic field affects most low-temperature thermometers, including carbon resistors [4.18]. Carbon resistors cannot be used for temperature measurements in which $\Delta T/T$ uncertainties must be kept less than 0.01%; they are acceptable as temperature sensors in 1.8-K cryostats provided that the sensors are located in a fringing magnetic field less than ~ 2 T. At 2.5 T, a typical value of $\Delta T/T$ for carbon resistors (47, 100, and 220 Ω) at 4.2 K is less than 1% or ΔT of no more than ~ 40 mK; this jumps to ~ 400 mK in a field of 14 T. Sensors having the smallest field effects are $SrTiO_3$ capacitors; their $\Delta T/T$ at 4.2 K is less than 1 mK in a field of 14 T [4.18].

References

[4.1] R.D. Cummings and J.L. Smith, Jr., "Boiling heat transfer to liquid helium," *Proc. Int'l Inst. of Refrigeration, Commission 1* (Pergamon Press, 1966), 85.

[4.2] D.N. Lyon, "Boiling heat transfer and peak nucleate boiling fluxes in saturated liquid helium," *Int'l Adv. in Cryogenic Eng.* (Plenum Press, 1965), 371.

[4.3] See, for example, C. Schmidt, "Transient heat transfer into a closed small volume of liquid or supercritical helium," *Cryogenics.* **28**, 585 (1988).

[4.4] G. Bon Mardion, G. Claudet, and P. Seyfert, "Practical data on steady state heat transport in superfluid helium at atmospheric pressure," *Cryogenics* **29**, 45 (1979).

[4.5] S. Van Sciver (personal communication, 1993).

[4.6] G. Claudet, C. Mwueia, J. Parain, and B. Turck, "Superfluid helium for stabilizing superconductors against local disturbances," *IEEE Trans. Magn.* **MAG-15**, 340 (1979).

[4.7] Steven W. Van Sciver, *Helium Cryogenics*, (Plenum Press, New York, 1986), 182.

[4.8] K.R. Efferson, "Helium vapor cooled current leads," *Rev. Sci. Instru.* **38**, 1776 (1967).

[4.9] J.L. Wu, J.T. Dederer, P.W. Eckels, S.K. Singh, J.R. Hull, R.B. Poepple, C.A Youngdahl, J.P. Singh, M.T. Lanagan, and U. Balachandran, "Design and testing of a high temperature superconducting current lead," *IEEE Trans. Mag.* **27**, 1861 (1991).

[4.10] Kazuo Ueda, Kiyoshi Takita, Toshio Uede, Masanao Mimura, Naoki Uno, Yasuzo Tanaka, "Thermal performance of a pair of current leads incorporating bismuth compound superconductors," *Advances in Superconductivity – V (ISS92)*, Eds. Y. Bando and H. Yamauchi (Springer-Verlag, Tokyo, 1993), 1235.

[4.11] See, for example, C. Taylor, R. Althaus, S. Caspi, W. Gilbert, W. Hassenzahl, R. Meuser, J. Reschen, R. Warren, "Design of epoxy-free superconducting dipole magnets and performance in both helium I and pressurized helium II," *IEEE Trans. Magn.* **MAG-17**, 1571 (1981).

[4.12] M.J. Leupold and Y. Iwasa, "A subcooled superfluid helium cryostat for a hybrid magnet system," *Cryogenics* **26**, 579 (1986).

[4.13] Gérard Claudet, Pierre Roubeau, and Jacques Verdier, "Method for the production of superfluid helium under pressure at very low temperature and an apparatus for carrying out said method," (U.S. Patent Application, February 11, 1975).

[4.14] Takashi Noguchi, "Vacuum insulation for a cryostat," *Cryogenic Engineering* (in Japanese) **28**, 355 (1993).

[4.15] Y. Iwasa, "Liquid hydrogen condenser and current lead designs for liquid hydrogen critical current measurements of superconductors," *Cryogenics* **31**, 174 (1991).

[4.16] Y. Iwasa and R.H. Bellis, "A three-lead test rig with a built-in neon condenser for field-dependent critical-current measurements of high-T_c superconductors at 27 K," *Cryogenics* **33**, 920 (1993).

[4.17] Sangkwon Jeong and Douglas H. Smith (an internal report, Cryogenic Engineering Laboratory, MIT, unpublished 1990).

[4.18] L.G. Rubin, B.L. Brandt, and H.H. Sample, "Some practical solutions to measurement problems encountered at low temperatures and high magnetic fields," *Adv. Cryogenic Eng.* **31**, 1221 (1986).

CHAPTER 5
MAGNETIZATION OF HARD SUPERCONDUCTORS

5.1 Introduction

This chapter discusses the magnetization of hard superconductors. When the first generation of superconducting magnets was built in the mid 1960s, many of them failed to reach their designed operating currents. These magnets quenched at currents far below their expected values because of a then unknown or little understood magnetic phenomenon called flux jumping. By the late 1960s, the reasons for flux jumping were elucidated, and solutions to eliminate it emerged soon afterwards. Today, well-established methods of producing superconducting wires and cables that minimize the problems of flux jumping are available. However, AC loss is another magnetization issue that directly affects magnet operation and that continues to challenge the magnet designer. This chapter discusses the fundamental theory behind the magnetization of hard superconductors; AC losses will then be addressed in more detail in Chapter 7. As in previous chapters, only those problems that can be treated analytically by simple closed-form mathematical expressions are considered.

5.2 Bean's Critical State Model

Of the two "mixed" state models to describe magnetic behaviors of hard (Type II) superconductors—the "lamina" by Goodman and the "vortex" by Abrikosov—Abrikosov's model, as remarked in Chapter 1, was verified experimentally. Nevertheless, we shall use Goodman's lamina model, which consists of superconducting and normal layers that permit penetration of field into the superconductor. In this model, each superconducting lamina that experiences a magnetic field carries an induced supercurrent density equivalent to the conductor's critical current density, J_c. Because each lamina that carries current does so at the critical value, we say the superconductor is in the "critical" state. This phenomenological model was first proposed in 1962 by Bean [5.1] and is known as Bean's Critical State model. Although J_c decreases with field in all real superconductors, the Bean model assumes J_c to be field-independent to simplify the mathematical treatment of the magnetization problem.

5.2.1 Magnetization of Hard Superconductors

Magnetization M is defined as $M = B/\mu_\circ - H$ (from Eq. 2.15); according to the Bean model, magnetic induction B is not zero within the interior of the hard superconductor and is equal to the superconductor's volumetric average of $\mu_\circ H_s$, where H_s is the magnetic field within the superconductor.

Figure 5.1 shows a hard superconducting slab infinitely high (in the y-direction), infinitely deep (in the z-direction), and $2a$ wide (in the x-direction). External field H_e, applied parallel to a slab previously unexposed to a magnetic field ("virgin" slab), creates $H_s(x)$ within the slab. With Ampere's law $\nabla \times \vec{H} = \vec{J} = \vec{J_c}$ applied to the slab, we obtain the magnetic field within the superconductor, $H_s(x)$:

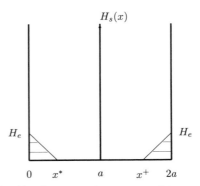

Fig. 5.1   Slab of hard superconductor exposed to an external field.

$$H_s(x) = \begin{cases} 0 & (x* \leq x \leq x\text{+}) & (5.1a) \\ H_e - J_c x & (0 \leq x \leq x*) & (5.1b) \\ H_e + J_c(x - 2a) & (x\text{+} \leq x \leq 2a) & (5.1c) \end{cases}$$

Note that the slope of the $H_s(x)$ is equal to J_c, positive where J_c is positive (z-directed, out of the paper) and negative where J_c is negative. $x*$ (and $2a - x\text{+}$) gives the extent of the field penetration in the slab; in terms of H_e and J_c:

$$x* = \frac{H_e}{J_c} \tag{5.2}$$

At $H_e = H_p \equiv J_c a$, $x* = x\text{+} = a$ and the entire slab is in the critical state; H_p is known as the penetration field.

The average magnetic induction within the slab, B_s, is thus given by

$$B_s = \frac{\mu_o}{2a} \int_0^{2a} H_s(x)\, dx = \frac{\mu_o}{2a} \times <\text{shaded area in Fig. 5.1}> \tag{5.3a}$$

$$= 2 \times \frac{\mu_o}{2a} \times \frac{H_e x*}{2} = \frac{\mu_o H_e^2}{2aJ_c} \tag{5.3b}$$

$$= \frac{\mu_o H_e^2}{2H_p} \tag{5.3c}$$

From the definition, $M = B_s/\mu_o - H_e$, we have:

$$-M = H_e - \frac{H_e^2}{2H_p} \qquad (0 \leq H_e \leq H_p) \tag{5.4}$$

Because the superconductor is diamagnetic, $-M$ is usually used to express its magnetization.

As the external field is increased further, the field finally penetrates the slab completely ($H_e \geq H_p$), and $B_s = H_e - H_p/2$, and thus:

$$-M = \frac{H_p}{2} \qquad (H_e \geq H_p) \tag{5.5}$$

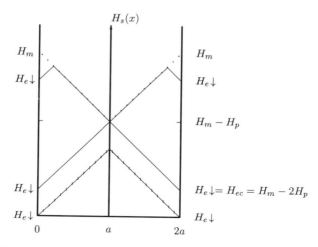

Fig. 5.2 $H_s(x)$ during a return sequence of H_e from H_m to zero.

The dotted lines in Fig. 5.2 represent $H_s(x)$ at $H_e = H_m > H_p$, where H_m is the maximum external field applied in this particular field sweep sequence.

As H_e is reduced from H_m towards 0, $H_s(x)$ changes as indicated by the solid lines in Fig. 5.2. When $H_{ec} = H_m - 2H_p$, as may be inferred from Fig. 5.2, because $H_s(x)$ is a complete mirror image of that at H_m, $-M$ becomes $-H_p/2$. That is, magnetization is now positive. It can be shown that for the return field sweep from H_m to H_{ec}, $-M(H_e)$ is given by:

$$-M(H_e) = \frac{H_p}{2} - H_m + \frac{H_m^2}{4H_p} + \left(1 - \frac{H_m}{2H_p}\right) H_e + \frac{H_e^2}{4H_p} \tag{5.6}$$

Note that $-M$ is a quadratic function of H_e as is the case when an external field is applied to the virgin slab. Note also that $-M(H_m) = H_p/2$ and $-M(H_{ec}) = -H_p/2$. For the range $H_e = H_{ec} \to 0$, $-M(H_e)$ is constant and remains $-H_p/2$ even when H_e returns to 0. $H_s(x)$ corresponding to this "remanent" magnetization is indicated by the hatched lines in Fig. 5.2. Once exposed to an external magnetic field, a hard superconductor will thus become magnetized. This remanent magnetization cannot be removed by means of external field; one way to remove it is by heating the superconductor and raising its temperature to above T_c.

Figure 5.3 gives $-M$ vs H_e plots for the entire positive field sweep sequence from 0 to $H_m = H_{c2}$ and back to 0. H_{c2} is the upper critical field. The solid curve is based on Eqs. 5.4~5.6 derived with Bean's assumption that J_c is field-independent. The dotted curve qualitatively corrects for the more realistic case in which J_c is a decreasing function of field, specifically becoming 0 at H_{c2}. Note that magnetization is hysteretic, and its magnitude at $H_p < H_e < H_m - 2H_p$, $\Delta M = -M(H_e\uparrow) + M(H_e\downarrow)$, is equal to $H_p = J_c a$. Magnetization measurement is thus sometimes performed to obtain $J_c(H_e)$ data, as discussed in Problem 5.4.

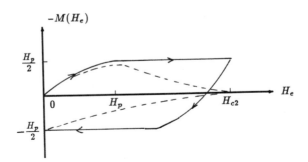

Fig. 5.3   Magnetization *vs* field traces for a hard superconducting slab subjected to an external field sequence of $0 \rightarrow H_{c2} \rightarrow 0$. The solid curve presents the case $J_c =$ constant; the dotted curve qualitatively presents the case $J_c(H_e)$, with $J_c = 0$ at H_{c2}.

Figure 5.4 shows the current distribution within the slab corresponding to the field distribution given in Fig. 5.1. The net current per unit length in the y-direction [A/m] flowing through the slab in the z-direction is given by:

$$I = \int_0^{2a} J(x)\,dx = 0 \tag{5.7}$$

As expected, $I = 0$ in the absence of transport current.

5.2.2 Effect of Transport Current on Magnetization

When a transport current I_t (per unit length in the y-direction) is applied uniformly over the slab in the $+z$-direction (out of the paper), we see an increase in magnetic field of $I_t/2$ at $x = 2a$ and a decrease in magnetic field of $I_t/2$ at $x = 0$.

Because the shielding current within the slab builds up from each surface into the interior, the field distribution $H_s(x)$ within the slab will be that shown in Fig. 5.5. In Fig. 5.5, $x*$ and $x+$ are given by:

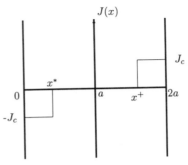

Fig. 5.4 $J(x)$ corresponding to $H_s(x)$ given in Fig. 5.1.

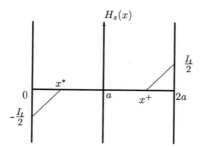

Fig. 5.5 $H_s(x)$ with transport current I_t in the slab.

$$-\frac{I_t}{2} + J_c x* = 0 \qquad (5.8a)$$

$$J_c(x^+ - 2a) + \frac{I_t}{2} = 0 \qquad (5.8b)$$

$$x* = \frac{I_t}{2J_c} \quad \text{and} \quad x^+ = 2a - \frac{I_t}{2J_c} \qquad (5.8c)$$

Figure 5.6 shows current distribution $J(x)$ in the slab. By integrating $J(x)$ across the slab width, we can show that the net current flowing in the slab is indeed I_t:

$$I = \int_0^{2a} J(x)\,dx = J_c x* + J_c(2a - x^+) \qquad (5.9a)$$

$$= \frac{I_t}{2} + \frac{I_t}{2} = I_t \qquad (5.9b)$$

As expected, the net current in the slab is the current supplied by the external source. Note that the presence of an external field $H_e\vec{i}_y$, when it is applied *after* I_t has been applied, does not fundamentally change the distributions shown in Figs. 5.5 and 5.6; but if $H_e\vec{i}_y$ is applied *before* I_t, different $H_s(x)$ and $J(x)$ would emerge. Problems 5.1~5.3 treat the effects of transport current on magnetization in detail.

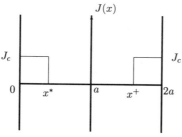

Fig. 5.6 $J(x)$ corresponding to $H_s(x)$ given in Fig. 5.5.

5.3 Experimental Confirmation of Bean's Model

Experimental confirmation of Bean's model came quickly [5.2]. The effects of transport current, treated in Problems 5.1~5.3, were also tested and confirmed by experiment [5.3, 5.4].

Perhaps the most direct and beautiful confirmation of the Bean model was by Coffey [5.5], who mapped field distributions in a test sample of a hard superconductor. Figure 5.7 presents tracings of the field distribution in a long test sample consisting of two Nb-Ti rods, each 60 mm long and 8.3 mm in diameter, joined together with a gap of 0.5 mm between two ends. Because the gap is much smaller than the diameter, the field distribution in the gap, measured by a Hall probe of thickness 5~10 μm, is a good approximation to that of a long single rod. Two aspects of field distribution evident in Fig. 5.7 have already been discussed:1) H_s decreases from surface to interior with nearly constant slope; and 2) the slope (J_c) decreases with increasing field. Another feature yet to be discussed is flux jumping, a thermal instability phenomenon that occurs in hard superconductors. (Flux jumping is treated in Problems 5.6 and 5.7.) Figure 5.7 shows four incidences of flux jumping, each evident from the flat field distribution across the rod.

When shown Coffey's paper, Bean is reported to have exclaimed, "I'm right!" Perhaps the model should be called ... "Coffey-Bean's!"

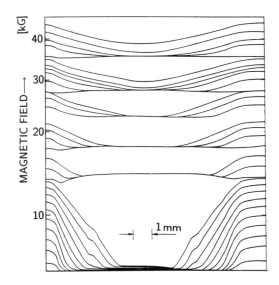

Fig. 5.7 Detailed tracings of field distribution data across a 0.5-mm gap between two Nb-Ti rods, each 60 mm long and 8.3 mm in diameter. Each of the four flat field lines running across the rod represents the field distribution immediately after a flux jump [5.5]. (Courtesy of H.T. Coffey.)

5.4 A Magnetization Measurement Technique

We describe here the technique most widely used to measure magnetization. It applies Faraday's law to analyze a voltage induced across the terminals of a loop coupled to a time-varying magnetic induction.

Figure 5.8 presents the key components of this technique [5.6]: 1) a primary search coil; 2) a secondary search coil; and 3) a balancing potentiometer. Not shown in the figure but equally essential is an integrator that converts the bridge output voltage, $V_{bg}(t)$, to a voltage that is directly proportional to $M(H_e)$. The test sample is placed within the primary search coil set. When the primary and secondary search coils are subjected to a time-varying external magnetic field $H_e(t)$ which is nearly uniform over the space occupied by each search coil, voltages, $V_{pc}(t)$ and $V_{sc}(t)$, are induced across the terminals of each search coil:

$$V_{pc}(t) = \mu_o N_{pc} A_{pc} \left[\frac{dM}{dt} + \left(\frac{d\tilde{H}_e}{dt} \right)_{pc} \right] \tag{5.10a}$$

$$V_{sc}(t) = \mu_o N_{sc} A_{sc} \left(\frac{d\tilde{H}_e}{dt} \right)_{sc} \tag{5.10b}$$

The subscripts pc and sc refer to the primary and secondary search coils, respectively. N is the number of turns of each search coil and A is the effective area of each turn in the coil through which $H_e(t)$ is coupled. $\tilde{H}_e$ is the space-averaged field over each search coil.

The bridge output voltage $V_{bg}(t)$ is given by:

$$V_{bg}(t) = (k - 1)V_{pc}(t) + kV_{sc}(t) \tag{5.11}$$

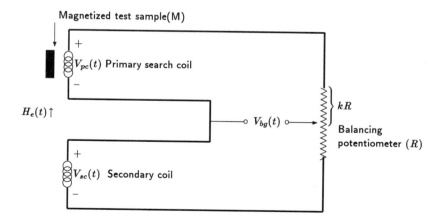

Fig. 5.8   Schematic of a magnetization measurement technique.

Combining Eqs. 5.10 and 5.11, we obtain:

$$V_{bg}(t) = (k-1)\mu_o N_{pc} A_{pc} \frac{dM}{dt}$$
$$+ (k-1)\mu_o N_{pc} A_{pc} \left(\frac{d\tilde{H}_e}{dt}\right)_{pc} + k\mu_o N_{sc} A_{sc} \left(\frac{d\tilde{H}_e}{dt}\right)_{sc} \qquad (5.12)$$

Now, it is possible to adjust k of the potentiometer to satisfy the following condition and make $V_{bg}(t)$ proportional only to dM/dt:

$$(k-1)\mu_o N_{pc} A_{pc} \left(\frac{d\tilde{H}_e}{dt}\right)_{pc} + k\mu_o N_{sc} A_{sc} \left(\frac{d\tilde{H}_e}{dt}\right)_{sc} = 0 \qquad (5.13a)$$

$$V_{bg}(t) = (k-1)\mu_o N_{pc} A_{pc} \frac{dM}{dt} \qquad (5.13b)$$

Although in practice the condition required by Eq. 5.13a is not always satisfied over a wide frequency range, Eq. 5.13b is a good approximation for most cases. Generally k is close to 0.5. $V_{bg}(t)$ is fed into an integrator and its output, $V_{mz}(t)$, is proportional to M. Specifically, if the test sample is at the virgin state ($M = 0$) and $H_e(t)$ is *increased* ($H_e\uparrow$) from 0 (at $t = 0$) to H_e (at $t = t_1$), then we have:

$$V_{mz}(H_e\uparrow) = \frac{1}{\tau_{it}} \int_0^{t_1} V_{bg}(t)\, dt = \frac{(k-1)\mu_o N_{pc} A_{pc}}{\tau_{it}} M(H_e) \qquad (5.14)$$

where τ_{it} is an effective integrator time constant. If $H_e > H_p$, then $M(H_e) = -H_p/2 = -J_c a/2$, and Eq. 5.14 simplifies to:

$$V_{mz}(H_e\uparrow > H_p) = -f_m \frac{(k-1)\mu_o N_{pc} A_{pc}}{\tau_{it}} \left(\frac{J_c a}{2}\right) \qquad (5.15a)$$

where the factor f_m is the ratio of magnetic material volume to the total test sample volume. This factor is needed because generally a test sample does not consist entirely of magnetic material for which magnetization is measured; in the case of a multifilamentary conductor, for example, the test sample consists not only of superconducting filaments but also of a matrix metal and other nonmagnetic materials, e.g. insulator. If an external field excursion is $0 \to H_m \gg H_p \to H_e\downarrow > H_p$, then we have:

$$V_{mz}(H_e\downarrow > H_p) = f_m \frac{(k-1)\mu_o N_{pc} A_{pc}}{\tau_{it}} \left(\frac{J_c a}{2}\right) \qquad (5.15b)$$

$\Delta V_{mz} = V_{mz}(H_e \uparrow > H_p) - V_{mz}(H_e \downarrow > H_p)$ is thus proportional to the "width" of magnetization curve at H_e:

$$\Delta V_{mz} = -f_m \frac{(k-1)\mu_o N_{pc} A_{pc}}{\tau_{it}} J_c a \qquad (5.16)$$

From Eq. 5.16 we note that ΔV_{mz} is directly proportional to J_c and a.

Figure 5.9 shows typical magnetization vs field traces for a Nb-Ti conductor. The solid curve corresponds to the case when the primary and secondary coils are well balanced (Eq. 5.13a is satisfied), while the dotted curve corresponds to a tilted magnetization trace that results when Eq. 5.13a is not quite satisfied.

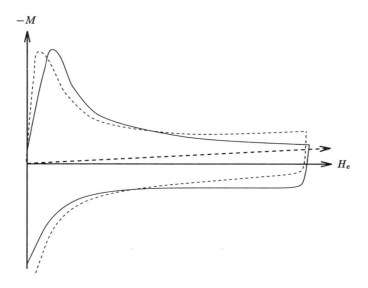

Fig. 5.9   Magnetization vs field traces for a Nb-Ti conductor. The solid curve: the primary and secondary coils are balanced (Eq. 5.13a is satisfied), the dotted curve: the coils are not quite balanced.

"Through measurement to knowledge." —Heike Kamerlingh Onnes

Problem 5.1: Magnetization with transport current
1. Field and then transport current

In this and the next two problems, we examine the effects of transport current on magnetization. As remarked briefly in the introductory section, magnetization in the presence of transport current depends on the order in which external field and transport current are applied. In this problem, an external field $H_e \vec{i}_y$ is already present when I_t is applied to the Bean slab of width $2a$.

a) With external field $H_e = 2.5 H_p = 2.5 J_c a$ applied to the slab, I_t is introduced. Plot $H_s(x)$ distributions for $I_t = 0$, $I_t = J_c a$, and $I_t = 2 J_c a$. Note that $2 J_c a = I_c$ [A/m] is the critical current of the slab. Also note that $I_t = I_c i = 2 H_p i$, where $i \equiv I_t/I_c$. As indicated in Fig. 5.10, take the slab's left-hand side to be $x = 0$ and its right-hand side to be $x = 2a$. In the presence of a transport current, field profiles are no longer symmetric with respect to the slab's midplane and the entire slab must be considered.

b) Show that an expression for $-M(i)$, the magnetization as a function of a normalized transport current i for $H_e \geq H_p$, is given by:

$$-M(i) = \frac{H_p}{2}(1 - i) \qquad (5.17a)$$

$$= -M(0)f_1(i) \qquad (5.17b)$$

where $f_1(i) = 1 - i$.

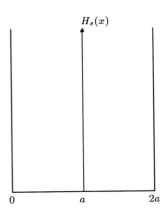

Fig. 5.10 Bean slab, $2a$ wide, for computation of magnetization.

Solution to Problem 5.1

a) At $H_e = 2.5H_p$, $H_s(x)$ for $I_t = 0$ (dotted), $H_s(x)$ for $I_t = J_c a$ (solid), and $H_s(x)$ for $I_t = 2J_c a$ (hatched) are shown in Fig. 5.11.

For $I_t = J_c a$, $H_s(x)$ (solid) is a piece-wise linear function consisting of $H_{s1}(x)$ for $0 \leq x \leq x*$, $H_{s2}(x)$ for $x* \leq x \leq x+$, and $H_{s3}(x)$ for $x+ \leq x \leq 2a$.

For $I_t = 2J_c a$, $H_s(x)$ (hatched) is a straight line, going from $1.5H_p$ at $x = 0$ to $3.5H_p$ at $x = 2a$.

$H_{s1}(x)$, $H_{s2}(x)$, and $H_{s3}(x)$ are given by:

$$H_{s1}(x) = 2H_p + J_c x = 2J_c a + J_c x \qquad (S1.1a)$$

$$H_{s2}(x) = 2.5H_p - J_c x = 2.5J_c a - J_c x \qquad (S1.1b)$$

$$H_{s3}(x) = H_p + J_c x = J_c a + J_c x \qquad (S1.1c)$$

$x*$ and $x+$ may be found:

$$H_{s1}(x*) = H_{s2}(x*) \quad \Longrightarrow \quad x* = 0.25a$$

$$H_{s2}(x+) = H_{s3}(x+) \quad \Longrightarrow \quad x+ = 0.75a$$

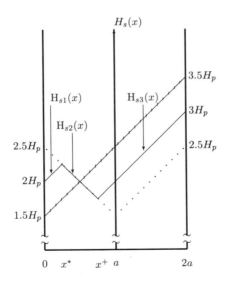

Fig. 5.11   Field profiles, with $I_t = 0$ (dotted), $J_c a$ (solid), and $2J_c a$ (hatched).

Solution to Problem 5.1

b) $H_s(x)$ for the general case in the presence of I_t is plotted in Fig. 5.12 with the solid lines, $H_{s1}(x)$, $H_{s2}(x)$, and $H_{s3}(x)$. For $\int H_s(x)\,dx$ integration, the slab is divided into three regions, I, II, and III.

$H_{s1}(x)$, $H_{s2}(x)$, and $H_{s3}(x)$ are given by:

$$H_{s1}(x) = \left(H_e - \frac{I_t}{2}\right) + J_c x \qquad\qquad (S1.2a)$$

$$H_{s2}(x) = H_e - J_c x \qquad\qquad (S1.2b)$$

$$H_{s3}(x) = \left(H_e + \frac{I_t}{2}\right) + J_c(x - 2a) \qquad\qquad (S1.2c)$$

We solve for $x*$ and $x+$, and determine $H_{s2}(x*)$ and $H_{s2}(x+)$.

$$H_{s1}(x*) = H_{s2}(x*) \qquad\qquad (S1.3a)$$

$$H_e - H_p i + J_c x* = H_e - J_c x*$$

$$x* = \frac{H_p}{2J_c}i = \frac{a}{2}i \qquad\qquad (S1.3b)$$

$$H_{s2}(x*) = H_e - \frac{aJ_c i}{2} = H_e - \frac{H_p i}{2} \qquad\qquad (S1.3c)$$

$$H_{s2}(x+) = H_{s3}(x+) \qquad\qquad (S1.4a)$$

$$H_e - J_c x+ = H_e + H_p i + J_c(x+ - 2a)$$

$$x+ = a\left(1 - \frac{i}{2}\right) \qquad\qquad (S1.4b)$$

$$H_{s2}(x+) = H_e - H_p + \frac{H_p i}{2} \qquad\qquad (S1.4c)$$

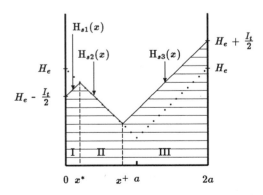

Fig. 5.12   Field profile with transport current (solid) for computation of magnetization.

Solution to Problem 5.1

M is proportional to the "shaded area," shown in Fig. 5.12, which is the sum of areas I (AI), II (AII), and III (AIII). As in earlier integral computations, a simple method to compute the area of a trapezoid—(base)×(height$_1$+height$_2$)/2—is used to compute AI, AII, and AIII.

$$AI = \frac{x*}{2}[H_{s1}(0) + H_{s2}(x*)] = \frac{ai}{4}[(H_e - H_pi) + (H_e - \tfrac{1}{2}H_pi)] \quad (S1.5a)$$

$$= \frac{ai}{4}(2H_e - \tfrac{3}{2}H_pi)$$

$$= a(\tfrac{1}{2}H_ei - \tfrac{3}{8}H_pi^2) \quad (S1.5b)$$

$$AII = \frac{(x+ - x*)}{2}[H_{s2}(x*) + H_{s2}(x+)] \quad (S1.6a)$$

$$= \tfrac{1}{2}(a - ai)(H_e - \tfrac{1}{2}H_pi + H_e - H_p + \tfrac{1}{2}H_pi)$$

$$= \frac{a(1 - i)}{2}(2H_e - H_p)$$

$$= a(H_e - H_ei - \tfrac{1}{2}H_p + \tfrac{1}{2}H_pi) \quad (S1.6b)$$

$$AIII = \frac{(2a - x+)}{2}[H_{s2}(x+) + H_{s3}(2a)] \quad (S1.7a)$$

$$= \tfrac{1}{2}\left(a + \frac{ai}{2}\right)(H_e - H_p + \tfrac{1}{2}H_pi + H_e + H_pi)$$

$$= a\left(1 + \frac{i}{2}\right)(H_e - \tfrac{1}{2}H_p + \tfrac{3}{4}H_pi)$$

$$= a(H_e + \tfrac{1}{2}H_ei - \tfrac{1}{2}H_p - \tfrac{1}{4}H_pi + \tfrac{3}{4}H_pi + \tfrac{3}{8}H_pi^2) \quad (S1.7b)$$

$$\text{Shaded area} = a(\tfrac{1}{2}H_ei - \tfrac{3}{8}H_pi^2 + H_e - H_ei - \tfrac{1}{2}H_p + \tfrac{1}{2}H_pi$$

$$+ H_e + \tfrac{1}{2}H_ei - \tfrac{1}{2}H_p - \tfrac{1}{4}H_pi + \tfrac{3}{4}H_pi + \tfrac{3}{8}H_pi^2) \quad (S1.8a)$$

$$= a(2H_e - H_p + H_pi) \quad (S1.8b)$$

Once the shaded area is known, M can be computed quickly:

$$-M(i) = H_e - \frac{1}{2a} \times (\text{Shaded area}) \quad (S1.9)$$

$$= H_e - H_e + \tfrac{1}{2}H_p - \tfrac{1}{2}H_pi$$

$$= \frac{H_p}{2}(1 - i) \quad (5.17a)$$

$$= -M(0)f_1(i) \quad (5.17b)$$

where $f_1(i) = 1 - i$. $-M(i)$ decreases linearly with i, becoming 0 at $i = 1$.

Problem 5.2: Magnetization with transport current

2. Transport current and then field

In this problem, we reverse the order in which external magnetic field and transport current are introduced to the slab. Specifically, a transport current of $J_c a = (I_c/2)\vec{i}_z$ (out of the paper) is introduced to the slab initially at the virgin state. While I_t is held at $I_c/2$, an external magnetic field is now applied to the slab in the $+y$-direction, increasing from 0 to $2H_p$.

a) Sketch $H_s(x)$ after the transport current of $J_c a$ $(= H_p)$ but before H_e of $2H_p$ is introduced. Also sketch $H_s(x)$ after H_e of $2H_p$ is introduced.

b) Using appropriate $H_s(x)$ profiles obtained in a), show that $I_t = J_c a$ before and after the application of H_e.

c) Show that an expression for $-M(i)$ valid for $H_e \geq H_p$ is given by:

$$-M(i) = \frac{H_p}{2}(1 - i^2) \tag{5.18a}$$

$$= -M(0)f_2(i) \tag{5.18b}$$

where $f_2(i) = 1 - i^2$. Again, it is easier to place the $x = 0$ point at the left end of the slab.

Use of SQUID for Magnetization Measurement

A SQUID (Superconducting Quantum Interference Device), based on the principle of Josephson effect, is an electronic device that can be used to measure a change in magnetic field with an extremely high resolution—in the unit of a quantized flux density, equal to 2.0×10^{-15} Wb (T/m^2).

A typical SQUID magnetization measurement setup consists of a test sample, at a constant temperature, placed in a uniform field. The test sample is moved back and forth in the uniform field; during each cycle it cuts through measurement coils, one located at one end of the test sample and the other located at the other end. The induced current in each measurement coil is measured by the SQUID in terms of a field generated by the current, which, in turn, is a measure of the test sample's magnetization. Because SQUIDs operate only in a low-field environment (perhaps no higher than ~100 oersted or ~0.01 T), they must be well shielded from the high-field environment of the test sample.

> *To do is to be.*   —Imanuel Kant
> *To be is to do.*   —Jean-Paul Sartre
> *Do be do be do.*   —Frank Sinatra

Solution to Problem 5.2

a) In Fig. 5.13, shown below, the dotted lines present $H_s(x)$ after a transport current of $J_c a$ $(H_p/2)$ but before H_e of $2H_p$ is introduced; the solid lines present $H_s(x)$ after H_e of $2H_p$ is introduced.

b) In both cases the net current in the slab is $J_c a$, as demonstrated below:

$H_e = 0$:

$$I_t = \int_0^{2a} J(x)\, dx = J_c \frac{a}{2} + J_c \left(2a - \frac{3a}{2}\right) = J_c a \qquad (S2.1)$$

$H_e = 2H_p$:

$$I_t = \int_0^{2a} J(x)\, dx = -J_c \frac{a}{2} + J_c \left(2a - \frac{a}{2}\right) = J_c a \qquad (S2.2)$$

c) To determine the magnetization in the slab, we must first find $x*$ (Fig. 5.14), which can be determined from $H_{s1}(x)$ and $H_{s2}(x)$.

$$H_{s1}(x) = (H_e - H_p i) - J_c x \qquad (S2.3a)$$

$$H_{s2}(x) = (H_e + H_p i) + J_c(x - 2a) \qquad (S2.3b)$$

$$H_{s1}(x*) = H_{s2}(x*) \qquad (S2.4a)$$

$$H_e - H_p i - J_c x* = H_e + H_p i + J_c x * -2J_c a$$

$$x* = a - ai = a(1 - i) \qquad (S2.4b)$$

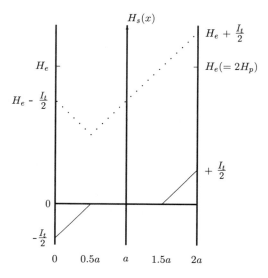

Fig. 5.13   Field profiles, with transport current only (solid) and with transport current and field (dotted).

Solution to Problem 5.2

Once $x*$ is determined, we can compute $H_{s1}(x*)$:

$$H_{s1}(x*) = H_e - H_p i - J_c a(1-i) = H_e - H_p \qquad (S2.5)$$

We can now compute the shaded area, which is the sum of AI and AII (Fig. 5.14).

$$\text{AI} = \frac{a(1-i)}{2}(H_e - H_p i + H_e - H_p) \qquad (S2.6a)$$

$$= a(1-i)(H_e - \tfrac{1}{2}H_p - \tfrac{1}{2}H_p i)$$

$$= a(H_e - H_e i - \tfrac{1}{2}H_p + \tfrac{1}{2}H_p i^2) \qquad (S2.6b)$$

$$\text{AII} = \frac{2a - a + ai}{2}(H_e + H_p i + H_e - H_p) \qquad (S2.7a)$$

$$= a(1+i)(H_e - \tfrac{1}{2}H_p + \tfrac{1}{2}H_p i)$$

$$= a(H_e + H_e i - \tfrac{1}{2}H_p + \tfrac{1}{2}H_p i^2) \qquad (S2.7b)$$

$$\text{Shaded area} = a(2H_e - H_p + H_p i^2) \qquad (S2.8)$$

Once the shaded area is known, we have M:

$$-M(i) = H_e - \tfrac{1}{2}(2H_e - H_p + H_p i^2) \qquad (S2.9)$$

$$= \frac{H_p}{2}(1 - i^2) \qquad (5.18a)$$

$$= -M(0)f_2(i) \qquad (5.18b)$$

where $f_2(i) = 1 - i^2$. This magnetization is a parabolic function of i.

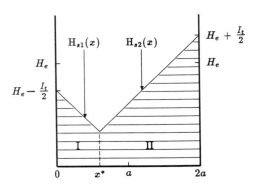

Fig. 5.14   Field profile with transport current
and field for computation of magnetization.

Problem 5.3: Magnetization with transport current
3. Field and then current changes

Finally, we shall consider $H_s(x)$ and $-M(i)$ for the slab when the following sequence of field and transport current is applied.

Step 1: Starting with a virgin state and $I_t = 0$ initially, the external magnetic field, $H_e = 2H_p$, is applied in the $+y$-direction.

Step 2: While H_e remains at $2H_p$, a transport current $I_t = J_c a$ is introduced into the slab in the z-direction (out of the paper).

Step 3: With H_e remaining at $2H_p$, I_t is reduced to zero.

Step 4: I_t is now reversed and $|J_c a|$ is introduced into the slab in the $-z$-direction (into the paper).

Step 5: I_t is again reduced to zero; H_e still remains at $2H_p$.

The field profile $H_s(x)$ after Step 5 is shown with solid lines in Fig. 5.15.

By generalizing the $H_s(x)$ profile shown in Fig. 5.15 for any $H_e(\geq H_p)$ and $I_t(\leq I_c)$, show that an expression for $-M(i)$ is given by:

$$-M(i) = -M(0)(1 - i)^2 \tag{5.19a}$$
$$= -M(0)f_3(i) \tag{5.19b}$$

where $f_3(i) = (1 - i)^2$.

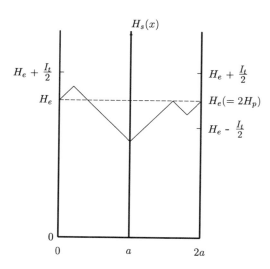

Fig. 5.15   Field profile after Step 5.

Solution to Problem 5.3

The two "dotted" areas in Fig. 5.16 are equal in magnitude but have "opposite" signs, hence they cancel out when we perform the area integral.

$$\text{Shaded area} = 2aH_e - \text{"crossed" area} \tag{S3.1a}$$

$$\text{crossed area} = \tfrac{1}{2}(\text{base})\times(\text{height}) \tag{S3.1b}$$

$$\text{base} = x\text{+}- x* \qquad \text{height} = H_e - H_{s1}(a) \tag{S3.1c}$$

$$H_{s1}(x) = H_e + H_pi - J_cx \tag{S3.2a}$$

$$H_{s3}(x) = H_e + H_pi + J_c(x - 2a) \tag{S3.2b}$$

$$H_{s1}(x*) = H_e \implies H_e + H_pi - J_cx* = H_e \tag{S3.3a}$$

$$x* = \frac{H_pi}{J_c} = ai \tag{S3.3b}$$

$$H_{s3}(x\text{+}) = H_e \implies H_e + H_pi + J_c(x\text{+}- 2a) = H_e \tag{S3.4a}$$

$$x\text{+} = 2a - \frac{H_p}{J_c}i = 2a - ai \tag{S3.4b}$$

$$\text{base} = x\text{+}- x* = (2a - ai) - ai = 2a(1 - i) \tag{S3.5a}$$

$$\text{height} = H_e - H_{s1}(a) = H_e - (H_e + H_pi - J_ca)$$

$$= J_ca - H_pi = H_p(1 - i) \tag{S3.5b}$$

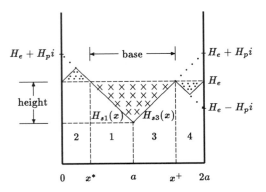

Fig. 5.16   Field profile (after Step 5) for computation of magnetization.

Solution to Problem 5.3

$$\text{Shaded area} = 2aH_e - \tfrac{1}{2}2a(1-i)H_p(1-i)$$

$$= 2aH_e - aH_p(1-i)^2 \qquad (S3.6)$$

$$-M = H_e - \frac{1}{2a}[2aH_e - aH_p(1-i)^2] = \frac{H_p}{2}(1-i)^2 \qquad (S3.7)$$

$$-M(i) = -M(0)(1-i)^2 \qquad (5.19a)$$

$$= -M(0)f_3(i) \qquad (5.19b)$$

where $f_3(i) = (1-i)^2$.

Magnetization Functions – Summary

Figure 5.17 presents three normalized magnetization functions, $f_1(i)$, $f_2(i)$, and $f_3(i)$, where $i = I_t/I_c$. It is interesting to note how different sequences of transport current and external field applications affect $M(i)$. These functions of magnetization *vs* normalized transport current were actually observed experimentally [5.3, 5.4] and shown to agree with analytical results based on Bean's model, thereby helping to quickly confirm its validity.

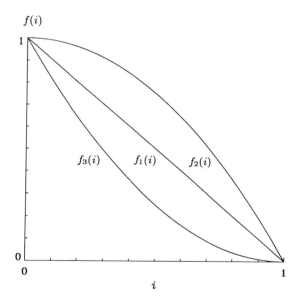

Fig. 5.17   Three normalized magnetization *vs* normalized transport current functions studied in Problems 5.1, 5.2, and 5.3.

Problem 5.4: Critical current density from magnetization

This problem demonstrates how to derive critical current density (J_c) data from the magnetization measurement of a typical HTS.

The most widely used and most direct technique to determine the J_c of a super-conductor is the "critical current measurement." A "long" test sample is prepared from the superconductor, and a direct current is slowly increased through it under given test conditions (T, H_e) until a resistive voltage across the sample is detected. The test sample must be "long," because if the sample is too short, two problems arise: 1) it becomes too difficult to obtain voltage resolution sufficient to detect a resistive electric field used to indicate the superconducting-to-normal transition (the typical criterion is between 0.1 to $1\,\mu V/cm$); 2) the contact resistance to the lead wires at each end of the test sample becomes too large, causing excessive heating at its ends and, therefore, a premature normal transition. The test sam-ples should normally be at least 10 mm long; perhaps under certain circumstances they can be as short as 5 mm, but not much shorter than this. It really depends to a large degree on the level of critical current.

During the early period (1987~1989) following the discovery of HTS, it was difficult to produce an HTS of sufficient length to perform a critical current measurement either because there was an insufficient quantity to manufacture a 10-mm long sample, or because it was nearly impossible to form the HTS into a conductor of any length due to its extreme brittleness, or both. Consequently, J_c data were often extracted from magnetization data through the Bean model.

Figure 5.18 presents the magnetization vs applied (external) field data at 4.5 K for a single crystal of $Y_1Ba_2Cu_3O_{7-x}$ (HTS) [5.7]. Compute J_c for this crystal at an applied field of 20 kilogauss. Assume the crystal is 300 μm high, 300 μm thick, and 180 μm wide. Treat the crystal as a Bean slab of thickness $2a = 180\,\mu$m.

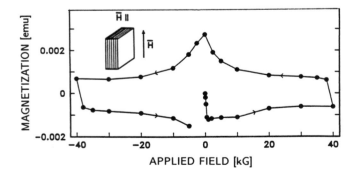

Fig. 5.18   Magnetization vs applied field hysteresis curves for an HTS single crystal at 4.5 K [5.7]. (Courtesy of W.J. Gallagher.)

Solution to Problem 5.4

First, we must divide magnetization by sample volume to express it in terms of a magnetization density, emu/cm^3. Then, we convert emu/cm^3 into the SI unit equivalent, A/m, by multiplying it by 1000. (See Appendix I.)

At $\mu_o H_e = 20$ kilogauss, ΔM, from Fig. 5.18, is 0.0015 emu, which when divided by the test sample volume of 1.62×10^{-5} cm^3, gives magnetization density of 92.6 emu/cm^3 or 9.26×10^4 A/m. With $2a = 180 \times 10^{-6}$ m and $\Delta M = J_c a$, we obtain: $J_c = 1.0 \times 10^9$ A/m^2. Note that because ΔM remains nearly constant for the range of $\mu_o H_e$ between 2 and 4 T, we may conclude that J_c is also nearly constant over this field range. At zero field $-M = 0.0027$ emu or 166.7 emu/cm^3, which is equal to $J_c a/2$. Thus, we obtain $J_c = 3.7 \times 10^9$ A/m^2 at zero field.

Contact-Resistance Heating at Test Sample Ends

As remarked above, excessive heating at the test sample's ends can cause an error in critical current measurements. Figure 5.19 shows temperature plots along a silver-sheathed bismuth-based HTS tape at selected transport currents, with the tape immersed in a bath of liquid nitrogen boiling at 77 K [5.8]. Transport currents are expressed as ratios to the tape's critical current (I_c) at 77 K in zero field. The plot labeled 1.00 corresponds to a transport current equal to I_c. Overheating is evident at the right-hand end, most likely because the tape at this end was improperly soldered to its electrode. Asymmetry in the temperature distribution increases with transport current. Although I_c measurement in this particular example was unaffected, due chiefly to its long length (60 mm between electrodes), it could certainly have been affected were the tape shorter than ~30 mm.

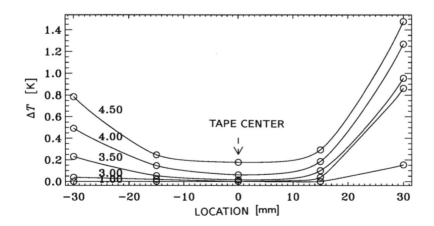

Fig. 5.19 Asymmetric temperature distributions of a silver-sheathed bismuth-based HTS tape at selected transport currents with the tape in a bath of liquid nitrogen [5.8].

Problem 5.5: Magnetization measurement

This problem presents an example of the magnetization measurement technique discussed in Section 5.4; the technique was applied for magnetization measurements of one of the four superconductors used in the Hybrid III SCM. The measurement was necessary to make sure that there would be no flux jumping. The absence of flux jumping is one of the necessary conditions for magnets that are not "cryostable"—this point will be discussed in more detail in Chapter 6.

Table 5.1 presents specifications of the superconductor, a bare Nb-Ti composite strip with overall dimensions of 9.2 mm width and 2.6 mm thickness. (Not all parameters in the table, *e.g.* twist pitch, are relevant for this problem.)

The test sample consists of 52 (13×4) 100-mm long strips assembled in a rectangular solid of square cross section, 38 mm×38mm, as shown in Fig. 5.20. Each bare strip was electrically insulated with a thin tape. In the orientation shown in Fig. 5.20a, each strip's broad surface is parallel to the direction of external magnetic induction B_e; in the orientation shown in Fig. 5.20b, it is normal to B_e. The test sample assembly was placed inside a rectangular-bore (cross section 107 mm×42mm) search coil set containing a primary search coil and two secondary coils (Fig. 5.20c). The test assembly midplane was placed to coincide with that of the primary search coil, whose midplane was in turn placed to coincide with that of a magnet providing B_e. The midplane-to-midplane distance between the primary and one of the secondary coils is 70 mm. The primary coil contains 500 turns of fine copper wire over an axial distance of 40 mm about its midplane; each secondary search coil contains 280 turns, extending an axial distance of 20 mm about its midplane. The turn density in the axial direction in each search coil may be assumed uniform.

When an external magnetic induction B_e was swept at a rate of 0.05 T/s between 0 and 5 T with the test sample at 4.2 K and oriented as shown in Fig. 5.20a (broad face *parallel* to external field), the V_{mz} *vs* B_e plot sketched in Fig. 5.21 was obtained. $+V_{mz}$ is the integrator output proportional to $-M$, the negative of the test sample magnetization. The effective integration time, τ_{it}, was 1 s; the balancing potentiometer's constant k was 0.5. Assume that the voltage drift of the integrator was zero during the time the magnetization plot was traced.

Table 5.1: Specifications for a Hybrid III Nb-Ti Conductor

Overall width, a	[mm]	9.2	Cu/Sc ratio, $\gamma_{c/s}$		3
Overall thickness, b	[mm]	2.6	T_c @ 10 T	[K]	4.7
Filament diameter*	[μm]	100	I_c @ 1.8 K, 10 T	[A]	6000
Twist pitch length, ℓ_p	[mm]	100	J_c @ 4.2 K, 5 T	[GA/m²]	2.0
Insulation		none	—		—

* Computed value for circular cross sectioned filaments.

Problem 5.5: Magnetization measurement

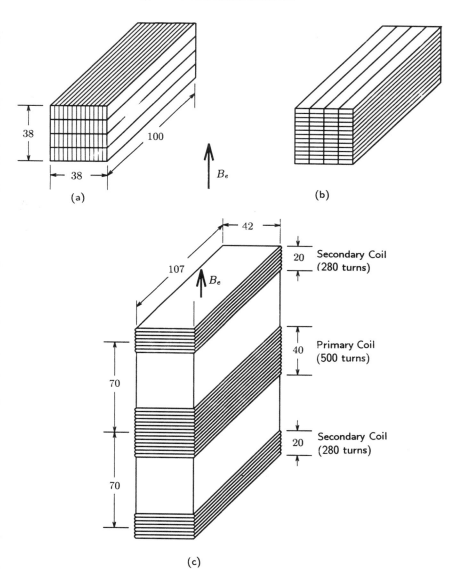

Fig. 5.20   Magnetization measurement details, dimensions in mm.
(a) Each conductor strip is orientated with its broad face parallel
to the external magnetic induction, B_e; (b) Each conductor strip is
orientated with its broad face normal to B_e; (c) Search coil setup.

Problem 5.5: Magnetization measurement

a) Make a *ballpark* estimate of ΔV_{mz} at $B_e \sim 2.5\,\mathrm{T}$ (Fig. 5.21). Note that $\tau_{it} = 1\,\mathrm{s}$ and $k = 0.5$. Assume $d_f = 2a$, where d_f is the filament diameter and $2a$ is the width of the Bean slab.

b) The 1.8-K measurement was performed by pumping on the cryostat and reducing the liquid helium bath pressure to 12.6 torr. During the 1.8-K measurement, the technician who controlled the cryostat pressure noticed that pressure control was more difficult, because of an increased liquid boil-off rate, when the test sample orientation was as in Fig. 5.20b rather than when it was as in Fig. 5.20a. Is this an aberration or does his observation make sense? Explain.

c) The z-component of the external induction B_e over the radial space occupied by the search coil may be approximated to vary as:

$$B_e(z) \simeq B_e(0) \left[1 - c \left(\frac{z}{z_o} \right)^2 \right] \tag{5.20}$$

where $z_o = 75\,\mathrm{mm}$. Based on information you have, compute the value of c.

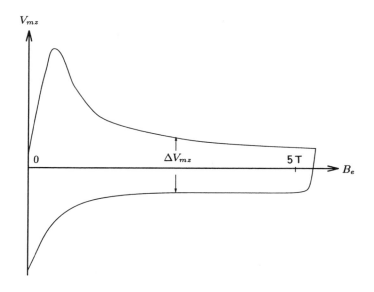

Fig. 5.21 V_{mz} vs B_e plot with the test sample at 4.2 K and oriented with respect to B_e as in Fig. 5.20a. B_e was swept at a |rate| of 0.05 T/s.

Solution to Problem 5.5

a) Equation 5.12 indicates that if the search coils were not balanced, a term proportional to the applied field would contribute to the magnetization. Since the V_{mz} vs B_e trace in Fig. 5.21 is not tilted, the search coils are balanced, similar to the solid trace of Fig. 5.9. From Eq. 5.13b:

$$V_{bg}(t) = (k-1)\mu_o N_{pc} A_{pc} \frac{dM}{dt} \tag{5.13b}$$

We have from Eq. 5.16:

$$\Delta V_{mz} = -f_m \frac{(k-1)\mu_o N_{pc} A_{pc}}{\tau_{it}} J_c a \tag{5.16}$$

We have: $k = 0.5$; $\tau_{it} = 1\,\text{s}$; $N_{pc} = 500$; $A_{pc} = (13)(0.1\,\text{m})(2.6 \times 10^{-3}\,\text{m}) = 3.38{\times}10^{-3}\,\text{m}^2$ [$(0.1\,\text{m}){\times}(38{\times}10^{-3}\,\text{m}) = 3.8{\times}10^{-3}\,\text{m}^2$ is also acceptable]; $f_m =$(Nb-Ti volume)/total composite volume)$= 1/(\gamma_{c/s} + 1) = 0.25$.

Estimate of J_c (4.2 K, 2.5 T):

From Table 5.1 we have J_c at $4.2\,\text{K}$ and $5\,\text{T}$ of $2.0 \times 10^9\,\text{A/m}^2$. It is generally accepted that for a given temperature, $J_c(B_e)$ may be given by the following approximate expression (based on Eq. 1.5):

$$J_c = \frac{J_0 B_0}{B_e + B_0} \tag{S5.1}$$

where for Nb-Ti, $B_0 \sim 0.3\,\text{T}$. J_0 is the zero-field critical current density, which is usually difficult to measure. Thus from the J_c value at $5\,\text{T}$ and $B_0 = 0.3\,\text{T}$, we can first solve for $J_0 B_0$:

$$2.0 \times 10^9\,\text{A/m}^2 = \frac{J_0 B_0}{5\,\text{T} + 0.3\,\text{T}} \implies J_0 B_0 = 10.6 \times 10^9\,\text{A T/m}^2$$

Once $J_0 B_0$ is known, then J_c may be solved at $2.5\,\text{T}$. Thus:

$$J_c(2.5\,\text{T}) = \frac{10.6 \times 10^9\,\text{A T/m}^2}{2.8\,\text{T}} = 3.8 \times 10^9\,\text{A/m}^2 \tag{S5.2}$$

Inserting appropriate values into Eq. 5.16, we have:

$$\Delta V_{mz} = -0.25 \frac{(-0.5)(4\pi \times 10^{-7}\,\text{H/m})(500)(3.38 \times 10^{-3}\,\text{m}^2)}{1\,\text{s}}$$
$$\times (3.8 \times 10^9\,\text{A/m}^2)(50 \times 10^{-6}\,\text{m}) \tag{S5.3}$$
$$\simeq 50\,\text{mV}$$

Solution to Problem 5.5

Because the strip is made from a round conductor by a rolling process, the projected diameter of filaments in the direction parallel to B_e would be actually slightly less than the equivalent circular-area radius, $a = 50\,\mu m$, which is used in the above computation for ΔV_{mz}. If a diameter less than $50\,\mu m$ is used, ΔV_{mz} would be less than $50\,mV$.

b) The anisotropic shape of the Nb-Ti filaments makes magnetization in the orientation shown in Fig. 5.20b, greater than that in the orientation of Fig. 5.20a—both J_c and the "effective" a are greater. Thus there will be more magnetization loss.

Eddy current loss is proportional to $(a\dot{H}_e)^2$ in the orientation of Fig. 5.20b, while it is proportional to $(b\dot{H}_e)^2$ in the orientation Fig. 5.20a—review Problem 2.7. Thus eddy current loss is increased by a factor of $(9.2/2.6)^2 = 12.5$ in Fig. 5.20b from that in Fig. 5.20a.

The increased heat load on the helium due to higher magnetization and eddy current losses causes a higher boil-off rate; thus his observation makes sense.

c) The $-M$ vs B plot shown in Fig. 5.21 suggests, as remarked above, that the search coils are well balanced. Thus:

$$N_{pc}A_{pc}\left(\frac{d\tilde{B}_e}{dt}\right)_{pc} = N_{sc}A_{sc}\left(\frac{d\tilde{B}_e}{dt}\right)_{sc} \tag{S5.4}$$

Because $A_{pc} = A_{sc}$, we have: $N_{pc}[\tilde{B}_e]_{pc} = N_{sc}[\tilde{B}_e]_{sc}$. From symmetry, we consider only the upper half (the unit mm is neglected in the following equations):

$$[\tilde{B}_e]_{pc} = \frac{B_e(0)}{20}\int_0^{20}\left[1 - c\left(\frac{z}{z_0}\right)^2\right]dz \tag{S5.5a}$$

$$[\tilde{B}_e]_{sc} = \frac{B_e(0)}{20}\int_{60}^{80}\left[1 - c\left(\frac{z}{z_0}\right)^2\right]dz \tag{S5.5b}$$

The $N_{pc}[\tilde{B}_e]_{pc} = N_{sc}[\tilde{B}_e]_{sc}$ equality gives:

$$\frac{250}{20}\int_0^{20}\left[1 - c\left(\frac{z}{z_0}\right)^2\right]dz = \frac{280}{20}\int_{60}^{80}\left[1 - c\left(\frac{z}{z_0}\right)^2\right]dz \tag{S5.6}$$

$$250\left[20 - \frac{c\,(20)^3}{3\,(75)^2}\right] = 280\left[80 - \frac{c\,(80)^3}{3\,(75)^2} - 60 + \frac{c\,(60)^3}{3\,(75)^2}\right]$$

$$5000 - 118.5c = 5600 - 8495.4c + 3584c$$

$$c \simeq \frac{600}{4793} \simeq 0.125$$

Problem 5.6: Criterion for flux jumping

This problem deals with the derivation of the critical conductor size above which flux jumping will occur. Specifically, if a slab's width, $2a$, exceeds a critical size, $2a_c$, the slab will be thermally unstable against a flux jump.

As mentioned earlier, flux jumping was the most important source of frustration for those who designed the first superconducting magnets of engineering significance in the early 1960s [5.9]. Flux jumping is a thermal instability peculiar to Type II superconductors that permits the magnetic field to penetrate far inside their interiors. A rapidly-changing magnetic field, $\dot{H}_e$, at the surface of a Bean slab induces an electric field $\vec{E}$ within the slab, which interacts with the supercurrent of critical current density, J_c. This $\vec{E} \cdot \vec{J}_c$ interaction generates heat within the slab. Because J_c decreases with temperature, the field (flux) penetrates further into the slab, generating more heat, which further decreases J_c, and so on. If the dissipation density associated with the field penetration exceeds the slab's heat capacity, the field will penetrate unchecked and the temperature will rise uncontrollably until the slab loses its superconductivity. Such an event is called a flux jump.

a) Using the Bean model and computing the $\vec{E} \cdot \vec{J}_c$ interaction over the positive half ($0 \leq x \leq a$) of the slab, show that an expression for the dissipative energy density, e_ϕ [J/m^3], generated within the slab when the critical current density J_c is suddenly *decreased* by $|\Delta J_c|$ is given by:

$$e_\phi = \frac{\mu_\circ J_c |\Delta J_c| a^2}{3} \tag{5.21}$$

Note that the entire slab is in the critical state with field at its surface ($\pm a$) exposed to external field of $H_e \vec{i}_y$.

b) Derive Eq. 5.21 by computing the Poynting energy flow into the slab at $x = a$ and equating it with change in magnetic energy storage and dissipation energy $\mathcal{E}_\phi$ in the positive half of the slab.

c) To relate ΔJ_c to an equivalent temperature rise in the conductor, we may assume a linear temperature dependence for $J_c(T)$:

$$J_c(T) = J_{c_\circ}\left(\frac{T_c - T}{T_c - T_{op}}\right) \tag{5.22}$$

where $J_{c_\circ}$ is the critical current density at operating temperature T_{op}. T_c is the critical temperature at a given magnetic induction $B_\circ$. From Eq. 5.22, ΔJ_c in Eq. 5.21 may be related to an equivalent temperature rise ΔT:

$$\Delta J_c = -J_{c_\circ}\left(\frac{\Delta T}{T_c - T_{op}}\right) \tag{5.23}$$

Now, by requiring that $\Delta T_s = e_\phi/C_s \leq \Delta T$, where C_s is the superconductor's heat capacity, show that an expression for thermal stability leads to the critical slab half width a_c given by:

$$a_c = \sqrt{\frac{3C_s(T_c - T_{op})}{\mu_\circ J_{c_\circ}^2}} \tag{5.24}$$

Solution to Problem 5.6

a) Because of symmetry about $x = 0$, we shall consider only one half of the slab, between $x = 0$ and $x = a$. As illustrated in Fig. 5.22, the solid line corresponds to $H_{s1}(x)$, which gives the field distribution within the slab carrying J_c. The dotted line corresponds to $H_{s2}(x)$ for the slab carrying $J_c - |\Delta J_c|$. Note that the field at the surface is H_e for each case. We thus have:

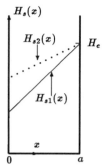

$$H_{s1}(x) = H_e + J_c(x - a) \qquad (S6.1a)$$
$$H_{s2}(x) = H_e + (J_c - |\Delta J_c|)(x - a) \quad (S6.1b)$$

Fig. 5.22   Field profiles.

Because there is a change in magnetic field within the slab, an electric field $\vec{E}$ is generated, which from Faraday's law of induction, is given by:

$$\oint_C \vec{E} \cdot d\vec{s} = -\mu_\circ \int_S \frac{\Delta H_s(x)\, \vec{i}_y \cdot d\vec{A}}{\Delta t} \qquad (S6.2)$$

From symmetry we have $\vec{E}(x = 0) = 0$ and $\vec{E}$ points in the z-direction. $\Delta H_s(x)$ is given by:

$$\Delta H_s(x) = H_{s1}(x) - H_{s2}(x)$$
$$= |\Delta J_c|(a - x) \qquad (S6.3)$$

Combining Eqs. $S6.2$ and $S6.3$, we obtain:

$$E_z(x) = \mu_\circ \frac{|\Delta J_c|}{\Delta t} \int_0^x (a - x)\, dx$$
$$= \mu_\circ \frac{|\Delta J_c|}{\Delta t} \left(ax - \frac{x^2}{2} \right) \qquad (S6.4)$$

Dissipation power density, $p(x)$, is given by $E_z(x)J_c$ and total energy density dissipated in the slab, $\mathcal{E}_\phi$ [W/m^2], per unit slab surface area in the y-z plane is given by:

$$\mathcal{E}_\phi = \int_0^a p(x)\Delta t\, dx$$
$$= \mu_\circ J_c |\Delta J_c| \int_0^a \left(ax - \frac{x^2}{2} \right)\, dx = \frac{\mu_\circ J_c |\Delta J_c| a^3}{3} \qquad (S6.5)$$

Dissipation energy density, e_ϕ, is given by $\mathcal{E}_\phi / a$:

$$e_\phi = \frac{\mu_\circ J_c |\Delta J_c| a^2}{3} \qquad (5.21)$$

Solution to Problem 5.6

b) The Poynting energy flux [W/m^2] in the y-z plane into the slab (in the $-x$-direction) at $x = a$ is equal to the change in magnetic energy storage flux ΔE_m [J/m^2] and dissipation energy flux $\mathcal{E}_\phi$ in the slab. Thus:

$$\int S_x(a)\,dt = \Delta E_m + \mathcal{E}_\phi \qquad (S6.6)$$

We can verify the direction of $\vec{S}$ by computing $\vec{S} = \vec{E} \times \vec{H}$ at $x = a$. At $x = a$, $\vec{H} = H_e\,\vec{i}_y$; from $E_z(x)$ derived in Eq. $S6.4$:

$$E_z(a) = \mu_o\frac{|\Delta J_c|a^2}{2\Delta t} \qquad (S6.7)$$

Thus:

$$\vec{S}(a) = \mu_o\frac{|\Delta J_c|a^2}{2\Delta t}\,\vec{i}_z \times H_e\,\vec{i}_y = -\mu_o\frac{H_e|\Delta J_c|a^2}{2\Delta t}\,\vec{i}_x \qquad (S6.8)$$

As expected, $\vec{S}(a)$ points in the $-x$-direction or energy flux indeed flows into the slab. Thus:

$$\int S_x(a)\,dt = \mu_o\frac{H_e|\Delta J_c|a^2}{2} \qquad (S6.9)$$

The difference in magnetic energy flux ΔE_m in the slab is given by:

$$\Delta E_m = \frac{\mu_o}{2}\int_0^a [H_{s2}^2(x) - H_{s1}^2(x)]\,dx \qquad (S6.10)$$

$$= \frac{\mu_o}{2}\int_0^a \{[H_e + (J_c - |\Delta J_c|)(x - a)]^2 - [H_e + J_c(x - a)]^2\}\,dx$$

$$= \frac{\mu_o}{2}\int_0^2 [-2H_e|\Delta J_c|(x - a) - 2J_c|\Delta J_c|(x - a)^2 + |\Delta J_c|^2(x - a)^2]\,dx$$

Neglecting the $|\Delta J_c|^2$ term in the above integral, we obtain:

$$\Delta E_m = \mu_o\left(\frac{H_e|\Delta J_c|a^2}{2} - \frac{J_c|\Delta J_c|a^3}{3}\right) \qquad (S6.11)$$

From Eq. $S6.6$, we have:

$$\mathcal{E}_\phi = \int S_x(a)\,dt - \Delta E_m \qquad (S6.12)$$

Combining Eqs. $S6.9$, $S6.11$, and $S6.12$, we obtain:

$$\mathcal{E}_\phi = \mu_o\frac{H_e|\Delta J_c|a^2}{2} - \mu_o\left(\frac{H_e|\Delta J_c|a^2}{2} - \frac{J_c|\Delta J_c|a^3}{3}\right)$$

$$= \mu_o\frac{J_c|\Delta J_c|a^3}{3} \qquad (S6.13)$$

Equation $S6.13$ leads directly to Eq. 5.21:

$$e_\phi = \frac{\mathcal{E}_\phi}{a} = \frac{\mu_o J_c|\Delta J_c|a^2}{3} \qquad (5.21)$$

Solution to Problem 5.6

c) As given by Eq. 5.22, $J_c(T)$ is a decreasing function of temperature. We thus have:

$$\Delta J_c = -J_{c_o} \left(\frac{\Delta T}{T_c - T_{op}} \right) \tag{5.23}$$

From Eq. 5.23, we have:

$$|\Delta J_c| = \frac{J_{c_o} \Delta T}{T_c - T_{op}} \tag{S6.14}$$

Replacing J_c with J_{c_o} in Eq. 5.21 and combining it with Eq. S6.14, we obtain:

$$e_\phi = \frac{\mu_o J_{c_o}^2 \Delta T a^2}{3(T_c - T_{op})} \tag{S6.15}$$

Note that e_ϕ is not only proportional to ΔT but more importantly varies with a^2. Under adiabatic conditions, the dissipation energy density e_ϕ increases the superconductor's temperature by ΔT_s, given by:

$$\Delta T_s = \frac{e_\phi}{C_s} > 0 \tag{S6.16}$$

where C_s is the superconductor's heat capacity $[J/m^3\,K]$. Combining Eqs. S6.15 and S6.16 and requiring $\Delta T_s < \Delta T$ for thermal stability, we have:

$$\frac{\Delta T_s}{\Delta T} < \frac{3C_s(T_c - T_{op})}{\mu_o J_{c_o}^2 a^2} \tag{S6.17}$$

For a given superconducting material and operating temperature, the only parameter that can be varied by the magnet designer to satisfy Eq. S6.17 is a. That is, thermal stability can only be satisfied if the slab half width a is less than the critical size a_c, given by:

$$a_c = \sqrt{\frac{3C_s(T_c - T_{op})}{\mu_o J_{c_o}^2}} \tag{5.24}$$

For a Nb-Ti conductor at $T_{op} = 4.2\,K$ and at a magnetic induction of $5\,T$, we have the following set of data (all approximate): $J_{c_o} = 2.0 \times 10^9\,A/m^2$; $C_s = 6 \times 10^3\,J/m^3\,K$; $T_c = 7.1\,K$. With these values inserted into Eq. 5.24, we obtain:

$$a_c = \sqrt{\frac{3(6 \times 10^3\,J/m^3\,K)(7.1\,K - 4.2\,K)}{(4\pi \times 10^{-7}\,H/m)(2.0 \times 10^9\,A/m^2)^2}} \tag{S6.18}$$

$$= 1.0 \times 10^{-4}\,m = 100\,\mu m$$

For a circular filament, $a_c = 100\,\mu m$ means a critical diameter of $\sim 200\,\mu m$.

Problem 5.7: Flux jumps

The magnetization *vs* ambient field trace shown in Fig. 5.23 was obtained with a monofilament of Nb-Zr wire (0.5 mm dia.) at 4.2 K carrying *no* transport current. [In the early 1960s, superconductors based on alloys of niobium and zirconium (Nb-Zr) preceeded Nb-Ti. Shortly after a composite superconductor became the standard form for magnet-grade superconductors in the mid 1960s, Nb-Ti replaced Nb-Zr; it is much easier to co-process Nb-Ti with copper than Nb-Zr with copper.] Note that both ordinate (magnetization) and abscissa (field) are given in non-SI units. You may use Bean's model and treat a single wire of diameter d_f as a slab of thickness $2a$.

a) Show that the field interval, ΔH_f, indicated in the trace is consistent with the measured magnitude of magnetization.

b) What is an estimated value of dissipation energy density [J/m³] resulting from one flux jumping event, say the one labeled A in Fig. 5.23?

c) Give an estimate of the temperature rise for flux jump A. Assume the heat capacity of Nb-Zr to be independent of temperature and equal to 6 kJ/m³ K.

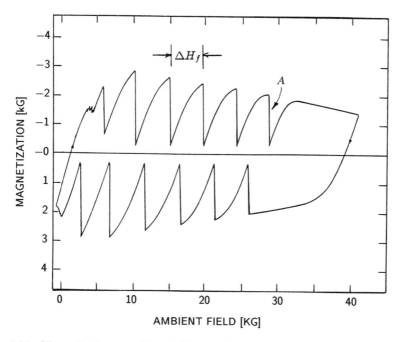

Fig. 5.23   Magnetization *vs* ambient field trace for a 0.5-mm dia. Nb-Zr monofilament.

Solution to Problem 5.7

a)   From Bean's model, flux jumping can occur every H_p. Clearly, $H_p = \Delta H_f$, where ΔH_f is indicated in Fig. 5.23. We also know full magnetization is $H_p/2$.

From Fig. 5.23, $\Delta H_f \simeq 5$ kilogauss, $\mu_\circ \Delta H_f = 0.5$ T. Also from Fig. 5.23, $H_p/2 \simeq$ 2.5 kilogauss, which is $(1/2)\,\Delta H_f$. They are consistent.

b)   We can derive the flux jump energy density, e_ϕ, using the Poynting energy balance: $e_s = e_\phi + \Delta e_m$, where e_s is the Poynting energy density entering the superconductor at $x = a$ and Δe_m is its change in stored magnetic energy density. Let's consider only $0 \leq x \leq a$. $\Delta H(x)$ within the slab is given by:

$$\Delta H(x) = H_p \frac{(a - x)}{a} \tag{S7.1}$$

From Eq. $S7.1$, we have:

$$E(x) = \mu_\circ \frac{H_p}{\Delta t} \int_0^x \frac{a - x}{a}\, dx = \frac{\mu_\circ H_p}{a\Delta t}\left(ax - \frac{x^2}{2}\right) \tag{S7.2}$$

$\vec{S}$ at $x = a$ is thus given by:

$$\vec{S}(a) = -\frac{\mu_\circ}{2\Delta t} H_p a H_e\, \vec{i}_x \tag{S7.3}$$

$\vec{S}(a)$ is directed towards the slab and the Poynting energy density e_s is given by:

$$e_s = \frac{\int S_x(a)\, dt}{a} = \frac{\mu_\circ}{2} H_p H_e \tag{S7.4}$$

The stored magnetic energy after flux jumping (e_{m2}) is $\mu_\circ H_e^2/2$. The stored magnetic energy before flux (e_{m1}) is given by:

$$e_{m1} = \frac{\mu_\circ}{2a} \int_0^a [H_e + J_c(x - a)]^2\, dx \tag{S7.5}$$

$$= \frac{\mu_\circ}{2a} \int_0^a [H_e^2 + 2H_e J_c(x - a) + J_c^2(x - a)^2]\, dx$$

$$= \frac{\mu_\circ}{2a}\left[H_e^2 a - H_e J_c a^2 + \frac{J_c^2 a^3}{3}\right] = \frac{\mu_\circ}{2} H_e^2 - \frac{\mu_\circ}{2} H_e H_p + \frac{\mu_\circ}{6} H_p^2 \tag{S7.6}$$

$$\Delta e_m = e_{m2} - e_{m1} = \frac{\mu_\circ}{2} H_e H_p - \frac{\mu_\circ}{6} H_p^2 \tag{S7.7}$$

Because $e_\phi = e_s - \Delta e_m$, we obtain:

$$e_\phi = \frac{\mu_\circ}{2} H_p H_e - \frac{\mu_\circ}{2} H_e H_p + \frac{\mu_\circ}{6} H_p^2 = \frac{\mu_\circ}{6} H_p^2$$

$$= \frac{1}{3\mu_\circ}\left(\frac{\mu_\circ H_p}{2}\right)(\mu_\circ H_p) = \frac{1}{3\mu_\circ}\left(\frac{\mu_\circ \Delta H_f}{2}\right)(\mu_\circ \Delta H_f) \tag{S7.8}$$

$$\simeq \frac{1}{(3)(4\pi \times 10^{-7}\,\text{H/m})}(0.25\,\text{T})(0.5\,\text{T}) \simeq 33 \times 10^3\,\text{J/m}^3$$

c) $e_\phi = C_s \Delta T_s$; $33 \times 10^3 = 6 \times 10^3 \Delta T_s$. Solving for ΔT_s, we obtain: $\Delta T_s = 5.5$ K, sufficient to drive the superconductor normal.

Problem 5.8: Filament twisting

As discussed in Problem 5.6, the criterion on flux jumping requires conductor diameters to be less than $2a_c$, which for Nb-Ti is ~200 μm. With J_{c_o} typically 2×10^9 A/m² (at 4.2 K and 5 T), a 200-μm diameter Nb-Ti filament has a critical current of only ~60 A—insufficient for most magnet applications if used alone. The idea of using many filaments, each meeting the diameter criterion of flux jumping, in a matrix of normal metal, emerged in the late 1960s to build conductors with critical currents as high as 1000 A. Today, 50-kA conductors are available.

In early (c. 1969) "multifilamentary" conductors, filaments were still untwisted, violating, it turned out, the size criterion. Problem 5.9 deals with such conductors. Results of a thorough study of multifilamentary conductors, both analytical and experimental, carried out by Wilson, Walters, Lewin, and Smith of the Superconducting Application Group, Rutherford Laboratory, and published in 1970 launched a new era of multifilamentary conductors [5.10].

Simply stated, when filaments are embedded in a conductive metal such as copper and subjected to a time-varying magnetic field, the filaments are electrically coupled according to Faraday's law. Thus they act as a single entity with an effective conductor diameter that is virtually equal to that of the whole conductor. The basic premise of the flux jumping criterion for isolated filaments is therefore violated in such an coupled filamentary conductor. In order to eliminate flux jumping in multifilamentary conductors, filaments must be decoupled. Twisting of the filaments does the trick.

Consider a two-dimensional conductor model comprised of two Bean slabs, each d_f wide, separated by a copper slab of width $2w$ and electrical resistivity ρ_{cu}. Figure 5.24 shows a representation of the conductor as seen looking down the z-axis. Note that unlike the one-dimensional Bean slab which extends into infinity in both the y- and z-directions, this conductor is 2ℓ long in the y-direction.

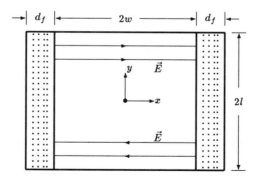

Fig. 5.24   Two-dimensional conductor comprising of a normal metal slab sandwiched between two Bean slabs.

Problem 5.8: Filament twisting

Suppose the conductor is subjected to a spatially uniform, time-varying magnetic field pointed in the z-direction, $\dot{H}_{0z}\vec{i}_z$.

a) Show that the x-directed electric field within the copper slab is given by:

$$E_{1x} = \mu_\circ \dot{H}_{0z} y \qquad (5.25)$$

Assume the electric field in each superconducting slab to be zero—strictly speaking it is not, but compared with that in the copper, it is extremely small hence the zero $\vec{E}$ field assumption is valid. Also assume that the quasi-static assumption applies. Under these assumptions, it is apparent that the $\vec{E}$ field in the copper, as indicated in Eq. 5.25, has only an x component.

b) Show that an expression for the net current flowing through the copper (per unit conductor depth in the z-direction), I_{cp} [A/m], from one superconducting slab to the other superconducting slab, over one half conductor length (from $y = 0$ to $y = \ell$), is given by:

$$I_{cp} = \int_0^\ell J_{cu}\,dy = \frac{\mu_\circ \dot{H}_{0z}}{\rho_{cu}} \int_0^\ell y\,dy = \frac{\mu_\circ \dot{H}_{0z}\ell^2}{2\rho_{cu}} \qquad (5.26)$$

c) At a critical length ℓ_c, the net current I_{cp} given by Eq. 5.26, becomes equal to $J_c d_f$, the slab's critical current (per unit conductor depth). Show that an expression for ℓ_c in terms of appropriate parameters is given by:

$$\ell_c = \sqrt{\frac{2\rho_{cu} J_c d_f}{\mu_\circ \dot{H}_{0z}}} \qquad (5.27)$$

d) Multifilamentary superconductors for 60-Hz power applications must have filaments size (d_f) that is extremely small, in the range 0.1~0.5 μm, which is even smaller than the wavelength of visible light (~0.7 μm). This extremely small size is required to keep hysteresis energy, generated within each filament every time a magnetic field is cycled, "manageable." (As will be discussed in Chapter 6, hysteresis loss per cycle of field excitation is proportional to filament diameter.)

Compute ℓ_c for a typical "submicron" superconductor having the following parameters: $\rho_m = 30\,\text{n}\Omega\,\text{m}$; $J_c = 2\,\text{GA/m}^2$; $d_f = 0.2\,\mu\text{m}$; $\mu_\circ \dot{H}_{0z} = 2\,\text{kT/s}$ (equivalent to a sinusoidal excitation of 5-T amplitude magnetic induction at 60 Hz). Note that ρ_m represents the matrix resistivity, which is generally of copper-nickel alloys.

e) Compute the number of filaments required for a submicron multifilamentary conductor with a filament diameter of 0.2 μm having a critical current of 100 A. Use the same values of parameters given in d).

Solution to Problem 5.8

a)   From Faraday's law, applied under the quasi-static assumption, we have:

$$\frac{\partial E_{1y}}{\partial x} - \frac{\partial E_{1x}}{\partial y} = -\mu_0 \dot{H}_{0z} \qquad (S8.1)$$

Because $\vec{E}$ is zero in the superconducting slabs, $E_y = 0$ at $x = \pm w$, forcing $E_{1y} = 0$ everywhere in the copper slab. Thus:

$$E_{1x} = \mu_0 \dot{H}_{0z} y \qquad (5.25)$$

b)   Once the E field is known, the current density J_{cu} in the copper slab is given by: $J_{cu} = E_{1x}/\rho_{cu}$. The net current flowing in the copper from one superconducting slab to the other over half the conductor length is given by:

$$I_{cp} = \int_0^\ell J_{cu}\, dy = \frac{\mu_0 \dot{H}_{0z}}{\rho_{cu}} \int_0^\ell y\, dy = \frac{\mu_0 \dot{H}_{0z} \ell^2}{2\rho_{cu}} \qquad (5.26)$$

c)   Equating I_{cp} given by Eq. 5.26 with $J_c d_f$ and solving for ℓ_c, we have:

$$\ell_c = \sqrt{\frac{2\rho_{cu} J_c d_f}{\mu_0 \dot{H}_{0z}}} \qquad (5.27)$$

d)   Inserting appropriate values into Eq. 5.27, we obtain:

$$\ell_c = \sqrt{\frac{2(3\times10^{-8}\,\Omega\,\mathrm{m})(2\times10^9\,\mathrm{A/m^2})(0.2\times10^{-6}\,\mathrm{m})}{2\times10^3\,\mathrm{T/s}}} \qquad (S8.2)$$

$$\simeq 10\,\mathrm{mm}$$

In typical submicron strands, the twist pitch length is ~10 mm! This means that the diameter of such strands, by mechanical requirements, should be ~1 mm; actually a thermal-magnetic stability criterion, similar to the flux jump criterion, requires it to be even smaller than 1 mm. This is because the strand, to reduce coupling losses, use Cu-Ni alloys as the matrix materials, resulting in a magnetic diffusion time constant that is smaller than its thermal diffusion time constant.

e)   Critical current (I_c), critical current density (J_c), the number of filaments (N_f) and filament diameter (d_f) in a multifilamentary conductor is related by:

$$I_c = N_f \frac{\pi d_f^2}{4} J_c \qquad (S8.3)$$

Solving for N_f from Eq. $S8.3$ with appropriate values of parameters, we obtain:

$$N_f = \frac{4I_c}{\pi d_f^2 J_c} = \frac{4(100\,\mathrm{A})}{\pi(0.2\times10^{-6}\,\mathrm{m})^2(2\times10^9\,\mathrm{A/m^2})} \qquad (S8.4)$$

$$= 3.2\times10^6$$

In submicron strands, the number of filaments may reach one million.

Problem 5.9: Magnetization of conductors

This problem illustrates the effect of filament size and twisting on magnetization. In the late 1960s, three Nb-Ti composite superconductors of equal volume were subjected to magnetization measurements [5.11]. Conductor 1 was twisted multifilamentary wire with a twist pitch length ℓ_{p1}. Conductor 2 was also twisted multifilamentary wire with a twist pitch length $\ell_{p2} > \ell_{p1}$, while Conductor 3 was a monofilament.

Figure 5.25 presents three magnetization curves, labeled A, B, and C, for the three Nb-Ti composite conductors. Each conductor was subjected to field pulses indicated by arrows in the figure. Traces A, B, and C do not necessarily correspond to Conductors 1, 2, and 3 respectively. Note that Traces B (B_1, B_2, B_3) show a dependence on field sweep rate; trace C is independent of field sweep rate; trace A is also independent of field sweep rate, but shows "partial" flux jumps induced by the field pulses.

a) Identify which magnetization trace corresponds to which conductor.

b) Estimate the ratio of filament diameter in the monofilament conductor to that in the multifilament conductors.

c) Estimate a value of ℓ_{p2}. Take $J_c d_f = 4 \times 10^4 \,\mathrm{A/m}$ for Conductors 1 and 2. Also comment on ℓ_{p1}.

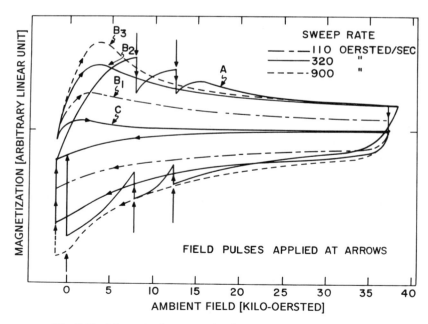

Fig. 5.25   Magnetization traces for Conductors 1, 2, and 3 [5.11].

Solution to Problem 5.9

a) Note that Traces A and C are field sweep-rate independent and the corresponding magnetization—an indication of filament diameter—is much greater for Trace A than that for Trace C. We therefore conclude that Trace A is for Conductor 3 (monofilament) and that Trace C is for Conductor 1 (ℓ_{p1}). That leaves Trace B for Conductor 2 (ℓ_{p2}). (Note that each conductor, whether monofilament or multifilament, has the same volume of Nb-Ti superconductor and thus its measured magnetization should be directly proportional to filament diameter.)

b) The ratio of magnetization width of Conductor 3 (monofilament, trace A) to that of Conductor 1 (trace C), is roughly 10 for $\mu_\circ H_e$ below ~1 T (10 kilo-oersted). Therefore, we conclude that the filament ratio is roughly 10.

c) Because a field sweep-rate of 900 oersted/sec ($\mu_\circ \dot{H}_{0z} = 0.09$ T/s) makes the magnetization of Conductor 2 (Trace B_3) nearly equal to that of Conductor 3 (Trace A), we may conclude that this sweep rate makes Conductor 2's filament twist pitch length ℓ_{p2} critical. Thus from Eq. 5.27:

$$\ell_{p2} = 2\sqrt{\frac{2\rho_{cu}J_c d_f}{\mu_\circ \dot{H}_{0z}}} \qquad (S9.1)$$

With $\rho_{cu} = 2 \times 10^{-10} \, \Omega \, \text{m}$; $J_c d_f = 4 \times 10^4 \, \text{A/m}$; and $\mu_\circ \dot{H}_{0z} = 0.09$ T/s, we obtain:

$$\ell_{p2} = 2\sqrt{\frac{(2)(2 \times 10^{-10} \, \Omega \, \text{m})(4 \times 10^4 \, \text{A/m})}{0.09 \, \text{T/s}}} \qquad (S9.2)$$

$$= 27 \, \text{mm}$$

This value is close enough to the actual twist pitch of 10 mm. Because magnetization of Conductor 1 (trace C) at a sweep rate of 320 oersted/sec is considerably smaller than that of Conductor 2 for the same field sweep rate, we conclude that ℓ_{p1} is significantly shorter than ℓ_{p2}.

Filament Twisting in Composite Superconductors

An important implication of the condition $I_{cp} = J_c d_f$, used to derive Eq. 5.27 (Problem 5.7), is that the two superconducting slabs are electrically coupled. Were the conductor length substantially shorter than $2\ell_c$, on the other hand, the two would be decoupled. In reality, these slabs may be decoupled, even if each is much longer than $2\ell_c$, if they are alternated in their position with a pitch length less than $2\ell_c$. Thus in multifilamentary conductors, we achieve partial decoupling by twisting the filaments with a pitch length $\ell_p \ll 2\ell_c$. Note that in a twisted conductors each filament remains at a fixed radial distance from the strand axis. By contrast, in a cable comprised of *transposed* strands, more complete decoupling is possible because each strand is made to occupy every radial position across the cable diameter as it spirals along the cable's twist pitch length. In multifilamentary conductors, because filaments are arranged in fixed radial locations in the matrix at the billet-making stage, transposition is not possible; the conductor is simply twisted during the production stage.

Problem 5.10: Flux jump criterion for HTS tapes

Although measurements of magnetization as comprehensive as that performed on
LTS have not yet been carried out on HTS, it is assumed that Bean's critical state
model can be used to describe the macroscopic magnetic behavior of HTS. (As
demonstrated in Problem 5.4 for an YBCO crystal, the model may be used to
extract J_c values from magnetization data.) To date, a series of flux jumps such
as the one studied in Problem 5.7 has not been observed in HTS. In this problem
we apply the flux jumping criterion derived in Problem 5.6 (Eq. 5.24) to HTS,
specifically a BiPbSrCaCuO (2223) HTS, by studying the effect of temperature on
the parameters appearing in Eq. 5.24.

a) Comment on the vulnerability of HTS to flux jumping by examining the
   temperature dependence of each parameter appearing in Eq. 5.24. Here, we
   are discussing the intrinsic stability and it is not necessary to include the
   effect of silver, the standard matrix metal used in bismuth-based HTS tapes
   presently available. For the heat capacity C_s appearing in Eq. 5.24, you may
   use copper's heat capacity.

b) Compute the values of a_c for BiPbSrCaCuO (2223) superconductor in zero
   magnetic field at 4, 10, 20, 30, 40, 50, 60, 70, and 80 K. Use the zero-field
   (0 T) values of $I_c(T)$ corresponding to Tape 1 (open circle data) shown in
   Fig. 5.26 [5.12]. The cross sectional area occupied by BiPbSrCaCuO alone
   is $0.118\,mm^2$.

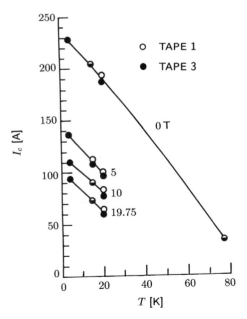

Fig. 5.26 I_c vs T data at 0, 5, 10, and 19.75 T [5.12].

Solution to Problem 5.10

a) The first important point to note is that in each of the HTS discovered so far J_c *decreases* with temperature, *i.e.* the HTS are in principle vulnerable to thermal instability. Their critical filament sizes should be considerably greater than those of LTS because of two parameters in Eq. 5.24: $(T_c - T_{op})$ and C_s. For LTS, $(T_c - T_{op})$ ranges 1∼5 K, while for HTS it ranges 10∼50 K, a factor of 10 increase. An increase in C_s (for copper) from $T_{op} \sim 4$ K to $T_{op} \sim 40$ K is ∼700; this increase in C_s alone can increase the critical size by a factor of over 25.

From these observations, we state that when HTS are operated at temperatures above ∼20 K, they are unlikely to suffer flux jump instability. Quantitative results are presented below.

b) Table 5.2 presents the values of a_c for the temperature range 4∼80 K. It also includes values of I_c extracted from Fig. 5.26 and corresponding J_c used for computation.

Table 5.2: Critical Size *vs* Temperature

T [K]	I_c [A]	J_c [MA/m²]	C_s* [kJ/m³K]	a_c [mm]
4	228	1932	0.8	0.2
10	217	1839	7.7	0.7
20	194	1644	68.5	2.4
30	163	1381	240	5.5
40	135	1144	534	9.8
50	108	915	881	16
60	80	678	1219	25
70	53	449	1540	43
80	24	203	1825	103

* Copper heat capacity.

The HTS tape used to compute a_c is of "single-core," meaning each tape contains a central strip 3∼5 mm wide ($2a$), processed with a silver matrix which occupies 60∼70% of the total conductor volume. Tape 1, including its silver matrix, is 4.09 mm wide and 0.153 mm thick. Values of a_c listed in Table 5.2 indicate that even these "single-core" tapes are stable against flux jumping when they are operated above ∼20 K. Within a few years these silver-sheathed tapes will likely be multifilamentary, consisting of many (up to ∼100) "mini strips" [5.13, 5.14]. The need to process tapes in the multifilamentary form, however, is driven not so much to satisfy the flux jumping criterion but to make them more strain resistant. Also making strips thin inherently improves the critical current density of these HTS tapes [5.15].

References

[5.1] C.P. Bean, "Magnetization of hard superconductors," *Phys. Rev. Lett.* **8**, 250 (1962).

[5.2] Y.B. Kim, C.F. Hempstead, and A.R. Strnad, "Magnetization and critical super-currents," *Phys. Rev.* **129**, 528 (1963).

[5.3] M.A.R. LeBlanc, "Influence of transport current on the magnetization of a hard superconductor," *Phys. Rev. Letts.* **11**, 149 (1963).

[5.4] Kō Yasukōchi, Takeshi Ogasawara, Nobumitsu Usui, and Shintaro Ushio, "Magnetic behavior and effect of transport current on it in superconducting Nb-Zr wire," *J. Phys. Soc. Jpn.* **19**, 1649 (1964).

Kō Yasukōchi, Takeshi Ogasawara, Nobumitsu Usui, Hisayasu Kobayashi, and Shintaro Ushio, "Effect of external current on the magnetization of non-ideal Type II superconductors," *J. Phys. Soc. Jpn.* **21**, 89 (1966).

[5.5] H.T. Coffey, "Distribution of magnetic fields and currents in Type II superconductors," *Cryogenics* **7**, 73 (1967).

[5.6] W.A. Fietz, "Electronic integration technique for measuring magnetization of hysteretic superconducting materials," *Rev. Sci. Instrum.* **36**, 1621 (1965).

[5.7] T.R. Dinger, T.K. Worthington, W.J. Gallagher, and R.L. Sandstrom, "Direct observation of electronic anisotropy in single-crystal $Y_1Ba_2Cu_3O_{7-x}$," *Phys. Rev. Lett.* **58**, 2687 (1987).

[5.8] M. Yunus and Y. Iwasa (preliminary results, FBNML, 1994).

[5.9] See, for example, M.S. Lubell, B.S. Chandrasekhar, and G.T. Mallick, "Degradation and flux jumping in solenoids of heat-treated Nb-25% Zr wire," *Appl. Phys. Lett.* **3**, 79 (1963).

[5.10] Superconducting Applications Group (Rutherford Laboratory), "Experimental and theoretical studies of filamentary superconducting composites," *J. Phys.* **D3**, 1517 (1970).

[5.11] Y. Iwasa, "Magnetization of single-core, multi-strand, and twisted multi-strand superconducting composite wires," *Appl. Phys. Lett.* **14**, 200 (1969).

[5.12] K. Sato, T. Hikata, and Y. Iwasa, "Critical currents of superconducting BiPb-SrCaCuO tapes in the magnetic flux density range 0 to 19.75 T at 4.2, 15, and 20 K," *Appl. Phys. Lett.* **57**, 1928 (1990).

[5.13] K. Sato, N. Shibutani, H. Maki, T. Hikata, M. Ueyama, T. Kato, and J. Fujikami, "Bismuth superconducting wires and their applications," *Cryogenics* **33**, 243 (1993).

[5.14] A. Otto, L.J. Masur, J. Gannon, E. Podtburg, D. Daly, G.J. Yurek, and A.P. Malozemoff, "Multifilamentary Bi-2223 composite tapes made by a metallic precursor route," *IEEE Trans. Appl. Superconduc.* **3**, 915 (1993).

[5.15] H. Ogiwara, M. Satou, Y. Yamada, T. Kitamura, T. Hasegawa, "Induced critical current density limit of Ag sheathed Bi-2223 tape conductor," *IEEE Trans. Magn.* **30**, 2399 (1994).

CHAPTER 6
STABILITY

6.1 Introduction

Reliability is one requirement that all devices must meet; superconducting magnets are not excepted. Historically, reliability has been one of the most troubling (and at the same time most challenging) aspects of superconducting magnet technology. An early failure of a magnet of significant size was that of a solenoidal magnet designed and built at MIT in 1961 [6.1]. Part of a thermonuclear fusion research project, the magnet, 20-cm bore and 1.2 m long, was wound with a 250-μm diameter single-core Nb-Zr superconductor. It may be noted that conductors like this—of single core exceeding the flux jumping criterion size $2a_c$ (Chapter 5) and processed without any normal metal matrix—were the only type available at that time; remarkably—a more appropriate expression would be *misleadingly*—these conductors worked well in "small" magnets. When it was finally tested in 1962, the magnet, instead of reaching a designed operating current of 20 A and generating a central field of 3 T, prematurely quenched at ~1 A; repeated tries did not improve its performance. No doubt there were many magnet failures of this nature—mostly unreported—during this early era of antiquity. Further discussion of premature quenches (quenches occurring at currents below designed operating current) and "training" in "adiabatic" magnets will be given in Chapter 7.

6.2 Stability Theories and Criteria

As stated in Chapter 1, superconductivity is confined within the phase surface bounded by magnetic field, temperature, and current. Of these three parameters, the temperature is neither completely controllable nor predictable within the winding of a superconducting magnet, because the energy stored in the magnet, both magnetic and mechanical, can easily be converted into heat, upsetting the thermal equilibrium of the winding and raising the conductor temperature to above its critical value. Therefore, in magnet operation the temperature stability or more generally thermal behavior of the conductor is crucial. We may discuss the thermal stability by examining the power equation operating on a unit conductor volume, one of the three basic constituents of the magnet winding.

The three winding constituents, as shown schematically in Fig. 6.1, are: 1) conductor; 2) structural material; and 3) coolant. The conductor temperature, T, is governed by the following power density equation:

$$\dot{e}_h = g_k + g_j + g_d - g_q \tag{6.1}$$

$\dot{e}_h$ represents the time rate of change of thermal energy density of the conductor unit, $\dot{e}_h = C_{cd}\partial T/\partial t$. C_{cd} is the heat capacity of the conductor, which, after the development in 1964 by Stekly [6.2] of composite conductors, consists of the superconductor and normal-metal (usually copper) matrix.

g_k represents the conduction heat flow into the conductor element, *i.e.* $g_k = \nabla \cdot (k_{cd}\nabla T)$. k_{cd} is the conductor element's thermal conductivity.

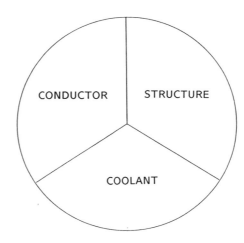

Fig. 6.1   Winding pack cross section comprising the three basic constituents.

g_j represents the Joule heating, *i.e.* $g_j = \rho_{cd}(T)j_{cd}^2(t)$. $\rho_{cd}(T)$ is the conductor's normal-state electrical resistivity and j_{cd} its current density. (This term is zero when the conductor is superconducting.)

g_d represents heat generation, primarily magnetic and mechanical in origin; this term is important in the so-called "adiabatic" magnets in which the cooling, represented by g_q, is absent in the winding.

g_q represents the cooling on the conductor element; it plays a prominent role in the so-called "cryostable" magnets and none at all in adiabatic magnets.

The history of the development of theories and concepts for stability (and protection to be discussed in Chapter 8) may be said to have evolved around specific solutions to Eq. 6.1. Table 6.1 lists various concepts derived from Eq. 6.1 under special conditions. In the table, the parameter with 0 signifies that the parameter is negligible or not considered in the equation; $\sqrt{}$ signifies it is included. Each case is briefly discussed below [6.3].

6.2.1 Flux Jumping

Flux jumping is a thermal runaway instability unique among superconductors. As discussed in Chapter 5, it occurs primarily at or near 4 K, the temperature at which LTS is operated, because the superconductor (as with everything else except helium) has an extremely small specific heat. The magnetic energy density stored in the superconductor in the form of the supercurrent can thus easily be converted into heat, exceeding the superconductor's critical temperature. As remarked in Chapter 5, flux jumping is no longer a serious problem for magnet operation.

Table 6.1: Concepts Derived from Eq. 6.1

$\dot{e}_h$	g_k	g_j	g_d	g_q	Application
$\checkmark$	0	0	$\checkmark$	0	Flux jump
0	0	$\checkmark$	0	$\checkmark$	Cryostability
$\checkmark$	$\checkmark$	$\checkmark$	0	$\checkmark$	Dynamic stability
0	$\checkmark$	$\checkmark$	0	$\checkmark$	"Equal area"
0	$\checkmark$	$\checkmark$	0	0	MPZ*
$\checkmark$	0	$\checkmark$	0	0	Protection
$\checkmark$	$\checkmark$	$\checkmark$	0	0	Adiabatic NZP†

* Minimum propagation zone.

† Normal zone propagation.

6.2.2 Cryostability

By the mid-1960s, the basic concept for an engineering solution to achieve reliable magnet operation was developed, chiefly by Stekly [6.2], who proposed composite conductors comprising a superconductor core—at the beginning still a single core— that is co-processed with a highly conductive copper. The copper reduces the normal-state Joule heating density, g_j, to a level sufficiently low so that it can be balanced by cooling, g_q, operating on a surface of the unit conductor volume. The theory based on this concept is known as cryostability and magnets designed according to this theory are cryostable. Problems are presented to discuss and study cryostability in more detail.

6.2.3 Dynamic Stability

In the adiabatic treatment of flux motion within the superconductor, discussed in Chapter 5, it is implicitly assumed that the superconductor's magnetic time constant, τ_{mg}, is much shorter than its thermal time constant, τ_{th}. Obviously, as confirmed by experiments, this assumption was valid in many cases. As a result flux jumping occurred in virtually all conductors through the mid-1960s. Hart [6.4] treated magnetic instabilities in which this assumption was no longer valid. His analyses led to what has become known as dynamic stability. Its practical applicability was mostly confined to Nb$_3$Sn superconductors that were then available only in the tape form. Since the termination of Nb$_3$Sn tape production in the mid-1970s and the availability henceforth of Nb$_3$Sn conductors in wire and mono- lithic strip forms, the dynamic stability theory has become dormant. However, because the only conductor form available in promising high-T_c superconductors has been tape, the dynamic stability theory was recently revived by Ogasawara for application to HTS magnets [6.5]. Because the standard operational mode of HTS magnets is to have them coupled to cryocoolers and operated in a vacuum ("adiabatic") environment, the relevancy of dynamic stability for HTS magnets is questionable. Nevertheless, a problem on dynamic stability as applied to tapes is presented.

6.2.4 "Equal Area"

The "equal-area" criterion is one of those rare instances in engineering where its formulation, though undoubtedly inspired by engineering need, seems to have been propelled more by mathematical logic. It is a special case of cryostability theory in that thermal conduction within the winding, not included in cryostability, is neatly used. The result is an improved magnet performance through an enhancement in the overall current density. A problem is devoted to tracing the criterion's original derivation in 1969, by Maddock, James, and Norris [6.6].

6.2.5 MPZ

The concept of MPZ (minimum propagating zone) was introduced in 1972 by Martinelli and Wipf [6.7] to consider the effect on coil performance of a quench-causing disturbance (the term expressed by g_d in Eq. 6.1) occurring locally, like a point source, within the coil winding. An epoxy-cracking event in epoxy-impregnated windings would be such a point source disturbance. The MPZ concept shows that it is possible to sustain "superconductivity" in a magnet even in the presence of a small normal-state region in its winding; the normal-zone volume must be smaller than a critical size defined by the MPZ theory. For typical winding parameters, however, this critical size is so small that its practical importance on cryostable magnets is virtually nil. Its importance in adiabatic magnets was recognized by Wilson in the late 1970s [6.8], and it has since become an indispensable concept for analyzing the stability of adiabatic magnets.

6.2.6 Nonsteady Cases

The last two cases in Table 6.1 concern the non-steady state thermal behavior of the winding. They are treated in Chapter 8.

6.3 Cable-in-Conduit (CIC) Conductors

It was through consideration of stability that the concept of cable-in-conduit (CIC) conductors was proposed by Hoenig and Montgomery in the early 1970s [6.9]. The concept is to encase a cable of superconducting strands in a leak-tight conduit through which supercritical helium is forced to maintain the strands' superconductivity. The conduit not only holds the strands and helium but also provides strength. Figure 6.2 presents sketches of CIC conductors.

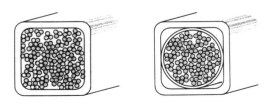

Fig. 6.2 Examples of CIC conductors.

The conduit makes CIC conductors suitable for Class 1 magnets and unsuitable for Class 2 magnets. Generally magnets require strong windings and the conduit is an effective strengthener. However, because the conduit itself occupies a significant fraction of the conductor cross sectional area, a CIC magnet must operate at high currents (above ~10 kA) to maintain reasonable overall current densities. Both of these features make CIC conductors desirable for Class 1 magnets, while neither *extra* strength nor high-current operation is needed or desirable for Class 2 magnets. Indeed, magnet engineers engaged in the fusion programs have been the most enthusiastic advocates of CIC conductors and have made significant contributions to the development of CIC conductors from the very beginning. Key issue still face designers of CIC conductors in the following areas.

- The conductor itself: encasing a bundle of cabled strands into a tight-fitting conduit is difficult, particularly when the strands are of composite Nb_3Sn, which are brittle and must not be strained above ~0.3%. This means that in the production of a CIC conductor, the final heat treatment for Nb_3Sn reaction must be performed after the unreacted strands are encased within a conduit and the CIC conductor, save heat treatment, is processed into its final form. The heat treatment temperature of ~700°C makes the manufacturing process of Nb_3Sn magnets at least one order of magnitude more complex than that required for Nb-Ti magnets. The strands, bundled quite tightly within the conduit, for example, should not sinter, because sintering dramatically increases AC losses in the conductor. Another important point is that the conduit must not lose its strength during heat treatment, resulting in the development of a special material for the conduit: a nickel-iron based superalloy, Incoloy 908. Also troubling is the differential thermal contraction among the conductor constituents—Nb_3Sn, bronze, copper, and other metals—which when the conductor is eventually cooled from ~700°C down to 4.2 K, may, as studied in Problem 3.18, strain Nb_3Sn close to its permissible limit.

- Stability: this is discussed below.

- AC losses: because superconducting magnets in fusion reactors are subjected to time-varying magnetic fields, AC losses occur within CIC conductors. How to deal with AC losses is a major topic in the design of CIC conductors. AC losses are discussed more fully in the next chapter.

- Other issues include: conductor joints; coolant flow circulation and distribution; and quench-induced pressure rise in the conductor. Conductor splicing and pressure rise are discussed, respectively, in Chapters 7 and 8.

6.3.1 Stability

From the very beginning in the early 1970s, much attention was given to the stability of CIC conductors. One of the important results from the early era was the observation in 1977 of "recovery" in a CIC conductor even in the absence of a net coolant flow through the conductor [6.10]. Apparently, heating-induced high-velocity local coolant flow in the heated region is responsible for supplying the cooling necessary for recovery.

An important milestone early in the stability work for CIC conductors is the discovery by Lue, Miller, and Dresner in 1979 of multivalued stability margins that can exist under certain operating conditions in CIC conductors [6.11]. Here the energy margin, Δe_h, is defined as the maximum dissipative energy density pulse (per unit strand volume) to which a conductor can be exposed and still remain superconducting when it is carrying a given transport current. Figure 6.3 presents a typical Δe_h vs I_t/I_{c_0} for constant values of operating temperature (T_{op}), field (B_0), and coolant flow rate. Here, I_t is the transport current and $I_{c_0}(T_{op}, B_0)$ is the critical current. The "dual stability" regime, characterized by multivalued stability margins, occurs near $I_t/I_{c_0} \sim 0.5$. The regime below the dual stability regime is referred to as "well-cooled" and that above as "ill-cooled" [6.12]. Because the greater the transport current, the lower will be the "current sharing" temperature (T_{cs}) beyond which partial Joule heating takes place, Δe_h is expected to decrease with increasing I_t/I_{c_0}. (The current sharing temperature is discussed in the Problem Section of this chapter.)

Thus in this dual stability regime, the conductor has multiple values of energy margin. Stability is maintained for energy density inputs up to Δe_{h1}. Beyond Δe_{h1}, the conductor is unstable until Δe_{h2} is reached. Then, for the next range of energy, from Δe_{h2} until Δe_{h3}, the conductor again remains stable. Dresner explained the stable range from Δe_{h2} to Δe_{h3} in terms of the same heating-induced local coolant flow used to explain the results of the earlier experiment [6.10]. The higher the heating, the higher will be the induced local cooling. Under the right circumstances, this induced cooling overtakes the heating, creating multivalued energy margins; this dual stability regime must necessarily be confined to a small range of transport current, as is suggested in Fig. 6.3.

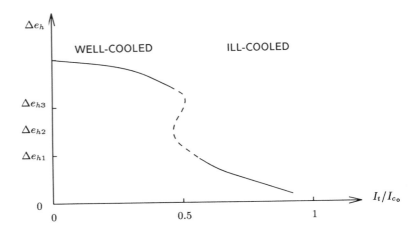

Fig. 6.3 General shape of an energy margin vs normalized transport current plot for a CIC conductor [6.12].

In the face of the requirement to *virtually* guarantee stable operation of large fusion superconducting magnets, the fusion community, which, as remarked in Chapter 3, was internationalized recently in the form of the ITER consortium, has decided to take a conservative route, ITER magnets will be designed to operate in the well-cooled regime. Despite intensive and numerous efforts to understand and improve the stability of CIC conductors, carried out analytically and experimentally during the 1980s [6.13∼6.16], CIC magnets for ITER, as remarked above, will be designed to satisfy a criterion similar to the classic Stekly criterion.

The major difference between the two criteria is cooling, or more precisely the spatial and temporal extent to which cooling is assumed to be available. In the Stekly criterion, to be studied in Problems 6.3 and 6.4, it is assumed that cooling by liquid helium is available to conductor situated at every region of the winding at all times and as long as required to achieve recovery; the winding designs of early magnets were filled with cooling channels to satisfy the Stekly criterion "unconditionally."

In a criterion applied to CIC windings ("pseudo-Stekly criterion"), which will be studied in Problem 6.11, conductor parameters are chosen to satisfy the Stekly criterion; however, cooling available within the conduit is not sufficient to guarantee unconditional stability against a disturbance that drives the *whole* winding normal nor a normal zone that remains normal for a *long* time period. It is for this reason that energy margin Δe_h and AC losses are considered critical parameters by the ITER magnet design team; these two issues are among the primary research targets of ITER's CIC conductor design effort. The concept of energy margin assumes *a priori* that disturbances, other than AC losses, are localized and transient: against a localized and transient disturbance the Stekly criterion's "infinite" cooling supply requirement may be relaxed. AC losses are also critical in CIC windings because their impact extends globally throughout the winding; it is absolutely necessary to avoid a global quench, an event in fusion magnets that can be induced by AC heating in the CIC windings.

This concludes a brief introduction to the stability of superconducting magnets [6.17]. Problems presented in the Problem Section examine some of the theories and concepts in greater detail to enhance the reader's understanding of them. The problems on the basic theories and concepts are presented first, with the topics appearing in the same order as in Table 6.1. Problems dealing with applications of these theories and concepts follow.

"Mendel's epoch-making discovery required little previous knowledge; what it needed was a life of elegant leisure spent in a garden." —Bertrand Russell

Problem 6.1: Cryostability

1. Circuit model

We shall study the theory of cryostability in two steps. In the first step, treated here, a circuit model is used to study the behavior of a composite superconductor comprised of superconducting filaments embedded in a matrix of copper.

Figure 6.4a shows an "ideal" R_s vs I plot for the superconducting filaments. Figure 6.4b shows an appropriate circuit model for the composite superconductor. R_s is the filaments' effective resistance and I_s is the total current through the filaments. The plot is ideal in the sense that for $I_s \leq I_c$, $R_s = 0$, where I_c is the superconductor's critical current at temperature T. (In Chapter 7 where the main topic is disturbances in magnets, a problem will deal with the effects on magnet operation of a more realistic R_s vs I_s curve, in which $R_s \neq 0$ even for $I_s < I_c$.) For $I_s \geq I_c$, $0 \leq R_s \leq R_n$, where R_n is the normal-state resistance. Note that at $I_s = I_c$, R_s is indeterminate; it satisfies conditions imposed by the circuit. R_m represents the total resistance of the copper matrix and generally $R_m \ll R_n$. Let us assume the conductor is at temperature T and carrying a net transport current I_t, which is varied slowly from 0 to I_c and beyond.

a) For $I_t < I_c$, derive expressions for V_{cd}, the voltage over the conductor length, and G_j, the total Joule dissipation in the conductor.

b) Repeat a) for $I_t \geq I_c$ and show:

$$G_j = R_m I_t (I_t - I_c) \qquad (6.2)$$

Assume that the conductor is well cooled and that its temperature remains constant at T. In reality, when G_j is nonzero, the conductor's temperature is actually raised slightly by ΔT even when it is well cooled; for this problem, assume $\Delta T = 0$.

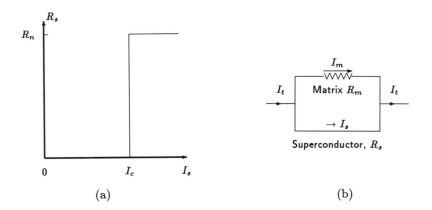

(a)   (b)

Fig. 6.4   (a) R_s vs I_s plot for superconducting filaments alone. (b) Circuit model for a composite superconductor.

Solution to Problem 6.1

a) We have $I_s = I_t \leq I_c$ and, from Fig. 6.4a, $R_s = 0$. Because $R_s = 0$, $V_{cd} = 0$ from Fig. 6.4b, and hence $G_j = 0$.

b) When $I_t \geq I_c$, any excess current greater than I_c now flows through the copper matrix because $R_m \ll R_n$. That is, $I_m = I_t - I_c$ and $I_s = I_c$. I_m is the current flowing through the copper matrix. We then have:

$$V_{cd} = R_m I_m = R_m(I_t - I_c) \qquad (S1.1)$$

Since the same voltage appears across R_s, we have:

$$V_{cd} = R_m(I_t - I_c) = R_s I_c \qquad (S1.2)$$

$$R_s = R_m\left(\frac{I_t}{I_c} - 1\right) \qquad (S1.3)$$

That is, R_s increases linearly with I_t starting with $R_s = 0$ at $I_t = I_c$. We have:

$$G_j = V_{cd} I_t \qquad (S1.4)$$

Combining Eqs. $S1.1$ and $S1.4$, we obtain:

$$G_j = R_m I_t(I_t - I_c) \qquad (6.2)$$

Note that G_j is temperature-independent as long as R_m and I_c remain independent of temperature. Because copper resistivity is nearly temperature-independent for the temperature range from 4 to $\sim$30 K (Appendix IV), R_m is always assumed constant in stability analyses of low-T_c superconductors. In the next problem, we let I_c be temperature-dependent and examine $G_j(T)$.

Peak Nucleate Boiling Heat Transfer Flux: Narrow Channels

As remarked briefly in Chapter 4, peak nucleate heat transfer flux (q_{pk}) is affected by many factors. For a cooling surface situated in a narrow channel, q_{pk} is smaller than that for an open surface. Because the cooling surfaces in most magnet winding geometries are situated in narrow channels, q_{pk} valid for narrow channel geometry should be used. Wilson's data [6.18] on q_{pk} [W/cm^2] vs channel width (w [cm]) in the range 0.01$\sim$0.15 cm, over the vertical channel height (z [cm]) range 2.5$\sim$20 cm, may be fit by the following simple expression:

$$q_{pk} = \frac{10w}{1.19 + 0.37z} \qquad (6.3)$$

Note here that q_{pk} is in W/cm^2, w and z are in cm.

Problem 6.2: Cryostability

2. Temperature dependence

We shall now investigate the temperature dependence of G_j, total power dissipation in the unit composite superconductor length considered in Problem 6.1 and given by Eq. 6.2. Figure 6.5 presents an often used linear approximation of the I_c vs T curve at a given magnetic field for superconductors. (Equation 5.22 gives the same linear approximation for critical current density.) Note that $I_c(T_{op}) = I_{c_o}$ and $I_c(T_c) = 0$. The net transport current through the composite conductor I_t remains constant as the conductor temperature is varied. T_{cs}, the current sharing temperature, is given by $I_t = I_c(T_{cs})$ and is indicated in the plot.

a) With $I_c(T)$ of the filaments approximated by:

$$I_c(T) = I_{c_o} \left(\frac{T_c - T}{T_c - T_{op}} \right) \qquad (T_{op} \leq T \leq T_c) \qquad (6.4)$$

show that an expression for $G_j(T)$, the total joule dissipation of the composite conductor as a function of temperature, is given by:

$$G_j(T) = 0 \qquad\qquad (T_{op} \leq T \leq T_{cs}) \qquad (6.5a)$$

$$G_j(T) = R_m I_t^2 \left(\frac{T - T_{cs}}{T_c - T_{cs}} \right) \qquad (T_{cs} \leq T \leq T_c) \qquad (6.5b)$$

Assume R_m to be temperature independent.

b) Make a plot of Eq. 6.5 for the temperature range, from T_{op} to $T > T_c$.

c) Give a physical explanation of $G_j(T)$ given by Eq. 6.5b.

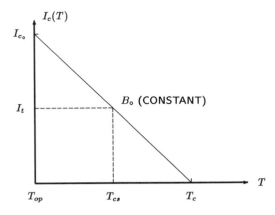

Fig. 6.5   Linear approximation of the I_c vs T curve (Eq. 6.4) for the superconducting filaments.

Solution to Problem 6.2

a) Because $I_c(T) > I_t$ for $T_{op} \leq T \leq T_{cs}$ (Fig. 6.5), we have:

$$G_j(T) = 0 \qquad (T_{op} \leq T \leq T_{cs}) \qquad (6.5a)$$

By inserting $I_c(T)$ given by Eq. 6.4 into G_j given by Eq. 6.2, we obtain:

$$G_j(T) = R_m I_t \left[I_t - I_{c_o} \left(\frac{T_c - T}{T_c - T_{op}} \right) \right] \qquad (T_{cs} \leq T \leq T_c) \qquad (S2.1)$$

Setting $I_t = I_c(T_{cs})$ and inserting this into Eq. 6.4, we can solve for I_{c_o}:

$$I_{c_o} = I_t \left(\frac{T_c - T_{op}}{T_c - T_{cs}} \right) \qquad (S2.2)$$

where $I_{c_o} \equiv I_c(T_{op})$. Combining Eqs. $S2.1$ and $S2.2$, we obtain:

$$G_j(T) = R_m I_t^2 \left(\frac{T - T_{cs}}{T_c - T_{cs}} \right) \qquad (T_{cs} \leq T \leq T_c) \qquad (6.5b)$$

b) Equation 6.5 is plotted in Fig. 6.6.

c) Clearly, as long as $I_t \leq I_c(T)$, all the transport current flows through the filaments and $V_{cd} = 0$, making $G_j(T) = 0$.

At T_{cs} when $I_t = I_c$, the filaments are carrying the maximum they can carry superconductively; beyond T_{cs} the current begins to "spill" over to the copper matrix, generating Joule dissipation in the composite. This spilling continues monotonically with T until T_c is reached, at which point all the transport current is now flowing through the matrix. Note that G_j varies linearly with T between T_{cs} and T_c. Beyond T_c, G_j remains constant as long as R_m remains constant; with copper as the matrix metal it does so for temperatures up to $\sim 30\,\mathrm{K}$.

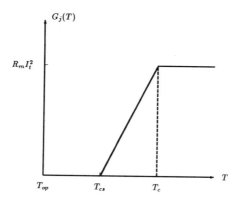

Fig. 6.6   Plot of $G_j(T)$ for the composite superconductor.

Problem 6.3:   Cryostability

3. Stekly criterion

We now proceed to derive an expression for the Stekly stability parameter α_{sk} and formulate the theory of cryostability. At the time Stekly formulated the theory (c. 1964), which was first applied to and tested with short lengths of crudely made composite superconductors, it was customary to assume that the superconducting magnet would always be operated at its conductor's critical current, I_c. Later on, it was realized that the magnet did not have to be operated at its utmost limit; a "safety margin" was introduced, *i.e.* $I_{op} < I_{c_o} = I_c(T_{op})$. With this safety margin came the current-sharing temperature T_{cs}.

a) Show that a new expression for $G_j(T)$ for the case $I_t = I_{c_o}$ is given by:

$$G_j(T) = R_m I_{c_o}^2 \left(\frac{T - T_{op}}{T_c - T_{op}}\right) \qquad (T_{op} \leq T \leq T_c) \qquad (6.6)$$

b) Show that $g_j(T)$, defined as $G_j(T)$ divided by the unit conductor volume, is given by:

$$g_j(T) = \frac{\rho_m I_{c_o}^2}{A_{cd} A_m} \left(\frac{T - T_{op}}{T_c - T_{op}}\right) \qquad (T_{op} \leq T \leq T_c) \qquad (6.7)$$

where ρ_m is the matrix metal electrical resistivity. A_{cd} is the total conductor cross sectional area. Note that $A_{cd} = A_m + A_s$, where A_m is the total matrix cross sectional area and A_s is that of the filaments.

c) Here, we shall use a constant heat transfer coefficient h_q for helium cooling. Under this assumption, cooling density, $g_q(T)$—the cooling per unit conductor volume—is given by:

$$g_q(T) = \frac{f_p P_{cd} h_q (T - T_b)}{A_{cd}} \simeq \frac{f_p P_{cd} h_q (T - T_{op})}{A_{cd}} \qquad (6.8)$$

T is the conductor surface temperature. T_b is the helium temperature, which may be assumed to be equal to the operating temperature, T_{op}: $T_b \simeq T_{op}$. P_{cd} is the total conductor perimeter. To introduce a variability to surface cooling, a constant f_p is used in Eq. 6.8 to quantify the fraction of P_{cd} exposed to liquid helium.

Stable conditions are achieved when cooling equals or exceeds Joule generation: $g_q(T) \geq g_j(T)$. By equating the two power densities, show that α_{sk}, known as the Stekly stability parameter, is given by:

$$\alpha_{sk} = \frac{f_p P_{cd} A_m h_q (T_c - T_{op})}{\rho_m I_{c_o}^2} \qquad (6.9)$$

Note that the conductor is stable when $\alpha_{sk} \geq 1$ and unstable when $\alpha_{sk} < 1$.

Solution to Problem 6.3

a) By inserting $I_t = I_{c_o}$ into Eq. 6.5b and noting that in this case $T_{cs} = T_{op}$ because $I_t = I_{c_o}$, we can immediately rewrite Eq. 6.5b as:

$$G_j(T) = R_m I_{c_o}^2 \left(\frac{T - T_{op}}{T_c - T_{op}}\right) \qquad (T_{op} \leq T \leq T_c) \qquad (6.6)$$

Note that $G_j(T)$ now increases linearly with T starting at T_{op}. Obviously, this is because the superconductor is now carrying its maximum allowable current, namely I_{c_o}, and $T_{cs} = T_{op}$.

b)   The matrix resistance over a unit conductor length is given by: $R_m = \rho_m/A_m$. The conductor volume over a unit conductor length is given by A_{cd}. Inserting $R_m = \rho_m/A_m$ into Eq. 6.6 and dividing it by A_{cd}, we have:

$$\frac{G_j(T)}{A_{cd}} \equiv g_j(T) = \frac{\rho_m I_{c_o}^2}{A_{cd} A_m}\left(\frac{T - T_{op}}{T_c - T_{op}}\right) \qquad (6.7)$$

c)   The total cooling operating on a unit conductor length, $G_q(T)$, is given by:

$$G_q(T) = f_p P_{cd} h_q(T - T_b) \qquad (S3.1)$$

Dividing $G_q(T)$ by the unit conductor volume and noting that $T_b \simeq T_{op}$, we obtain:

$$g_q(T) \simeq \frac{f_p P_{cd} h_q(T - T_{op})}{A_{cd}} \qquad (6.8)$$

The Stekly cryostability theory requires that $g_q(T) \geq g_j(T)$. Combining Eqs. 6.7 and 6.8, we obtain:

$$\frac{f_p P_{cd} h_q(T - T_{op})}{A_{cd}} \geq \frac{\rho_m I_{c_o}^2}{A_{cd} A_m}\left(\frac{T - T_{op}}{T_c - T_{op}}\right)$$

$$\frac{f_p P_{cd} A_m h_q(T_c - T_{op})}{\rho_m I_{c_o}^2} \geq 1 \qquad (S3.2)$$

The Stekly stability parameter, α_{sk}, is given by the left-hand side of Eq. S3.2:

$$\alpha_{sk} = \frac{f_p P_{cd} A_m h_q(T_c - T_{op})}{\rho_m I_{c_o}^2} \qquad (6.9)$$

Note that α_{sk}, a dimensionless number, expresses the ratio of cooling density to Joule dissipation density. The operation is thus stable when $\alpha_{sk} \geq 1$ (sufficient cooling) and unstable when $\alpha_{sk} < 1$ (insufficient cooling.)

Discussion of Stekly Cryostability Criterion

Stekly's cryostability criterion requires "composite superconductors" but not necessarily "multifilamentary superconductors;" all multifilamentary superconductors
are, however, composite superconductors. (The development of twisted multifilamentary superconductors came in the late 1960s, after the formulation of cryostability criterion.) With Stekly's cryostability established, the stage was set for
building in earnest—and with great confidence—large superconducting magnets.
Examples of large magnets built and operated reliably in the early era of this
new era of magnet building include, just to cite a few, those for bubble chambers
[6.19, 6.20], MHD [3.20, 3.21], and motors [6.21].

In magnet design, as discussed in Chapter 3, an important parameter is "overall
winding" current density, $\lambda J \equiv J_{ov}$, rather than actual operating current, I_{op}.
As noted in Chapter 3, λ (space factor), represents the fraction of the total cross
sectional area occupied by the conductor to that occupied by the winding; after
all it is only the conductor that carries current and contributes to the field. In
discussing "composite superconductors," an important parameter is their "overall
conductor" current density, J_{cd}, given by I_{op}/A_{cd}, which in this case would be
I_{c_o}/A_{cd}. Note that J_{cd} is always greater than J_{ov} because whereas J_{cd} only ac-
counts for A_{cd}, given by A_s and A_m, J_{ov} must in addition account for the areas
occupied by coolant, insulators, and reinforcing materials. Still, the greater the
J_{cd}, generally the greater will be J_{ov}.

Equation 6.9 states $\alpha_{sk} \propto A_m$ and implies that for a given cooling condition
stability (or reliability) is directly linked to A_m and vice versa:

$$A_m = \frac{\alpha_{sk}\rho_m I_{c_o}^2}{f_p P_{cd} h_q (T_c - T_{op})} \tag{6.10}$$

To achieve a greater degree of stability for given cooling conditions, therefore, it
is necessary to increase A_m, which is usually copper in LTS and most likely in
HTS. The conductor's overall critical current density $[J_{c_o}]_{cd}$, and the conductor's
"intrinsic" critical current density, $[J_{c_o}]_s$ are related as:

$$[J_{c_o}]_{cd} \equiv \frac{I_{c_o}}{A_{cd}} = \frac{I_{c_o}}{A_s + A_m} = \frac{[J_{c_o}]_s}{1 + \gamma_{c/s}} \tag{6.11}$$

where $\gamma_{c/s} \equiv A_m/A_s$ is commonly known as copper-to-superconductor ratio.

It is clear from Eq. 6.11 that $[J_{c_o}]_{cd}$ is always less than $[J_{c_o}]_s$ because $\gamma_{c/s} > 0$.
Equation 6.11 implies stability can be costly. The greater the α_{sk} is, which makes
magnet operate more stably, the greater must be A_m and hence $\gamma_{c/s}$, and the
smaller will be $[J_{c_o}]_{cd}$.

Values of α_{sk} selected for early large magnets were quite large, some exceeding 10
[6.22, 6.23]. Clearly, magnet reliability was unquestionably favored over magnet
efficiency. Right or wrong, this philosophy continues to this day, particularly with
Class 1 magnets.

Problem 6.4: Cryostability

4. Nonlinear cooling curves

The parameter α_{sk} derived in Problem 6.3 is based on the cooling characteristic that assumes h_q to be temperature independent. In reality, cooling curves are quite nonlinear, even in the nucleate boiling heat transfer regime where these cryostable magnets generally operate, see for example, Fig. 4.1. It is thus more accurate to incorporate the heat transfer curve, $q(T)$, directly in the derivation of the cryostability theory.

a) Show that an expression for $[J_{c_o}]_{cd}$ for cryostable operation at $I_{op} = I_{c_o}$ incorporating the heat transfer flux curve $q(T)$ [W/m²] may be given by:

$$[J_{c_o}]_{cd} = \sqrt{\frac{f_p P_{cd} A_m q_{fm}}{\rho_m (A_s + A_m)^2}} \qquad (6.12a)$$

$$\simeq \sqrt{\frac{f_p P_{cd} q_{fm}}{\rho_m A_m}} \qquad (A_m \gg A_s) \qquad (6.12b)$$

where q_{fm} is the minimum heat transfer flux in the film boiling regime.

b) Draw qualitatively, on the same plot, a $q(T)$ curve and a dimensionally consistent generation curve, and indicate on the plot the region of stable operation.

c) Generalize **b)** for the case $I_{op} < I_{c_o}$ on the same plot used in **b)**.

Composite Superconductors: "Monolithic" and "Built-up"

Magnet-grade superconductors are available in two types, one known as "monolithic" and the other known as "built-up."

Monolithic: As the name implies, the superconductor and the normal metal in a monolithic form one entity, achieved chiefly through metallurgical processes. By visual inspection from outside, it is impossible to distinguish, except through the conductor cross section, the existence of more than one constituent in a monolithic conductor. Virtually all round conductors are of this type. For values of $\gamma_{c/s}$ above ~10, however, it becomes difficult to manufacture monolithic conductors without breaking filaments in the metal forming processes, particularly for conductors with filament sizes less than ~100 μm. Early monolithic conductors [6.22, 6.23] had filaments with large diameters, *i.e.* 1~5 mm; filament breakage was hence rare. These sizes, however, clearly exceeded the critical size set by the flux jumping criterion.

Built-Up: A built-up conductor is comprised of a monolithic conductor having a $\gamma_{c/s}$ value close to 1 and normal-metal stabilizer parts that are generally soldered to the monolith, after the monolith has been prepared. Mechanical properties of the stabilizer parts are therefore unaffected by manufacturing processes of the monolith, making it sometimes easier to satisfy conductor specifications. The CIC conductor is a variant of built-up conductors.

Solution to Problem 6.4

a) In most applications where cryostability is applied, we must assume that the conductor may operate—even briefly—in the fully normal state. Under this assumption, it is safe to use the minimum heat flux (q_{fm} in Table 4.2) in the film boiling regime. We thus have:

$$\frac{\rho_m I_{c_o}^2}{A_m} = f_p P_{cd} q_{fm} \qquad (S4.1)$$

Combining Eqs. $S4.1$ and 6.11 and solving for $[J_{c_o}]_{cd}$, we obtain:

$$[J_{c_o}]_{cd} = \sqrt{\frac{f_p P_{cd} A_m q_{fm}}{\rho_m (A_s + A_m)^2}} \qquad (6.12a)$$

$$\simeq \sqrt{\frac{f_p P_{cd} q_{fm}}{\rho_m A_m}} \qquad (A_m \gg A_s) \qquad (6.12b)$$

Equation 6.12b indicates that increasing A_m to improve stability decreases $[J_{c_o}]_{cd}$.

b) Figure 6.7 presents a typical plot of $q(T)$ for liquid helium. Plotted also is a curve of $\hat{g}_j(T) \equiv (A_{cd}/f_p P_{cd}) g_j(T)$ [W/m^2]. Parameters are chosen to make $\hat{g}_j(T_c) = (A_{cd}/f_p P_{cd}) g_j(T_c)$ slightly less than q_{fm} so that operation is stable for the entire temperature range from T_{op} to temperatures even above T_c.

c) The dotted line in Fig. 6.7 presents the case in which $I_t < I_{c_o}$. Note that in the temperature range $T_{op} \leq T \leq T_{cs}$ the conductor is fully superconducting.

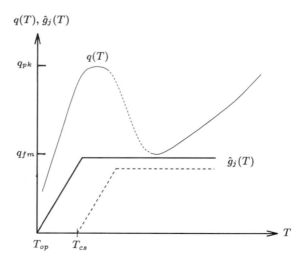

Fig. 6.7   Qualitative plots of $q(T)$ for liquid helium and $\hat{g}_j(T)$ for the case $I_{op} = I_{c_o}$ (solid straight line) and $I_{op} < I_{c_o}$ (dotted straight line).

Problem 6.5: Dynamic stability for tape conductors

1. Magnetic and thermal diffusion

Before formulating the dynamic stability theory applicable to superconducting tapes—the only conductor configuration for which this theory is considered in this book—we shall study basic equations of magnetic and thermal diffusion to identify two time constants in dynamic stability: magnetic, τ_{mg}, and thermal, τ_{th}. These time constants are, in turn, related to magnetic diffusivity, D_{mg}, and thermal diffusivity, D_{th}, respectively.

a) Starting with diffusion equations, both magnetic and thermal, show that expressions for magnetic diffusivity, D_{mg}, and thermal diffusivity, D_{th}, are given by:

$$D_{mg} = \frac{\rho_e}{\mu_\circ} \tag{6.13a}$$

$$D_{th} = \frac{k}{C} \tag{6.13b}$$

ρ_e is electrical resistivity, k is thermal conductivity, and C is heat capacity. As in Chapter 5, consider a one-dimensional slab of width $2a$, where magnetic field $\vec{H}(x,t)$, directed only in the y-direction, is given by: $\vec{H}(x,t) = H_y(x,t)\,\vec{i}_y$.

b) Compute D_{mg} and D_{th} for stainless steel and copper at temperature near 4.2 K. Discuss how differences in these two diffusivities may impact the magnetic and thermal behaviors of these metals. Stainless steel, an alloy, has electrical and thermal properties very much like those of normal-state low-T_c superconductors.

c) A solution to either one of these diffusion equations is generally in the form of an infinite series. For a one-dimensional slab of width $2a$, the first term in the series for magnetic field, $H_y(x,t)$, is given by:

$$H_y(x,t) = H_0 \cos\left(\frac{\pi x}{2a}\right) e^{-t/\tau_{mg}} \tag{6.14}$$

Inserting Eq. 6.14 into the magnetic diffusion equation, show that an expression for τ_{mg} in terms of D_{mg} is given by:

$$\tau_{mg} = \frac{1}{D_{mg}}\left(\frac{2a}{\pi}\right)^2 = \frac{\mu_\circ}{\rho_e}\left(\frac{2a}{\pi}\right)^2 \tag{6.15}$$

By analogy, an expression for τ_{th} is given by:

$$\tau_{th} = \frac{1}{D_{th}}\left(\frac{2a}{\pi}\right)^2 = \frac{C}{k}\left(\frac{2a}{\pi}\right)^2 \tag{6.16}$$

Solution to Problem 6.5

a) Two Maxwell's equations applicable to derive the magnetic diffusion equation are Ampere's law and Faraday's law, both given by differential forms:

$$\text{Ampere's law:} \qquad \nabla \times \vec{H} = \vec{J}_f \qquad\qquad (2.5)$$

$$\text{Faraday's law:} \qquad \nabla \times \vec{E} = -\frac{\partial \vec{B}}{\partial t} \qquad\qquad (2.8)$$

For the slab (width $2a$) geometry studied in Chapter 5, we can express Eqs. 2.5 and 2.8, respectively, as:

$$\text{Ampere's law:} \qquad \frac{\partial H_y}{\partial x} = J_z = \frac{E_z}{\rho_e} \qquad\qquad (S5.1)$$

$$\text{Faraday's law:} \qquad \frac{\partial E_z}{\partial x} = \frac{\partial B_y}{\partial t} = \mu_o \frac{\partial H_y}{\partial t} \qquad\qquad (S5.2)$$

Combining Eqs. $S5.1$ and $S5.2$, we obtain:

$$\rho_e \frac{\partial^2 H_y}{\partial x^2} = \mu_o \frac{\partial H_y}{\partial t}$$

$$\frac{\rho_e}{\mu_o} \frac{\partial^2 H_y}{\partial x^2} \equiv D_{mg} \frac{\partial^2 H_y}{\partial x^2} = \frac{\partial H_y}{\partial t} \qquad\qquad (S5.3)$$

Equation $S5.3$ is a magnetic diffusion equation and we obtain:

$$D_{mg} = \frac{\rho_e}{\mu_o} \qquad\qquad (6.13a)$$

Similarly, the one-dimensional thermal diffusion equation having constant thermal properties can be derived from Eq. 6.1 with g_j (Joule heating), g_d (other dissipations), and g_q (cooling) terms zero:

$$k \frac{\partial^2 T}{\partial x^2} = C \frac{\partial T}{\partial t} \qquad\qquad (S5.4)$$

Dividing both sides of Eq. $S5.4$ by C, we obtain:

$$\frac{k}{C} \frac{\partial^2 T}{\partial x^2} \equiv D_{th} \frac{\partial^2 T}{\partial x^2} = \frac{\partial T}{\partial t} \qquad\qquad (S5.5)$$

Equation $S5.5$ is a thermal diffusion equation and from the equation we have:

$$D_{th} = \frac{k}{C} \qquad\qquad (6.13b)$$

Solution to Problem 6.5

b) Table 6.2 presents *approximate* values of electrical and thermal properties and corresponding diffusivities (Eqs. 6.13a and 6.13b), all near 4 K, for stainless steel and copper.

From Table 6.2 we can clearly see that stainless steel, essentially representing normal-state superconductors, and copper have diffusivities that are asymmetric with respect to magnetic and thermal diffusivities. Specifically, changes in magnetic field propagate quickly through stainless steel, whereas temperature gradients are relatively slow to propagate; hence, large nonuniform temperature distributions can be created in stainless steel during changing magnetic fields. Physically, it means that magnetic heating happens quickly and uniformly within stainless steel just as it does in the hard superconductors studied in Chapter 5. In copper, the reverse is true: the magnetic field moves very slowly, while any nonuniformity in temperature is quickly "evened out." The presence of copper right next to hard superconductor should therefore alleviate field-motion induced instability in hard superconductors. This thinking is the essence of dynamic stability.

c) From Eq. 6.14 we obtain:

$$\frac{\partial^2 H_y}{\partial x^2} = -H_0 \left(\frac{\pi}{2a}\right)^2 \cos\left(\frac{\pi x}{2a}\right) e^{-t/\tau_{mg}} \qquad (S5.6)$$

$$\frac{\partial H_y}{\partial t} = -\frac{H_0}{\tau_{mg}} \cos\left(\frac{\pi x}{2a}\right) e^{-t/\tau_{mg}} \qquad (S5.7)$$

Combining Eqs. S5.3, S5.6, and S5.7, we have:

$$-D_{mg} H_0 \left(\frac{\pi}{2a}\right)^2 \cos\left(\frac{\pi x}{2a}\right) e^{-t/\tau_{mg}} = -\frac{H_0}{\tau_{mg}} \cos\left(\frac{\pi x}{2a}\right) e^{-t/\tau_{mg}} \qquad (S5.8)$$

From Eq. S5.8, we can solve for τ_{mg} and combining it with Eq. 6.13a, we have:

$$\tau_{mg} = \frac{1}{D_{mg}} \left(\frac{2a}{\pi}\right)^2 = \frac{\mu_o}{\rho_e} \left(\frac{2a}{\pi}\right)^2 \qquad (6.15)$$

Table 6.2: Diffusivities* of Stainless Steel and Copper at 4.2 K

Metal	ρ_e [nΩ m]	k [W/m K]	C [J/m³ K]	Diffusivity [m²/s]	
				D_{mg}	D_{th}
Stainless Steel	500	0.2	3000	0.5	7×10^{-5}
Copper	0.2	400	800	2×10^{-4}	0.5

* Properties values are approximate.

Problem 6.6: Dynamic stability for tape conductors
2. Criterion for edge-cooled tapes

This problem deals with the formulation of the dynamic stability criterion applicable to composite tapes. The criterion, as stated earlier, has been reexamined recently, because many promising HTS are available only in tape form.

Figure 6.8 shows the cross section of a composite tape $2a$ wide, comprised of a superconductor d thick sandwiched between copper layers, each $D/2$ thick. Because the winding is in the form of a stack of tapes, cooling for each tape is only at the edges ($x = \pm a$) and may be given in terms of a constant heat flux coefficient h_q.

An energy equation (per unit tape length) valid for the tape after its temperature was raised ΔT following the deposition of an energy (per unit tape length), G_d, is given by:

$$2a(d+D)C_{cd}\Delta T = 2ade_\phi + G_d - 2(d+D)h_q\tau_{mg}\Delta T \qquad (6.17)$$

where e_ϕ is the magnetization energy density (Eq. 5.21, p. 189) triggered and released in the tape. Defining $\lambda_s = d/(d+D)$, the volumetric fraction of the superconductor, we can derive an energy density equation:

$$C_{cd}\Delta T = \lambda_s e_\phi + g_d - \frac{h_q\tau_{mg}\Delta T}{a} \qquad (6.18)$$

Because of the presence of copper, the condition $\tau_{th} \ll \tau_{mg}$ is valid and the process time is given by τ_{mg}. Show that the critical tape half width a_c for edge-cooled tapes to satisfy the dynamic stability criterion is given by:

$$a_c = \sqrt{\frac{3C_{cd}(T_c - T_{op})}{\lambda_s\mu_0 J_{c_o}^2} + \frac{(1-\lambda_s)12a_c(T_c - T_{op})h_q}{\lambda_s\pi^2\rho_{cu}J_{c_o}^2}} \qquad (6.19)$$

where ρ_{cu} is the electrical resistivity of copper and J_{c_o} is the critical current density of the superconductor at operating temperature T_{op}. In arriving at Eq. 6.19, use τ_{mg} corresponding to *copper* because the copper layers in the composite tape are the ones slowing down the magnetic process and keeping the tape essentially at a uniform temperature. (Note that the second term in the right-hand side of Eq. 6.19 also contains a_c.)

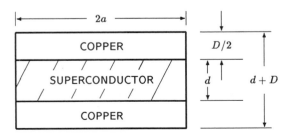

Fig. 6.8 Cross section of a composite superconducting tape.

Solution to Problem 6.6

From Eqs. 5.21 and 5.23, we have:

$$e_\phi = \frac{\mu_o J_{c_o}^2 \Delta T a^2}{3(T_c - T_{op})} \qquad (S6.1)$$

The composite's effective resistivity, ρ_{ef}, is a weighted average of both copper (ρ_{cu}) and normal-state superconductor (ρ_{ns}):

$$\frac{1}{\rho_{ef}} = \frac{1 - \lambda_s}{\rho_{cu}} + \frac{\lambda_s}{\rho_{ns}} \qquad (S6.2)$$

Because $\rho_{cu} \ll \rho_{ns}$, we have:

$$\rho_{ef} \simeq \frac{\rho_{cu}}{1 - \lambda_s} \qquad (S6.3)$$

Combining Eqs. 6.15 and S6.3, we obtain:

$$\tau_{mg} = (1 - \lambda_s) \frac{4\mu_o a^2}{\pi^2 \rho_{cu}} \qquad (S6.4)$$

By combining Eqs. 6.18, S6.1, and S6.4, we obtain:

$$C_{cd}\Delta T = \lambda_s \frac{\mu_o a^2 J_{c_o}^2}{3(T_c - T_{op})} \Delta T + g_d - (1 - \lambda_s) \frac{4\mu_o a h_q}{\pi^2 \rho_{cu}} \Delta T \qquad (S6.5)$$

The "effective" heat capacity of the composite, C_e, defined by $C_e \equiv g_d / \Delta T$, is thus given by:

$$C_e \equiv \frac{g_d}{\Delta T} = C_{cd} - \lambda_s \frac{\mu_o a^2 J_{c_o}^2}{3(T_c - T_{op})} + (1 - \lambda_s) \frac{4\mu_o a h_q}{\pi^2 \rho_{cu}} \qquad (S6.6)$$

For thermal stability, C_e must be positive. Solving Eq. S6.6 for a_c under this condition, we have:

$$a_c = \sqrt{\frac{3C_{cd}(T_c - T_{op})}{\lambda_s \mu_o J_{c_o}^2} + \frac{(1 - \lambda_s)12 a_c (T_c - T_{op})h_q}{\lambda_s \pi^2 \rho_{cu} J_{c_o}^2}} \qquad (6.19)$$

Except for λ_s which indicates the composite nature of the tape conductor, note that Eq. 6.19 is identical to the flux jumping ("adiabatic") criterion given by Eq. 5.24 if the cooling term represented by the second term on the right-hand side of Eq. 6.19 is set to 0 (which will be the case when the edge-cooling is absent).

The dynamic stability criterion as given by Eq. 6.19 is applicable to silver-sheathed HTS tapes, with ρ_{cu} replaced with ρ_{ag} the electrical resistivity of silver. The criterion could be useful for HTS magnets operated in a bath of cryogen or force-cooled by stream of cold gas (helium). For HTS magnets operated in a vacuum environment ($h_q = 0$) of cryocoolers, the criterion is not applicable.

Problem 6.7: "Equal-area" criterion

This problem concerns the "equal-area" criterion proposed by Maddock, James, and Norris in 1969. Following the work of Wilson [6.24], we shall extend the criterion, originally derived for one-dimensional geometry, to a special case of two-dimensional geometry.

a) Under steady-state conditions ($dT/dt = 0$), an expression for the 1-D power density (Eq. 6.1) with no dissipation other than Joule heating dissipation, *i.e.* $g_d = 0$, may be expressed:

$$g_k(T) = \frac{d}{dx}\left[k_{cd}(T)\frac{dT}{dx}\right] = g_q(T) - g_j(T) \qquad (6.20)$$

where $k_{cd}(T)$ is the composite conductor's temperature-dependent thermal conductivity. We now introduce a new variable $S(T)$ defined by:

$$S(T) = k_{cd}(T)\frac{dT}{dx} \qquad (6.21)$$

Inserting a new variable $S(T)$ given by Eq. 6.21 into Eq. 6.20 and being clever about mathematics, derive first the following equation:

$$S(T)\frac{dS(T)}{dT} = k_{cd}(T)[g_q(T) - g_j(T)] \qquad (6.22)$$

Equation 6.22 is an intermediate point on the way to the equal-area criterion.

b) Integrate Eq. 6.22 between the two points on the conductor where $S_1(T) = S_2(T)$ and where equilibrium is reached, *i.e.* $g_j(T) = g_q(T)$, and $dT/dx = 0$. Next, make $k_{cd}(T)$ temperature-independent and show that:

$$\int_{T_1}^{T_2}[g_q(T) - g_j(T)]\,dT = 0 \qquad (6.23)$$

T_1 is temperature at the center of normal zone and T_2 is conductor temperature far way from the normal zone.

c) Interpret Eq. 6.23 in terms of $g_q(T)$ and $g_j(T)$ plots. In plotting $g_j(T)$ consider two cases of interest: 1) a "long" normal zone; and 2) a "short" normal zone. The "long" normal zone corresponds nearly to that satisfied by the Stekly criterion.

d) Consider 2-D geometry such as pancake windings and derive a 2-D equivalent of Eq. 6.21 for the special case in which $dT/d\theta = 0$; explain why under this condition the 2-D windings may tolerate a greater value of g_j than that satisfying Eq. 6.23 for the 1-D geometry.

Solution to Problem 6.7

a) Defining a new variable $S(T) = k_{cd}(T) \, dT/dx$, we may express Eq. 6.20 as:

$$\frac{dS(T)}{dx} = g_q(T) - g_j(T) \qquad (S7.1)$$

Also from the definition of $S(T)$ given by Eq. 6.21, we have:

$$\frac{dS(T)}{dx} = \frac{dS(T)}{dT}\frac{dT}{dx}$$
$$= \frac{dS(T)}{dT}\frac{S(T)}{k_{cd}(T)} \qquad (S7.2)$$

Combining Eqs. $S7.1$ and $S7.2$, we obtain:

$$S(T)\frac{dS(T)}{dT} = k_{cd}(T)[g_q(T) - g_j(T)] \qquad (6.22)$$

b) Integrating Eq. 6.22 between point 1 (T_1) and point 2 (T_2), we have, when $k_{cd}(T) = k_0$, a constant:

$$\tfrac{1}{2}[S^2(T_2) - S^2(T_1)] = k_0 \int_{T_1}^{T_2} [g_q(T) - g_j(T)] \, dT \qquad (S7.3)$$

At both points, equilibrium conditions exist ($dT/dx = 0$): that is, $S(T_1) = S(T_2)$, and thus:

$$\int_{T_1}^{T_2} [g_q(T) - g_j(T)] \, dT = 0 \qquad (6.23)$$

c) As long as the net area under the $g_q(T) - g_j(T)$ curve is zero, Eq. 6.23 is satisfied. This is the "equal area" criterion. Here $g_q(T) = f_p P_{cd} q(T)/A_{cd}$, where f_p is the fraction of the conductor perimeter, P_{cd}, exposed to cooling; A_{cd} is the conductor cross sectional area. Figure 6.9 plots $g_q(T)$ and $g_j(T)$. For a given cooling density curve $g_q(T)$, the generation curve $g_{j1}(T)$ satisfies Eq. 6.23 because the net area under the $g_q(T) - g_{j1}(T)$ curve is zero. Wilson [6.24] also notes that for "short" normal zones, $S(T) = 0$ at the center (by symmetry) even though heating and cooling may not be equal at that point. Thus, the equal area criterion may still be applied to generation curves such as $g_{j2}(T)$ in the figure. The excess of heating over cooling in the center is conducted outwards along the conductor to the cooler regions, where it is transferred to the coolant.

Solution to Problem 6.7

d) In 2-D cylindrical coordinates with $dT/d\theta = 0$, the term $\nabla\cdot[k_{cd}(T)\nabla T]$, from Fourier's heat conduction equation, may be given by:

$$\nabla\cdot[k_{cd}(T)\nabla T] = \frac{d}{dr}\left[k_{cd}(T)\frac{dT}{dr}\right] + \frac{1}{r}k_{cd}(T)\frac{dT}{dr} \qquad (S7.4)$$

Thus, in 2-D cylindrical coordinates with $dT/d\theta = 0$, Eq. 6.20 may be modified to:

$$\frac{d}{dr}\left[k_{cd}(T)\frac{dT}{dr}\right] = g_q(T) - g_j(T) - \frac{1}{r}k_{cd}(T)\frac{dT}{dr} \qquad (S7.5)$$

That is, as long as $dT/dr < 0$—which is valid in most cases—the last term on the right-hand side of Eq. S7.5 has the same effect as an increase in the cooling density or a decrease in the heating density. Therefore, in this special 2-D case, the equal area criterion can be met with a higher conductor current density.

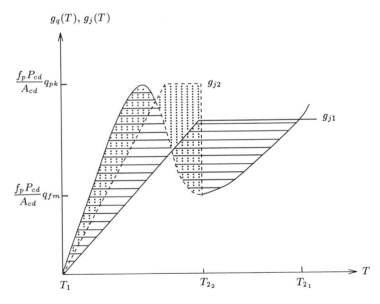

Fig. 6.9 Cooling density and heating density curves, showing the "equal area" criterion. The $g_{j1}(T)$ curve (the solid piece-wise line) corresponds to the generation curve with a "long" normal zone and $g_{j2}(T)$ (the dotted piece-wise line) corresponds to the generation curve with a "short" normal zone. The shaded areas below and above the cooling curve are equal; similarly the dotted areas below and above the cooling curve are equal.

Problem 6.8: The MPZ concept

As briefly stated in the introductory section, the MPZ (minimum propagation zone) concept, introduced by Martinelli and Wipf in 1972, has played a key role in advancing our understanding of "disturbances" that occur within the magnet winding and their effects on the performance of virtually every kind of magnet, adiabatic as well as cooled. The concept showed the minuteness of disturbances degrading the magnet performance. It also helped to launch a more thorough and systematic inquiry into detrimental mechanical disturbances taking place within the winding: microscopic conductor motions and cracking of the winding impregnants. These mechanical disturbances are discussed in Chapter 7.

This problem traces the formulation of MPZ size and computes an actual size for typical winding parameters.

a) Consider a "spherical" winding geometry of infinite radial extent as shown in Fig. 6.10, in which region 1 ($r \leq R_{mz}$) is fully normal and dissipating Joule heating density of g_j; region 2 ($r \geq R_{mz}$) is superconducting, and far away from R_{mz}, the winding temperature is T_{op}. Using spherical coordinates and assuming that the winding is characterized by thermal conductivity k_{wd}, show that under "adiabatic" and steady-state conditions ($g_q = 0$; $\dot{e}_h = 0$ in Eq. 6.1), the radial function of temperature in region 1, $T_1(r)$, with no other dissipation sources ($g_d = 0$), is given by:

$$T_1(r) = T_{op} + \frac{g_j R_{mz}^2}{2k_{wd}} \left[1 - \frac{1}{3} \left(\frac{r}{R_{mz}} \right)^2 \right] \qquad (6.24)$$

Assume also that both g_j and k_{wd} are temperature independent.

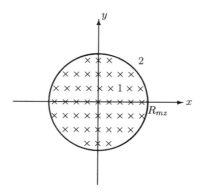

Fig. 6.10 "Spherical" winding where in region 1 ($r \leq R_{mz}$) the winding is normal and generating Joule heating and in region 2 ($r \geq R_{mz}$) the winding is superconducting.

Problem 6.8: The MPZ concept

b) Show that the total Joule heating generated within region 1 is matched exactly with that leaving the winding at R_{mp} by thermal conduction.

c) Equating $g_j = \rho_m J_m^2$ and $T_1(r = R_{mz}) = T_c$, the critical temperature of the conductor, show that R_{mz} is given by:

$$R_{mz} = \sqrt{\frac{3k_{wd}(T_c - T_{op})}{\rho_m J_m^2}} \tag{6.25}$$

ρ_m and J_m are, respectively, the electrical resistivity and current density of the matrix metal; $J_m = I/A_m$, where A_m is the matrix's cross sectional area.

d) Compute R_{mz1} for the following winding parameters: $k_{wd} = 400\,\text{W/m K}$ (thermal conductivity of copper at 4 K); $T_c = 6\,\text{K}$; $T_{op} = 4\,\text{K}$; $\rho_m = 2 \times 10^{-10}\,\Omega\,\text{m}$; $J_m = 300 \times 10^6\,\text{A/m}^2$. R_{mz1} gives the MPZ size along the conductor axis.

e) Compute R_{mz2} for the same set of winding parameters as in **d)** except $k_{wd} = 400\,\text{W/m K}$ is replaced with $k_{wd} = 0.1\,\text{W/m K}$, thermal conductivity of an epoxy at 4 K. R_{mz2} gives an approximate MPZ size transverse to the conductor axis of epoxy-impregnated windings.

f) Compute the total energy margin ΔE_h for the ellipsoid-shaped MPZ shown in Fig. 6.11 with major radius R_{mz1} computed in **d)** and minor radii R_{mz2} of **e)**.

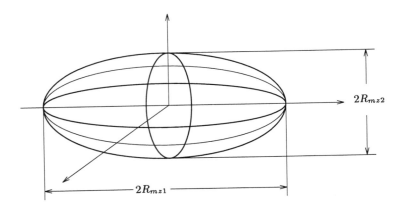

Fig. 6.11   Ellipsoid-shaped MPZ with major radius R_{mz1} and minor radii R_{mz2}.

Solution to Problem 6.8

a) Under adiabatic and steady-state conditions with Joule heating as the only source of dissipation, Eq. 6.1 becomes:

$$0 = g_k + g_j \tag{S8.1}$$

From symmetry, temperatures in both regions are dependent only on r. Thus, with k_{wd} assumed constant, we may express Eq. $S8.1$ for $T_1(r)$ in region 1 and $T_2(r)$ in region 2 in spherical coordinates:

$$\frac{k_{wd}}{r^2} \frac{d}{dr}\left(r^2 \frac{dT_1}{dr}\right) = -g_j \tag{S8.2a}$$

$$\frac{k_{wd}}{r^2} \frac{d}{dr}\left(r^2 \frac{dT_2}{dr}\right) = 0 \tag{S8.2b}$$

$T_1(r)$ and $T_2(r)$ have solutions of the form given below:

$$T_1(r) = -\frac{g_j}{6k_{wd}}r^2 - \frac{A}{r} + B \tag{S8.3a}$$

$$T_2(r) = -\frac{C}{r} + D \tag{S8.3b}$$

A, B, C, and D are constants of integration.

Boundary Conditions

At $r = \infty$, $T_2 = T_{op} \implies D = T_{op}$;

At $r = 0$, $T_1 \neq \infty \implies A = 0$;

At $r = R_{mz}$, $dT_1/dr = dT_2/dr$:

$$\left.\frac{dT_1}{dr}\right|_{r=R_{mz}} = -\frac{g_j R_{mz}}{3k_{wd}} = \left.\frac{dT_2}{dr}\right|_{r=R_{mz}} = \frac{C}{R_{mz}^2} \implies C = -\frac{g_j R_{mz}^3}{3k_{wd}}$$

At $r = R_{mz}$, $T_1 = T_2$.

Substituting A, C, and D found above into Eqs. $S8.3a$ and $S8.3b$, we obtain:

$$T_1(r) = B - \frac{g_j r^2}{6k_{wd}} \tag{S8.4a}$$

$$T_2(r) = T_{op} + \frac{g_j R_{mz}^3}{3k_{wd}r} \tag{S8.4b}$$

By equating T_1 and T_2 at $r = R_{mz}$, we obtain:

$$B = T_{op} + \frac{g_j R_{mz}^2}{2k_{wd}} \tag{S8.5}$$

Combining Eqs. $S8.4a$ and $S8.5$, we have:

$$T_1(r) = T_{op} + \frac{g_j R_{mz}^2}{2k_{wd}}\left[1 - \frac{1}{3}\left(\frac{r}{R_{mz}}\right)^2\right] \tag{6.24}$$

Solution to Problem 6.8

b) The total Joule heating generated within region 1, G_j, is given by the volume of region 1 times g_j:

$$G_j = \frac{4\pi R_{mz}^3}{3} g_j \qquad (S8.6)$$

Under steady-state conditions, G_j given by $S8.6$ is equal to the total heat leaving Region 1 at $r = R_{mz}$ by conduction. Thus,

$$G_j = -4\pi R_{mz}^2 k_{wd} \frac{dT_1}{dr}\bigg|_{r=R_{mz}}$$

$$= -4\pi R_{mz}^2 k_{wd} \left(\frac{g_j R_{mz}^2}{2k_{wd}}\right)\left(-\frac{2}{3R_{mz}}\right) = \frac{4\pi R_{mz}^3}{3} g_j$$

c) By noting that $g_j = \rho_m J_m^2$ and $T_1 = T_c$ at $r = R_{mz}$, we modify Eq. 6.24 to:

$$T_c = T_{op} + \frac{\rho_m J_m^2 R_{mz}^2}{2k_{wd}} \frac{2}{3} \qquad (S8.7)$$

Solving Eq. $S8.7$ for R_{mz}, we obtain:

$$R_{mz} = \sqrt{\frac{3k_{wd}(T_c - T_{op})}{\rho_m J_m^2}} \qquad (6.25)$$

d) For the first set of parameters ($k_{wd} = 400\,\text{W/m K}$; $T_c = 6\,\text{K}$; $T_{op} = 4\,\text{K}$; $\rho_m = 2\times10^{-10}\,\Omega\,\text{m}$; $J_m = 300\times10^6\,\text{A/m}^2$) we have:

$$R_{mz1} = \sqrt{\frac{3(400\,\text{W/m K})(6\,\text{K} - 4\,\text{K})}{(2\times10^{-10}\,\Omega\,\text{m})(300\times10^6\,\text{A/m}^2)^2}} \qquad (S8.8)$$

$$= \sqrt{\frac{4}{3}} \times 10^{-2}\,\text{m} \simeq 1.2\,\text{cm}$$

e) Because R_{mz} is proportional to the square root of k_{wd}, we have:

$$R_{mz2} = \sqrt{\frac{k_{wd2}}{k_{wd1}}}\, R_{mz1} \qquad (S8.9)$$

$$\simeq \sqrt{\frac{0.1\,\text{W/m K}}{400\,\text{W/m K}}}\,(1.2\,\text{cm}) \simeq 0.2\,\text{mm}$$

Solution to Problem 6.8

As Wilson points out in his textbook [6.3], the thermal conductivity is anisotropic in real windings, being essentially that of copper (matrix) along the conductor axis and that of a filler, usually an epoxy, in the directions transverse to the conductor axis, thus making the MPZ ellipsoidal rather than spherical as assumed in the present analysis. Such an ellipsoidal MPZ will have the major radius given by R_{mz1} and the minor radii given by R_{mz2}.

f)   The total energy margin ΔE_h is given by:

$$\Delta E_h = V_{mz} \int_{T_{op}}^{T_c} C_{wd} \, dT \qquad (S8.10)$$

where C_{wd} is the heat capacity of the winding, which we shall assume to be nearly equal to that of copper. (Volumetrically, the heat capacities of most solid materials are nearly equal.) Note that the upper limit of integration is T_c rather than $T_1(r)$ given by Eq. 6.24, in which case it is also necessary to perform the volume integral instead of simply multiplying the energy density term by V_{mz} as is done in Eq. $S8.10$. The reason for this simplification is that once the winding temperature within the MPZ is driven to T_c, Joule heating is generated in the MPZ which will supply the additional energy needed to create the stable temperature profile defined by $T_1(r)$.

The ellipsoidal MPZ volume, V_{mz}, having major radius R_{mz1} and minor radii R_{mz2}, is given by: $V_{mz} = 4\pi R_{mz2}^2 R_{mz1}/3$. We thus have:

$$\Delta E_h = \frac{4\pi R_{mz2}^2 R_{mz1}}{3} [h(T_c) - h(T_{op})]_{cu} \qquad (S8.11)$$

With $R_{mz2} \simeq 2 \times 10^{-4}\,\mathrm{m}$, $R_{mz1} = 1.2 \times 10^{-2}\,\mathrm{m}$, $[h(T_c)]_{cu} = 3.9\,\mathrm{kJ/m^3}$ and $[h(T_{op})]_{cu} = 1.3\,\mathrm{kJ/m^3}$, we obtain:

$$V_{mz} = \frac{4\pi(2 \times 10^{-4}\,\mathrm{m})^2(1.2 \times 10^{-2}\,\mathrm{m})}{3} \qquad (S8.12)$$

$$= \frac{4\pi(4.8 \times 10^{-10}\,\mathrm{m^3})}{3} \simeq 2 \times 10^{-9}\,\mathrm{m^3}$$

$$\Delta E_h = (2 \times 10^{-9}\,\mathrm{m^3})(3.9\,\mathrm{kJ/m^3} - 1.3\,\mathrm{kJ/m^3}) \qquad (S8.13)$$

$$= (2 \times 10^{-9}\,\mathrm{m^3})(2.6 \times 10^3\,\mathrm{J/m^3}) = 5.2 \times 10^{-6}\,\mathrm{J}$$

$$\sim 10\,\mu\mathrm{J}$$

Since $R_{mz2} \simeq 0.2$ mm is smaller than typical conductor diameters, V_{mz} and therefore ΔE_h in real windings would be greater than the values computed above.

In any event, only a minute amount of energy, in the range of microjoules, is needed to create an MPZ in typical windings, implying that adiabatic magnets are very susceptible to quench with tiny inputs of heat. This conclusion was indeed verified experimentally by Superczynski [6.25] and Scott [6.26].

Problem 6.9: V vs I traces of a cooled composite conductor

This problem investigates V vs I traces of a composite superconductor immersed in a bath of liquid helium boiling at 4.2 K; we will generate V vs I traces for three different cooling conditions. The conductor parameters are as follows: $I_{c_o} = 1000$ A, the critical current at $T_{op} = 4.2$ K; $\rho_m = 4 \times 10^{-10}$ Ω m, electrical resistivity of matrix metal; $A_m = 2 \times 10^{-5}$ m^2, the total matrix cross section; $P_{cd} = 2 \times 10^{-2}$ m, the total conductor perimeter exposed to liquid helium and it is varied as $f_p P_{cd}$; $h_q = 10^4$ W/m^2 K, heat transfer coefficient. V is measured across the conductor length $\ell = 0.1$ m. In deriving V vs I traces, assume also that $I_c(T)$ is given by Eq. 6.4.

a) For $I < I_{c_o}$, we have $V = 0$ V. For $I \geq I_{c_o}$, show that V is given by the following expression:

$$V = R_m(I - I_{c_o}) + \frac{R_m^2 I_{c_o} I(I - I_{c_o})}{f_p P_{cd}\ell h_q(T_c - T_{op}) - R_m I_{c_o} I} \qquad (6.26)$$

where $R_m = \rho_m \ell / A_m$. To derive this equation you must assume that the conductor is always in thermal equilibrium, with the resistive dissipation balanced by the cooling, which requires that the conductor temperature be at $T_{op} + \Delta T$. Also note that current in the matrix is given by $I - I_s$, where I_s, the current in the superconductor, is given by Eq. 6.4 with $T = T_{op} + \Delta T$.

b) By defining two additional dimensionless parameters, $v = V/R_m I_{c_o}$, $i = I/I_{c_o}$, and also using α_{sk} (Stekly parameter), show that dimensionless voltage in terms of dimensionless current, $v(i)$, and dimensionless current in terms of dimensionless voltage, $i(v)$ are given by:

$$v(i) = \frac{\alpha_{sk}(i - 1)}{\alpha_{sk} - i} \qquad (6.27a)$$

$$i(v) = \frac{\alpha_{sk}(v + 1)}{\alpha_{sk} + v} \qquad (6.27b)$$

c) *Condition 1:* $f_p = 1$. For $T_c = 5.2$ K (and $T_{op} = 4.2$ K), compute v at $i = 1$, 1.1, 1.5, and 2.

d) *Condition 2:* $f_p = 0.1$. Show that v is independent of i.

e) *Condition 3:* $f_p = 0.05$. Here, because most of the surface area is insulated from liquid helium, we expect the conductor to behave unstably. Specifically, solve i for $v = 0$, 0.125, 0.25, 0.5, and 0.625. Also compute v which lands on the $v = i$ line.

f) Plot $v(i)$ traces for the three conditions studied above. Plot $v = i$ with a solid line. Label $\alpha_{sk} = 10$ for the curve corresponding to Condition 1; $\alpha_{sk} = 1$ for the Condition 2 curve; $\alpha_{sk} = 0.5$ for the Condition 3 curve.

Solution to Problem 6.9

a) For $I > I_{c_o}$, V across the voltage taps is given by $R_m(I - I_s)$ where I_s is the current in the superconductor, *i.e.* $I_s = I_c(T)$. Joule heating generation $G_j(T_{op} + \Delta T)$ in the composite is thus given by:

$$G_j(T_{op} + \Delta T) = VI = R_m \left\{ I - I_{c_o} \left[\frac{T_c - (T_{op} + \Delta T)}{T_c - T_{op}} \right] \right\} I$$

$$= R_m I \left[(I - I_{c_o}) + \frac{I_{c_o} \Delta T}{T_c - T_{op}} \right] \tag{S9.1}$$

$G_j(T_{op} + \Delta T)$ is matched by the cooling, which is given by $f_p P_{cd} \ell h_q (T - T_{op}) = f_p P_{cd} \ell h_q \Delta T$. Equating these two powers and solving for ΔT, we obtain:

$$\Delta T = \frac{R_m I(I - I_{c_o})(T_c - T_{op})}{f_p P_{cd} \ell h_q (T_c - T_{op}) - R_m I_{c_o} I} \tag{S9.2}$$

Combining Eqs. S9.1 and S9.2 and solving for V, we obtain:

$$V = R_m \left\{ (I - I_{c_o}) + \frac{I_{c_o}}{(T_c - T_{op})} \left[\frac{R_m I(I - I_{c_o})(T_c - T_{op})}{f_p P_{cd} \ell h_q (T_c - T_{op}) - R_m I_{c_o} I} \right] \right\} \tag{S9.3}$$

$$V = R_m(I - I_{c_o}) + \frac{R_m^2 I_{c_o} I(I - I_{c_o})}{f_p P_{cd} \ell h_q (T_c - T_{op}) - R_m I_{c_o} I} \tag{6.26}$$

b) With $v = V/R_m I_{c_o}$, $i = I/I_{c_o}$, and $\alpha_{sk} = f_p P_{cd} A_m h_q (T_c - T_{op})/\rho_m I_{c_o}^2$ (Eq. 6.9) or $\alpha_{sk} = f_p P_{cd} \ell h_q (T_c - T_{op})/R_m I_{c_o}^2$, all dimensionless parameters, we can rewrite Eq. 6.26 as:

$$v(i) = (i - 1) + \frac{i(i-1)}{\alpha_{sk} - i}$$

$$= \frac{\alpha_{sk}(i-1)}{\alpha_{sk} - i} \tag{6.27a}$$

Solving Eq. 6.27a for i, we obtain:

$$i(v) = \frac{\alpha_{sk}(v+1)}{\alpha_{sk} + v} \tag{6.27b}$$

c) R_m: $\rho_m \ell / A_m = (4 \times 10^{-10}\,\Omega\,m)(0.1\,m)/(2 \times 10^{-5}\,m^2) = 2 \times 10^{-6}\,\Omega$. With $f_p = 1$, we have $\alpha_{sk} = (1)(2 \times 10^{-2}\,m)(0.1\,m)(10^4\,W/m^2\,K)(1\,K)/(2 \times 10^{-6}\,\Omega)(10^6\,A^2) = 10$, and Eq. 6.27a is given by: $v(i) = 10(i - 1)/(10 - i)$. Values of $v(i)$ at selected values of i are given in Table 6.3.

Table 6.3: v vs i for $\alpha_{sk} = 10$

i	1	1.1	1.5	2
v	0	0.11	0.59	1.25

Solution to Problem 6.9

Table 6.4: i vs v for $\alpha_{sk}=0.5$

v	0	0.125	0.25	0.5	0.625	0.707
i	1	0.9	0.833	0.75	0.722	0.707

d) When $f_p = 0.1$, α_{sk} becomes 1 and from Eq. 6.27a, $v(i) = -\alpha_{sk} = -1$. Equation 6.27b gives $i = 1$ when $\alpha_{sk} = 1$. Thus, $v(i)$ is really indeterminate; physically this means that v can be any point on the vertical line at $i = 1$.

e) Here, with $f_p = 0.05$, $\alpha_{sk} = 0.5$, Eq. 6.27b is solved for various values of i and presented in Table 6.4. To compute v that lands on the $v = i$ line, we set $\alpha_{sk} = 0.5$ in Eq. 6.27b and solve for v. Thus:

$$v = \frac{0.5(v+1)}{0.5+v} \qquad (S9.4)$$

From Eq. $S9.4$, we have: $v = \sqrt{0.5} = 0.707$, as indicated in Table 6.4; $i = 0.707$ is known as the "recovery" current (normalized) for $\alpha_{sk} = 0.5$. Once the conductor is driven normal, the transport current must be reduced to below $0.707I_{c_o}$ before superconductivity is restored.

f)

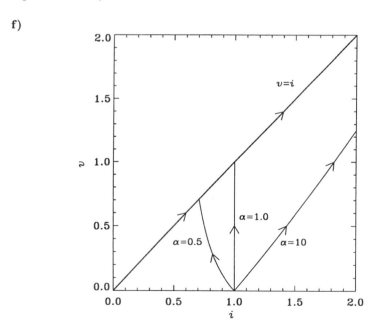

Fig. 6.12   Normalized voltage vs normalized current traces for $\alpha_{sk} = 10$, 1, and 0.5.

Problem 6.10: Stability analyses of Hybrid III SCM

This problem deals with the cryostability of the Hybrid III's Nb-Ti coil, whose conductor specifications are given in Table 5.1 of Problem 5.5 (p. 184). (Unlike the real Nb-Ti coil which is wound with two grades of Nb-Ti composite conductors, the coil in this problem is assumed to be wound with only one grade.)

The Nb-Ti coil, as described in Problem 3.3, consists of a stack of 32 double-pancake coils, with a total overall winding height of 640 mm. Each pancake coil has winding i.d. and o.d., respectively, of 658 mm and 907 mm (Table 3.1). Figure 6.13 shows important details of the winding. As is evident from the figure, each turn is separated by a thin insulator strip. The two pancakes in each double-pancake unit are separated by a sheet of 0.5-mm thick insulator bonded with epoxy resin. Between adjacent double pancakes are 1.0-mm thick insulating spacers extending radially from the inside radius to the outside radius to provide cooling. The spacers cover, on the average, 60% of the flat surface of each coil exposed to liquid helium. Note that in each double-pancake set, the top coil has its top surface (40%) exposed to liquid helium, while the bottom coil has its bottom surface (40%) exposed to liquid helium. Because the liquid helium is at 1.8 K (superfluid), there is no adverse effect on cooling for the "face-down" bottom coil.

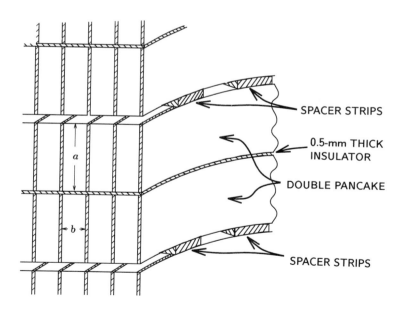

Fig. 6.13   Winding details for the Nb-Ti pancakes.

Problem 6.10: Stability analyses of Hybrid III SCM

The Nb-Ti coil (and Nb_3Sn coil) is designed to be operated while immersed in a bath of 1-atm subcooled 1.8-K superfluid helium. Assume the cooling to be dominated by Kapitza resistance; q_k given by Eq. 4.4 should hence be used for cooling, $q(T)$:

$$q_k = a_k(T_{cd}^{n_k} - T_b^{n_k}) \qquad (4.4)$$

T_{cd} [K] is the conductor temperature (really the temperature of the conductor surface, which is copper) and T_b [K] is the bath temperature. You may take $a_k = 0.02\,W/cm^2$ and $n_k = 4.0$. At an operating temperature T_{op} of 1.8 K ($T_{op} = T_b$), the coil carries a transport current I_{op} of 2100 A and is exposed to a maximum magnetic induction of ~10 T.

a) Make an appropriate $I_c(T)$ plot for this conductor covering the temperature range $1.8\,K\sim T_c(10\,T)$. Determine, from the plot, the current sharing temperature T_{cs} for a transport current I_t of 2100 A. Indicate T_{cs} in the plot.

b) Make, and label, power flux [W/cm^2] vs temperature [K] plots for both cooling and heat generation at 10 T when $T_{op} = T_b = 1.8\,K$, $I_{op} = 2100\,A$. Table 5.1 gives useful data. Based on the plots, state whether the pancakes are stable, and if so under which criterion. If not, explain why they are not stable. For the purpose of solving this question you may assume that: 1) $q(T)$ given above is valid over the *entire* temperature range of interest and that 2) heat generated within each pancake is transported freely through the 1-mm high radial channels.

c) In the pancake coils in previous Hybrids built at FBNML, each turn was separated by thin (~0.4 mm thick) spacers to make the coils cryostable. In the early phase of the Hybrid III project, pancake designs with such turn-to-turn cooling spacers were seriously considered but were abandoned in favor of a spacerless winding design, because turn-to-turn spacers reduce the radial stiffness of the winding. Assuming turn-to-turn cooling channels are present in the Hybrid III Nb-Ti pancakes, draw again neatly on another graph power flux [W/cm^2] vs temperature [K] plots for both cooling and heat generation at 10 T when $T_b = 1.8\,K$, $I_{op} = 2100\,A$. For this case, assume 50% of the total conductor perimeter is exposed directly to liquid helium. Again, based on these plots, comment on the stability of the pancakes.

"...there must be discipline. For many things are not as they appear. Discipline must come from trust and confidence." —Robert Jordan

Solution to Problem 6.10

a) Figure 6.14 presents the $I_c(T)$ plot for this conductor, constructed by joining two points: one at 6000 A at 1.8 K and 10 T (Table 5.1) and the other at 0 A at 4.7 K (T_c at 10 T; Table 5.1). The current sharing temperature T_{cs} is given by 2100 A $= I_c(T_{cs})$: it is 3.7 K.

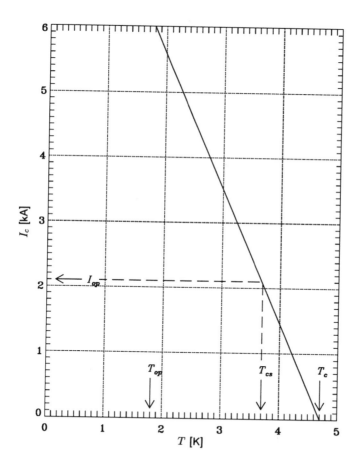

Fig. 6.14 I_c vs T plot (solid) for a Hybrid III Nb-Ti conductor at 10 T. The intersection of the line at $I_c = 0$ determines $T_c = 4.7$ K. $I_t = 2100$ A is given by the dotted line, which intersects the solid line at $T_{cs} = 3.7$ K.

Solution to Problem 6.10

b) Let us first compute $\hat{g}_j$ in the normal state valid for $T \geq T_c = 4.7\,\mathrm{K}$. As discussed in Problem 6.4, $\hat{g}_j(T_c)$, unlike $g_j(T_c) = \rho_m I_{op}^2/A_m A_{cd}$ [W/m^3], is the normal-state generation flux (per unit conductor surface exposed to liquid helium). That is, $\hat{g}_j(T_c) = (A_{cd}/f_p P_{cd})g_j(T)$, where $f_p P_{cd}$ is the conductor perimeter exposed to liquid helium. A_{cd} and A_m are the overall conductor cross sectional area and matrix metal cross sectional area, respectively. We have: $A_{cd} = ab$, where a and b are the overall conductor width and thickness, respectively; and $A_m = ab\gamma_{c/s}/(\gamma_{c/s} + 1)$, where $\gamma_{c/s}$ is the copper-to-superconductor ratio. We thus have: $\hat{g}_j(T_c) = \rho_m I_{op}^2(\gamma_{c/s} + 1)/\gamma_{c/s}abf_p P_{cd}$.

With $I_t = 2100\,\mathrm{A}$, $\rho_m = 4.5\times10^{-10}\,\Omega\,\mathrm{m}$, $a = 9.2\times10^{-3}\,\mathrm{m}$, $b = 2.6\times10^{-3}\,\mathrm{m}$, $\gamma_{c/s} = 3$, $f_p P_{cd} = (0.4)(2.6\times10^{-3}\,\mathrm{m}) = 1.04\times10^{-3}\,\mathrm{m}$, we obtain: $\hat{g}_j = 1.06\times10^5\,\mathrm{W/m^2}$.

$\hat{g}_j(T)$ is zero for $1.8\,\mathrm{K} \leq T \leq T_{cs}$ (3.7 K) and starting at T_{cs} it rises linearly with T until T_c (4.7 K), at which point $\hat{g}_j(T_c) = 1.06\times10^5\,\mathrm{W/m^2}$.

Figure 6.15 shows $\hat{g}_j(T)$ and $q(T)$ plots. From the plots, it is clear that the pancakes are almost cryostable; they certainly satisfy the equal area criterion.

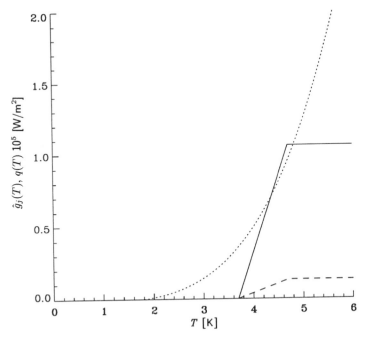

Fig. 6.15 $\hat{g}_j(T)$ plot (solid) and $q(T)$ plot (dotted); $\hat{g}_j(T)$ (dashed) corresponds to question **c)**.

Solution to Problem 6.10

c) Both $g_j(T)$ and $q(T)$ are the same as those computed in **b)**. The area exposed to helium per unit conductor length will of course be much greater in this case than in the spacerless winding. For this case, $f_p P_{cd} = (0.5)(23.6 \times 10^{-3}\,\mathrm{m}) = 11.8 \times 10^{-3}\,\mathrm{m}$ and $\hat{g}_j(T_c) = 0.13 \times 10^5\,\mathrm{W/m^2}$.

The dashed curve in Fig. 6.15 shows the $\hat{g}_j(T)$ plot; the same $q(T)$ used in the previous case is valid. From the plots, it is clear that the pancakes are cryostable.

Cryostable vs Quasi-Adiabatic (QA) Magnets

As described in Problem 6.10, the double pancakes of the Hybrid III Nb-Ti coil have no cooling channels between turns. Decision to make these double pancakes "quasi-adiabatic" (QA) was based more on stress considerations than on stability considerations. The term "quasi-adiabatic" was used because it was thought that the Nb-Ti coil, without cooling channels, would not be cryostable but would approximate adiabatic performance—later, when the stability analysis presented above was performed as an exercise for the students, the coil was found to be stable. (A stability study, examining the effects of mechanical disturbances on coil operation, was performed with satisfactory conclusions at the time of the decision to make the Nb-Ti coil quasi-adiabatic [6.27].)

Figure 6.16 shows hoop stress (σ_h) vs winding radius (r) plots of the Nb-Ti conductors (HF and LF) for cryostable and QA windings [3.7]. For cryostable windings, because of the presence of structurally "soft" spacer material, σ_h vs r traces essentially present ($r \times J \times B$) stresses. (A jump in the stress at the HF-LF transition is due to reduction in conductor cross section.) For the QA winding, because the winding is much more rigid in the radial direction (no spacers), radial expansion of the inner turns is supported by the outer turns, increasing stress at the outer turns. The net result is a more uniform stress distribution. (In the analysis it was assumed that the both sections had the same conductor cross sectional area.) Equally important is a substantial reduction in the overall size of the Nb-Ti coil, from a winding o.d. of over 1 m to an o.d. of ~0.9 m (Table 3.1, p. 59).

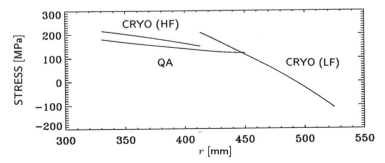

Fig. 6.16   Hoop stress vs radius plots for cryostable (Cryo) and quasi-adiabatic (QA) windings.

Problem 6.11: Stability of CIC conductors

This problem deals with the stability of CIC conductors; in particular, the amount of copper matrix needed in CIC conductors from stability and energy margin (Δe_h) considerations. We progress using a step-by-step approach similar to that used by Bottura in designing a 40-kA CIC conductor for the Next European Torus [NET] machine [6.28].

The cable consists of N_{st} numbers of circular strands, with f_{cu} defined as the ratio of A_{cu} to A_{cd}: $f_{cu} = A_{cu}/A_{cd}$. A_{cu} is the total copper cross sectional area in the conductor whose total cross sectional area is A_{cd}. A_{cd} includes areas occupied by copper, non-copper metals (superconductor, unreacted metals, etc.), and helium. Note that $f_{cu} + f_{nc} + f_{he} = 1$. f_{nc} is the fraction of the conductor cross section occupied by non-copper metals. f_{he} is commonly known as void fraction; in most CIC conductors it generally ranges from 33 to 40%. In the "well-cooled" region of operation [6.12], a CIC conductor may be treated like a bath-cooled conductor— except here cold helium, confined within the conduit, is not available in abundance. In this case, we have a stability criterion very much like that of Stekly's. For a *single* circular cross-section strand of diameter, d_{st}, operating current density in the *copper matrix*, $[j_{op}]_{cu}$ must satisfy the following condition to achieve cryostability:

$$[j_{op}]_{cu} \leq \sqrt{\frac{4f_p(f_{cu} + f_{nc})h_q(T_c - T_{op})}{f_{cu}\rho_{cu}d_{st}}} \qquad (6.28)$$

f_p is the fraction of conductor surface exposed directly to helium at $T_b = T_{op}$; h_q is the heat transfer coefficient; and ρ_{cu} is the copper resistivity.

a) For a CIC conductor containing N_{st} strands, each of diameter d_{st}, show that the $[J_{op}]_{cd}$, operating current density in the *conductor* containing N_{st} strands, must satisfy the following condition for cryostability:

$$[J_{op}]_{cd} \leq \sqrt{\frac{4f_p f_{cu}(f_{cu} + f_{nc})h_q(T_c - T_{op})}{\rho_{cu}d_{st}}} \qquad (6.29)$$

Note that $[J_{op}]_{cd} = I_{op}/A_{cd}$.

b) Energy margin density, Δe_h, is defined as the amount of dissipation density (per unit volume of *all* strands) imposed into the CIC conductor just sufficient to drive the conductor to the current sharing temperature, T_{cs}. Because the heat capacity of the conductor is essentially that of helium, Δe_h may be equated with the helium enthalpy available between T_{op} and T_{cs}. Show that Δe_h is given by the following expression:

$$\Delta e_h = \frac{f_{he}}{f_{cu} + f_{nc}}[h(T_{cs}) - h(T_{op})] \simeq \frac{f_{he}C_p(T_{cs} - T_{op})}{f_{cu} + f_{nc}} \qquad (6.30)$$

where h and C_p are, respectively, helium's volumetric enthalpy and heat capacity. The heating takes place with constant helium pressure and mass.

Problem 6.11: Stability of CIC conductors

c) By combining Eq. 6.30 and the usual linear dependance of $J_c(T)$, show that an expression that relates $[J_{op}]_{cd}$, f_{cu}, f_{nc}, and Δe_h is given by:

$$[J_{op}]_{cd} = [J_{co}]_{nc} f_{nc} \left\{ 1 - \frac{(f_{cu} + f_{nc})\Delta e_h}{[1 - (f_{cu} + f_{nc})]C_p(T_c - T_{op})} \right\} \quad (6.31)$$

where $[J_{co}]_{nc} = I_{co}/A_{nc}$ and $I_{co} = I_c(T_{op})$.

d) An equation relating f_{cu} and f_{nc} that satisfies both the stability criterion of Eq. 6.29 and the energy margin density criterion of Eq. 6.30 is given by:

$$4 f_p f_{cu} (f_{cu} + f_{nc}) h_q (T_c - T_{op})$$

$$= \rho_{cu} d_{st} [J_{co}]_{nc}^2 f_{nc}^2 \left\{ 1 - \frac{(f_{cu} + f_{nc})\Delta e_h}{[1 - (f_{cu} + f_{nc})]C_p(T_c - T_{op})} \right\}^2 \quad (6.32)$$

Make a f_{cu} vs f_{nc} plot of Eq. 6.32 on graph paper for the parameters presented in Table 6.5 applicable to a 40-kA CIC conductor proposed by the NET team [6.28]

Also on the same plot, draw a line given by $f_{cu} + f_{nc} = 1 - f_{he} = 0.6$; this line corresponds to the case for a helium void fraction of 40%. The intersection of the Eq. 6.32 curve and the line gives the desired set of f_{cu} and f_{nc} for this 40-kA CIC conductor, making it satisfy the stability criterion of Eq. 6.29 and the energy margin criterion of Eq. 6.30.

Table 6.5: Parameter Values for
An NET 40-kA CIC Conductors [6.28]

f_p		5/6
h_q	[W/m^2]	800
T_c @ 12 T	[K]	8.5
T_{op}	[K]	4.5
$[J_{co}]_{nc}$	[MA/m^2]	490
ρ_{cu}	[nΩ m]	0.7
d_{st}	[mm]	0.75
Δe_h	[kJ/m^3]	500
C_p	[MJ/m^3 K]	0.75

Solution to Problem 6.11

a) We may express the stability condition of Eq. 6.28 for the whole conductor by balancing the total normal-state Joule heating at T_c and total cooling. Thus:

$$\frac{\rho_{cu} I_{op}^2}{A_{cu}} \leq P_{cd} h_q (T_c - T_{op}) \tag{S11.1}$$

where P_{cd} is the total wetted perimeter of the cable. P_{cd} may be given by:

$$P_{cd} = f_p N_{st} \pi d_{st} \tag{S11.2}$$

Also, we have:

$$A_{cu} + A_{nc} = \frac{N_{st} \pi d_{st}^2}{4} \tag{S11.3}$$

Combining Eqs. $S11.2$ and $S11.3$, we obtain:

$$P_{cd} = \frac{4 f_p (A_{cu} + A_{nc})}{d_{st}} \tag{S11.4}$$

Now, by combining Eqs. $S11.1$ and $S11.4$, we obtain:

$$I_{op}^2 \leq A_{cu}(A_{cu} + A_{nc}) \frac{4 f_p h_q (T_c - T_{op})}{\rho_{cu} d_{st}} \tag{S11.5}$$

Dividing Eq. $S11.5$ by A_{cd}^2 and recognizing $[J_{op}]_{cd} = I_{op}/A_{cd}$, we have:

$$[J_{op}]_{cd} \leq \sqrt{\frac{4 f_p f_{cu}(f_{cu} + f_{nc}) h_q (T_c - T_{op})}{\rho_{cu} d_{st}}} \tag{6.29}$$

b) The *total* energy margin ΔE_h (per unit conductor length) required to raise the conductor to its current sharing temperature is transferred almost entirely to the surrounding helium and given by:

$$\Delta E_h = A_{he}[h(T_{cs}) - h(T_{op})] = A_{he} C_p (T_{cs} - T_{op}) \tag{S11.6}$$

Δe_h is given by ΔE_h over the total volume of strands. Thus:

$$\Delta e_h \equiv \frac{\Delta E_h}{A_{cu} + A_{nc}} = \frac{A_{he} C_p (T_{cs} - T_{op})}{A_{cu} + A_{nc}} \tag{S11.7}$$

By dividing the numerator and denominator of Eq. $S11.7$ by A_{cd}, we have:

$$\Delta e_h = \frac{f_{he} C_p (T_{cs} - T_{op})}{f_{cu} + f_{nc}} \tag{6.30}$$

Solution to Problem 6.11

c) To determine the value of $[J_{op}]_{cd}$ corresponding to Δe_h, the usual linear dependance of critical current, $I_c(T)$, first is given by Eq. 6.4:

$$I_c(T) = I_{c_o} \left(\frac{T_c - T}{T_c - T_{op}} \right) \tag{6.4}$$

Equation 6.4 may be rewritten as:

$$A_{cd}[J_c(T)]_{cd} = A_{nc}[J_{c_o}]_{nc} \left(\frac{T_c - T}{T_c - T_{op}} \right) \tag{S11.8}$$

Dividing the above expression by A_{cd}, we obtain:

$$[J_c(T)]_{cd} = f_{nc}[J_{c_o}]_{nc} \left(\frac{T_c - T}{T_c - T_{op}} \right) \tag{S11.9}$$

At T_{cs}, by definition, we have $[J_c(T_{cs})]_{cd} = [J_{op}]_{cd}$, and solving Eq. $S11.9$ for T_{cs} we obtain:

$$T_{cs} = T_c - \frac{[J_{op}]_{cd}(T_c - T_{op})}{f_{nc}[J_{c_o}]_{nc}} \tag{S11.10}$$

Combining Eqs. 6.30 and $S11.10$, we have:

$$\Delta e_h = \frac{f_{he}}{f_{cu} + f_{nc}} C_p \left[(T_c - T_{op}) - \frac{[J_{op}]_{cd}(T_c - T_{op})}{f_{nc}[J_{n_o}]_{nc}} \right]$$

$$= \frac{f_{he}}{f_{cu} + f_{nc}} \left(1 - \frac{[J_{op}]_{cd}}{f_{nc}[J_{c_o}]_{nc}} \right) C_p(T_c - T_{op}) \tag{S11.11}$$

Solving Eq. $S11.11$ for $[J_{op}]_{cd}$ and noting $f_{he} = 1 - (f_{cu} + f_{nc})$, we obtain:

$$[J_{op}]_{cd} = f_{nc}[J_{c_o}]_{nc} \left\{ 1 - \frac{(f_{cu} + f_{nc})\Delta e_h}{[1 - (f_{cu} + f_{nc})]C_p(T_c - T_{op})} \right\} \tag{6.31}$$

d) By inserting parameter values into Eq. 6.32, we have:

$$(4)(5/6)f_{cu}(f_{cu} + f_{nc})(800 \,\text{W/m}^2 \,\text{K})(4 \,\text{K})$$
$$= (7 \times 10^{-10} \,\Omega\,\text{m})(7.5 \times 10^{-4} \,\text{m})(4.9 \times 10^8 \,\text{A/m}^2)^2$$

$$\times f_{nc}^2 \left\{ 1 - \frac{(f_{cu} + f_{nc})(5 \times 10^5 \,\text{J/m}^2)}{[1 - (f_{cu} + f_{nc})](7.5 \times 10^5 \,\text{J/m}^3 \,\text{K})(4 \,\text{K})} \right\}^2 \tag{S11.12a}$$

$$f_{cu}(f_{cu} + f_{nc})(1.067 \times 10^4)$$

$$= (7)(7.5)(4.9)^2(100)f_{nc}^2 \left\{ 1 - \frac{(f_{cu} + f_{nc})}{[1 - (f_{cu} + f_{nc})]6} \right\}^2 \tag{S11.12b}$$

Solution to Problem 6.11

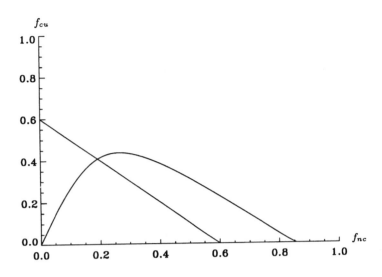

Fig. 6.17 f_{cu} vs f_{nc} plots for a 40-kA CIC conductor
proposed by the NET team [6.28].

$$f_{cu}(f_{cu} + f_{nc})(1 - f_{cu} - f_{nc})^2 = 0.328 f_{nc}^2 (6 - 7f_{cu} - 7f_{nc})^2 \qquad (S11.13)$$

Equation $S11.13$ is plotted in Fig. 6.17. From the figure we have: $f_{cu} \simeq 0.4$; $f_{nc} \simeq 0.2$. Note that $1 - f_{cu} - f_{nc} = 0.4 = f_{he}$, as expected.

"In this very spot there are whole forests which were buried millions of years ago; now they have turned to coal, and for me they are an inexhaustible mine." —Captain Nemo (c. 1870)

"The seas of this planet contain 100,000,000,000,000,000 tons of hydrogen and 20,000,000,000,000 tons of deuterium. Soon we will learn to use these simplest of all atoms to yield unlimited power." —Arthur C. Clarke (c. 1960)

Problem 6.12: "Ramp-rate-limitation" in CIC conductors

This problem deals with what has become known as the "ramp-rate limitation" phenomenon. Simply stated, a CIC conductor afflicted with ramp-rate-limitation quenches *prematurely* before reaching its design current when it is ramped too fast in current, background field, or a combination of the two. These premature quenches were first observed in the Demonstration Poloidal Coil (DPC) project experiments for large fusion experimental superconducting magnets [6.29,6.30]. The experiments were conducted at the Japan Atomic Energy Research Institute (JAERI), Tokai. The ramp-rate-limitation phenomenon studied in this problem was observed in a CIC coil built in the U.S. (known as the US-DPC coil)and tested at JAERI's DPC test facility .

Although the mechanism responsible for ramp-rate limitation has not been identified, a model proposed in 1993 by Takayasu [6.31] leads to an equation for quench currents that fits experimental data quite well. Figure 6.18 presents quench and nonquench currents (respectively, solid and open circles) *vs* ramping time data for the US-DPC coil [6.31]. For a given current reached, quenched or nonquenched, the shorter the time, the greater the ramp rate. The data cover the current ramp rate range from ~3 to ~75 kA/s. The solid curve is based on the Takayasu equation, the derivation of which is the focal point of this problem.

The Takayasu model, unlike Stekly's model which assumes continuous Joule heating, postulates that during a ramp, each strand is subjected to a periodic Joule heating. The mechanism for a periodic disturbance that induces this Joule heating has not been identified other than that it is apparently triggered by the ramping.

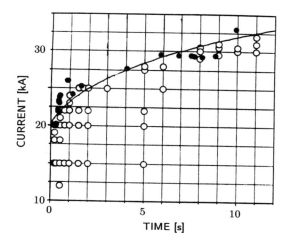

Fig. 6.18 Ramp-rate-limitation data for the US-DPC coil [6.31]. Solid circles indicate quenches and open circles nonquenches. The solid curve is based on Eq. 6.39 to be derived in this problem.

Problem 6.12: "Ramp-rate-limitation" in CIC conductors

We start with a single strand of a CIC conductor which contains N_{st} strands. The strands are cooled by helium, which is initially at T_{b_o}. Although the helium is forced through the conduit, for the purpose of the present analysis it may be assumed static, because in these current-ramping events, as can be inferred from Fig. 6.18, each event lasts at most several seconds, while helium "replenishment" time over the conductor length of interest is of the order of tens of seconds.

For a strand undergoing Joule heating, the model postulates a scenario which progresses through the following steps.

1. The strand is subjected to periodic (frequency f_h) "disturbance pulses," each pulse of constant amplitude and "negligible" duration that is repeated every t_p ($t_p = 1/f_h$). The pulsing is active during current ramp that starts at $t = 0$ and ends at $t = t_r$, at which time the strand current has reached i_q. Note that $n_p = t_r/t_p = f_h t_r$, where n_p is the number of pulses in one ramping sequence. The conductor quench current, I_q, is given by $I_q = N_{st} i_q$.

2. Each time the strand is "pulsed," its temperature is raised "immediately" to T_c, the critical temperature for a given magnetic field, and it generates Joule heating for a duration of t_d, after which, it is cooled to below T_{cs}, the current sharing temperature, and regains superconductivity. The bottom trace of Fig. 6.19 indicates this periodic Joule heating generation. Note that because $i_{st}(t)$ is ramping, the size of $g_j(t)$ increases as t^2; Fig. 6.19 indicates this quadratic dependence schematically.

3. The Joule heat is released to the helium and, because the helium is assumed static, the helium temperature, initially at T_{b_o}, increases, as indicated in the top trace. If the helium temperature reaches T_{cs} at the end of the ramping, the strand will not recover and quench ensues at $I_q = N_{st} i_q$.

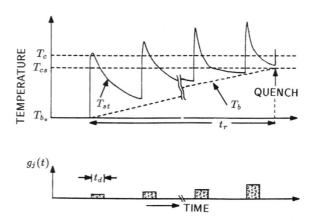

Fig. 6.19 Heating scenario according to the model proposed by Takayasu [6.31]. $g_j(t)$ function is modified from that shown in [6.31].

Problem 6.12: "Ramp-rate-limitation" in CIC conductors

a) Assume that the cooling power density g_q (Eq. 6.1) acting on the heat generating strand to be given by:

$$g_q = \frac{p_{st}}{a_{st}} h_q (T_{st} - T_b) \qquad (6.33)$$

where p_{st} is the strand perimeter, a_{st} is the strand cross sectional area, h_q is the heat transfer coefficient; T_{st} and T_b are, respectively, the strand temperature and helium temperature, both time-dependent during current ramping. Equation 6.33 implies that the entire strand surface is exposed to helium. Using Eq. 6.33, show that an expression for the energy density equation in integral form, valid over the ramping period (from $t = 0$ to $t = t_r$) for constant h_q, is given by:

$$\int_{T_{bo}}^{T_{cs}} C_{st} \, dT = \tilde{\zeta}_{n_p} \frac{\rho_{cu} i_q^2 t_d f_h t_r}{a_{cu} a_{st}} - \frac{p_{st}}{a_{st}} h_q \int_0^{t_r} [T_{st}(t) - T_b(t)] \, dt \qquad (6.34)$$

where C_{st} is the strand heat capacity, $\tilde{\zeta}_{n_p}$ is used during ramped current experiments to represent the average value of n_p Joule heating pulses in terms of the quench current i_q [strand current, $i_{st}(t)$, as noted above, is ramping with time and is not equal to i_q], ρ_{cu} is the copper resistivity, i_q is the quench current in the strand, and a_{cu} is the strand's copper cross sectional area. Note that $n_p = f_h t_r$.

b) The second term on the right-hand side of Eq. 6.34 represents cooling and may be divided into two terms, given by:

$$\frac{p_{st}}{a_{st}} h_q \int_0^{t_r} [T_{st}(t) - T_b(t)] \, dt = \frac{p_{st}}{a_{st}} h_q \sum_{m=0}^{n_p} \left\{ \int_{mt_p}^{mt_p + t_d} [T_{st}(t) - T_b(t)] \, dt \right.$$
$$\left. + \int_{mt_p + t_d}^{(m+1)t_p} [T_{st}(t) - T_b(t)] \, dt \right\} \qquad (6.35)$$

The first term on the right-hand side of Eq. 6.35 represents cooling when the strand is in the normal state: it is on n_p times, each time for a duration of t_d or the total duration of $t_d n_p = t_d f_h t_r$. The second term represents cooling when the strand is in the superconducting state: it is on n_p times, each time for a duration of $t_p - t_d$. For the first term, Takayasu simplifies the integral by assuming $T_{st}(t) \simeq T_c$ and $T_b(t) \simeq T_{bo}$, both constant; the assumption seems reasonable. Combining Eqs. 6.34 and 6.35, and making the assumption noted above, show that an expression for i_q^2 may be given by:

$$i_q^2 = \frac{a_{cu} p_{st} h_q (T_c - T_{bo})}{\tilde{\zeta}_{n_p} \rho_{cu}} + \frac{a_{cu}}{\tilde{\zeta}_{n_p} \rho_{cu} f_h} \chi \qquad (6.36)$$

where

$$\chi = \left(\frac{a_{st}}{t_r t_d} \int_{T_{bo}}^{T_{cs}} C_{st} \, dT + \frac{p_{st} h_q}{t_r t_d} \sum_{m=0}^{n_p} \left\{ \int_{mt_p + t_d}^{(m+1)t_p} [T_{st}(t) - T_b(t)] \, dt \right\} \right) \qquad (6.37)$$

Problem 6.12: "Ramp-rate-limitation" in CIC conductors

c) By recognizing that the conductor quench current, I_q, is N_{st} times i_q, i.e. $I_q = N_{st}i_q$, Eq. 6.36 applied to I_q may be written:

$$I_q^2 = \frac{N_{st}^2 a_{cu} p_{st} h_q (T_c - T_{b_o})}{\tilde{\zeta}_{n_p} \rho_{cu}} + \frac{N_{st}^2 a_{cu}}{\tilde{\zeta}_{n_p} \rho_{cu} f_h} \chi$$

$$= I_{sk}^2 + \frac{N_{st}^2 a_{cu}}{\tilde{\zeta}_{n_p} \rho_{cu} f_h} \chi \qquad (6.38)$$

where

$$I_{sk} = N_{st} \sqrt{\frac{a_{cu} p_{st} h_q (T_c - T_{b_o})}{\tilde{\zeta}_{n_p} \rho_{cu}}} \qquad (6.39)$$

I_{sk} is the Stekly stability current with $\alpha_{sk} = 1$, which is valid when Joule heating is on *continuously*; in this event, the helium must of course be replenished.

Now Takayasu makes additional postulates for the pulsing frequency, f_{cd}, for a conductor that contains N_{st} strands: 1) $f_{cd} = N_{st} f_h$ and 2) $f_{cd} = \xi N_{st}^n (dB^2/dt) = \xi N_{st}^n B\dot{B}$. Here B is the magnetic induction at the quench location; it can be generated by the test coil itself, the test coil and the background-field magnet, or the background-field magnet alone. The first postulate recognizes that disturbance frequency f_{cd} for the whole conductor is N_{st} times that for each strand—a plausible assumption. The second postulate, somewhat contrived, chooses a dB^2/dt dependence of f_{cd} because it makes the model fit the data best; among other candidates tried were dB/dt, $d(BI)/dt$, and dI/dt. None works as well as the $B\dot{B}$ dependence. ξ and n, both appearing in Postulate 2, are constants. Combining Eq. 6.38 and these two postulates, show that an expression for I_q is given by:

$$I_q = \sqrt{I_{sk}^2 + \frac{N_{st}^{3-n} a_{cu} \chi}{\xi \tilde{\zeta}_{n_p} \rho_{cu} B\dot{B}}} \qquad (6.40)$$

The solid curve in Fig. 6.18 is a plot of Eq. 6.40 for the US-DPC coil. In general Eq. 6.40 must be solved numerically because T_c and ρ_{cu} depend on B, which can be self field, background field, or a combination. The solid curve shown in Fig. 6.18 was calculated with the following values of parameters: $N_{st} = 225$; $a_{cu} = 2.6 \times 10^{-7}\,\mathrm{m}^2$; $p_{st} = 2.5\,\mathrm{mm}$; $h_q = 1400\,\mathrm{W/m}^2\,\mathrm{K}$; $T_c = 15.1 - 0.59B$ K (B in tesla); $T_{b_o} = 4.5\,\mathrm{K}$; $\tilde{\zeta}_{n_p} = 1$; $\rho_{cu} = 7.4 \times 10^{-10} + 4.8 \times 10^{-11}B\ \Omega\,\mathrm{m}$ (B in tesla); $\beta = 800\,\mathrm{W\,T}^2/\mathrm{ms}$; and $n = 1.25$.

d) Interpret I_{sk} and explain why I_q is always greater than I_{sk}. Discuss whether Eq. 6.40 is valid for very slow ramp rates.

Solution to Problem 6.12

a) An equation of energy balance (per unit strand length) valid over the entire ramping duration of t_r is given by:

$$a_{st}e_h = a_{st}e_j - p_{st}e_q \qquad (S12.1)$$

where e_h and e_j, both per unit strand volume, are the thermal energy density absorbed by and the Joule heating density generated in the strand, respectively. e_q is the energy flux (per unit strand surface) released into helium and p_{st} is the strand perimeter in contact with helium. Because the conductor's temperature rises from its operating temperature, which is equal to the initial bath temperature (T_{b_o}), to the strand's current sharing temperature (T_{cs}), e_h is given by:

$$e_h = \int_{T_{b_o}}^{T_{cs}} C_{st}\, dT \qquad (S12.2)$$

The Joule energy density generated by a single mth pulse of t_d, Δe_j, is given by:

$$\Delta e_j^m = \frac{\rho_{cu}}{a_{cu}a_{st}} \int_{mt_p}^{mt_p+t_d} i_{st}^2(t)\, dt = \zeta_m(i_{st})\frac{\rho_{cu}i_q^2 t_d}{a_{cu}a_{st}} \qquad (S12.3)$$

where $i_{st}(t)$ is strand current, which is being ramped from 0 to i_q, the quench current. $\zeta_m(i_{st})$ is a current-dependent parameter to make the second equality valid. Because this Joule heating density occurs n_p ($= f_h t_r$) times over one ramping period, we have:

$$e_j = \sum_{m=0}^{n_p} \Delta e_j^m = \tilde{\zeta}_{n_p} \frac{\rho_{cu}i_q^2 t_d f_h t_r}{a_{cu}a_{st}} \qquad (S12.4)$$

$\tilde{\zeta}_{n_p}$ is an average parameter value from $m=1$ to $m=n_p$. The energy flux (per unit strand surface) released into the helium over one ramping period is given by:

$$e_q = h_q \int_0^{t_r} [T_{st}(t) - T_b(t)]\, dt \qquad (S12.5)$$

Combining Eqs. $S12.1$, $S12.2$, $S12.4$, and $S12.5$, we obtain:

$$a_{st} \int_{T_{b_o}}^{T_{cs}} C_{st}\, dT = \tilde{\zeta}_{n_p} \frac{\rho_{cu}i_q^2 t_d f_h t_r}{a_{cu}} - p_{st}h_q \int_0^{t_r} [T_{st}(t) - T_b(t)]\, dt \qquad (S12.6)$$

By dividing $S12.6$ by a_{st}, we have:

$$\int_{T_{b_o}}^{T_{cs}} C_{st}\, dT = \tilde{\zeta}_{n_p} \frac{\rho_{cu}i_q^2 t_d f_h t_r}{a_{cu}a_{st}} - \frac{p_{st}}{a_{st}} h_q \int_0^{t_r} [T_{st}(t) - T_b(t)]\, dt \qquad (6.34)$$

Solution to Problem 6.12

b) We may simplify the first term of the right-hand side of Eq. 6.35 by substituting T_c for $T_{st}(t)$ and T_{b_o} for $T_b(t)$. Thus:

$$\frac{p_{st}}{a_{st}} h_q \sum_{m=0}^{n_p} \left\{ \int_{mt_p}^{mt_p+t_d} [T_{st}(t) - T_b(t)] dt \right\} \simeq \frac{p_{st}}{a_{st}} h_q (T_c - T_{b_o}) \sum_{m=0}^{n_p} \left(\int_{mt_p}^{mt_p+t_d} dt \right)$$

$$= \frac{p_{st}}{a_{st}} h_q (T_c - T_{b_o}) f_h t_r t_d \qquad (S12.7)$$

Substituting Eq. $S12.7$ into Eq. 6.35, we simplify Eq. 6.35 to:

$$\frac{p_{st}}{a_{st}} h_q \int_0^{t_r} [T_{st}(t) - T_b(t)] \, dt = \frac{p_{st}}{a_{st}} h_q (T_c - T_{b_o}) f_h t_r t_d$$

$$+ \frac{p_{st}}{a_{st}} h_q \sum_{m=0}^{n_p} \left\{ \int_{mt_p+t_d}^{(m+1)t_p} [T_{st}(t) - T_b(t)] \, dt \right\} \qquad (S12.8)$$

Combining Eqs. 6.34 and $S12.8$, we obtain:

$$\int_{T_{b_o}}^{T_{cs}} C_{st} \, dT = \tilde{\zeta}_{n_p} \frac{\rho_{cu} i_q^2 t_d f_h t_r}{a_{cu} a_{st}} - \frac{p_{st}}{a_{st}} h_q (T_c - T_{b_o}) f_h t_r t_d$$

$$- \frac{p_{st}}{a_{st}} h_q \sum_{m=0}^{n_p} \left\{ \int_{mt_p+t_d}^{(m+1)t_p} [T_{st}(t) - T_b(t)] \, dt \right\} \qquad (S12.9)$$

Solving Eq. $S12.9$ for i_q^2, we obtain:

$$i_q^2 = \frac{a_{cu} p_{st} h_q (T_c - T_{b_o})}{\tilde{\zeta}_{n_p} \rho_{cu}} + \frac{a_{cu}}{\tilde{\zeta}_{n_p} \rho_{cu} f_h} \chi \qquad (6.36)$$

where

$$\chi = \left(\frac{a_{st}}{t_r t_d} \int_{T_{b_o}}^{T_{cs}} C_{st} \, dT + \frac{p_{st} h_q}{t_r t_d} \sum_{m=0}^{n_p} \left\{ \int_{mt_p+t_d}^{(m+1)t_p} [T_{st}(t) - T_b(t)] \, dt \right\} \right) \qquad (6.37)$$

c) By substituting $f_h = f_{cd}/N_{st} = \xi N_{st}^{n-1} B \dot{B}$ into Eq. 6.38, we have:

$$I_q^2 = I_{sk}^2 + \frac{N_{st}^2 a_{cu} \chi}{\xi \tilde{\zeta}_{n_p} \rho_{cu} N_{st}^{n-1} B \dot{B}}$$

$$= I_{sk}^2 + \frac{N_{st}^{3-n} a_{cu} \chi}{\xi \tilde{\zeta}_{n_p} \rho_{cu} B \dot{B}} \qquad (S12.10)$$

Solution to Problem 6.12

From Eq. $S12.10$, we have:

$$I_q = \sqrt{I_{sk}^2 + \frac{N_{st}^{3-n} a_{cu} \chi}{\xi \tilde{\zeta}_{n_p} \rho_{cu} B \dot{B}}} \qquad (6.40)$$

d) As remarked above, I_{sk} is the Stekly stability current with $\alpha_{sk} = 1$, which is valid when Joule heating is on *continuously*. The reason I_q is greater than I_{sk} is because Joule heating is on only for a fraction of the time. In fact, Eq. 6.40 will break down for very slow ramp rates. This may be seen more clearly by rewriting the heat transfer term in Eq. 6.34 in terms of heat absorbed by the helium. In this model, in which the helium is essentially stagnant, heat transferred to the helium must be absorbed by it. Thus, on a per unit conductor length basis, we have:

$$a_{st} \int_{T_{b_o}}^{T_{cs}} C_{st}\, dT = \tilde{\zeta}_{n_p} \frac{\rho_{cu} i_q^2 t_d f_h t_r}{a_{cu}} - a_{he} \int_{T_{b_o}}^{T_{cs}} C_{he}\, dT \qquad (S12.11)$$

where a_{he} is the equivalent helium cross sectional area for a single strand. In terms of the entire conductor (N_{st} strands), Eq. $S12.11$ may be written:

$$A_{cd} \int_{T_{b_o}}^{T_{cs}} C_{st}\, dT = \tilde{\zeta}_{n_p} \frac{\rho_{cu}}{A_{cu}} I_q^2 t_d f_h t_r - A_{he} \int_{T_{b_o}}^{T_{cs}} C_{he}\, dT \qquad (S12.12)$$

where A_{cd} is the total conductor cross sectional area; A_{cu} is the total copper cross sectional area; and A_{he} is the total cross sectional area of the helium. Because $C_{he} \gg C_{st}$, we can approximate $S12.12$ and solve for I_q^2:

$$I_q^2 = \frac{A_{cu} A_{he}}{\tilde{\zeta}_{n_p} \rho_{cu} t_d f_h t_r} \int_{T_{b_o}}^{T_{cs}} C_{he}\, dT \qquad (S12.13)$$

Equation $S12.13$ implies that I_q can be arbitrarily large by making t_r small. However, above a critical power density (per unit conductor length), power flow into the helium will be heat transfer-limited, requiring Eq. $S12.13$ to be modified as:

$$I_q^2 < \frac{P_{cd} A_{cu} \breve{q}_{mx}}{\rho_{cu}} < \frac{A_{cu} A_{he}}{\tilde{\zeta}_{n_p} \rho_{cu} t_d f_h t_r} \int_{T_{b_o}}^{T_{cs}} C_{he}\, dT \qquad (S12.14)$$

where $\breve{q}_{mx}$ is the maximum nucleate heat transfer flux under transient conditions. Generally, $\breve{q}_{mx}$ can be as large as 10 times the maximum for steady-state nucleate heat transfer flux. Accordingly, Eq. 6.40 should be modified to:

$$I_q = \sqrt{I_{sk}^2 + \frac{N_{st}^{3-n} a_{cu} \chi}{\xi \tilde{\zeta}_{n_p} \rho_{cu} B \dot{B}}} < \sqrt{\frac{P_{cd} A_{cu} \breve{q}_{mx}}{\rho_{cu}}} \qquad (S12.15)$$

Problem 6.13: MPZ for a composite tape conductor

This problem studies the MPZ in "one-dimensional" composite tape superconductors. Figure 6.20 shows the tape configuration (width w, superconductor
thickness d, matrix thickness D). The tape, with a critical current of I_{c_o} at
operating temperature T_{op}, is carrying transport current $I_t = I_{c_o}$. In the MPZ
region, $-x_{mz} \leq x \leq +x_{mz}$, the tape is fully normal. Assume that the MPZ zone
($|x| \leq x_{mz}$) is insulated from cooling and that one of the tape's broad surfaces is
exposed to cooling for $|x| \geq x_{mz}$. The coolant temperature is equal to the tape's
operating temperature and surface cooling flux acting on *one* side of the broad
(w) tape surface is given by $h_q(T - T_{op})$, where h_q is a heat transfer coefficient,
assumed constant in this analysis. Because of symmetry about the origin ($x = 0$),
we shall consider only the $+x$ region. The steady-state power density equations
for region 1 ($0 \leq x \leq x_{mz}$) and region 2 ($x \geq x_{mz}$) are given by:

$$\text{(Region 1)} \quad k_m \frac{d^2 T_1}{dx^2} + \rho_m J_m^2 = 0 \tag{6.41a}$$

$$\text{(Region 2)} \quad k_m \frac{d^2 T_2}{dx^2} - \frac{w}{A_m} h_q (T_2 - T_{op}) = 0 \tag{6.41b}$$

where k_m and ρ_m are, respectively, the matrix thermal conductivity and electrical
resistivity. J_m is current density in the matrix: $J_m = I_t/wD = I_{c_o}/wD$.

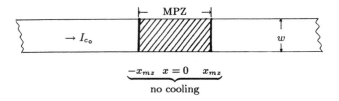

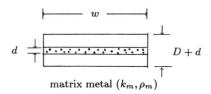

Fig. 6.20 One-dimensional MPZ in a composite superconducting tape.

Problem 6.13: MPZ for a composite tape conductor

a) Solve Eq. 6.41 and show that the temperature distribution, $T_1(x)$, within the $+x$ half of the MPZ (region 1) and the temperature distribution, $T_2(x)$, beyond $+x_{mz}$ (region 2) are given by:

$$T_1(x) = T_c + \frac{\rho_m J_m^2}{2k_m}(x_{mz}^2 - x^2) \qquad (6.42a)$$

$$T_2(x) = T_{op} + (T_c - T_{op})\exp[\alpha_{mz}(x_{mz} - x)] \qquad (6.42b)$$

where T_c is the tape's critical temperature and $\alpha_{mz} = \sqrt{wh_q/(k_m A_m)}$. In deriving Eqs. 6.42a and 6.42b, assume that heat is conducted in the tape only by the matrix metal because $k_m \gg k_s$ and that Joule heating is generated only in the matrix because $\rho_m \ll \rho_s$ (normal state). Assume also that k_m and ρ_m are temperature-independent.

b) Show that an expression for x_{mz} is given by:

$$x_{mz} = \left(\frac{D}{d}\right)^2 \frac{(T_c - T_{op})}{\rho_m [J_{co}]_s^2} \sqrt{\frac{h_q k_m}{D}} \qquad (6.43)$$

where $[J_{co}]_s$ is the critical current density in the superconductor at T_{op}.

c) Initially, the entire tape is at T_{op} and superconducting and it is insulated over the region $-x_{mz} \le x \le x_{mz}$. Compute the energy margin, ΔE_h [J], required to create this MPZ for the parameters given in Table 6.6.

d) Compute T_{max} within the MPZ for the parameter values of Table 6.6.

Table 6.6: Tape Parameters

T_c	[K]	12
T_{op}	[K]	10
ρ_m	[nΩ m]	0.2
k_m	[W/m K]	1200
h_q	[W/m^2 K]	250
w	[mm]	5
D	[μm]	50
d	[μm]	8
$[J_{co}]_s$	[GA/m^2]	2

"What you need at the outset is a high degree of uncertainty; otherwise it isn't likely to be an important problem." —Lewis Thomas

Solution to Problem 6.13

a) Let us solve Eq. 6.41a first.

$$\frac{d^2T_1}{dx^2} = -\frac{\rho_m J_m^2}{k_m} \implies \frac{dT_1}{dx} = -\frac{\rho_m J_m^2}{k_m}x + A \qquad (S13.1)$$

From symmetry, $dT_1/dx = 0$ at $x = 0$, thus $A = 0$. Integrating Eq. $S13.1$, we have:

$$T_1(x) = -\frac{\rho_m J_m^2}{2k_m}x^2 + B \qquad (S13.2)$$

At $x = x_{mz}$, $T_1 = T_c$. Thus:

$$T_c = -\frac{\rho_m J_m^2}{2k_m}x_{mz}^2 + B \implies B = T_c + \frac{\rho_m J_m^2}{2k_m}x_{mz}^2$$

$$T_1(x) = T_c + \frac{\rho_m J_m^2}{2k_m}(x_{mz}^2 - x^2) \qquad (6.42a)$$

Now, let us solve Eq. 6.41b with $\theta = T_2 - T_{op}$:

$$\frac{d^2\theta}{dx^2} - \frac{wh_q}{k_m A_m}\theta = 0 \implies \theta = \theta_0 \exp(-\alpha_{mz}x) \text{ where } \alpha_{mz} = \sqrt{\frac{wh_q}{k_m A_m}}$$

$$T_2(x) = T_{op} + \theta_0 \exp(-\alpha_{mz}x) \qquad (S13.3)$$

At $x = x_{mz}$, we have $T_2 = T_c$. Thus:

$$T_c = T_{op} + \theta_0 \exp(-\alpha_{mz}x_{mz}) \text{ where } \theta_0 \exp(-\alpha_{mz}x_{mz}) = (T_c - T_{op})$$

We thus have:

$$T_2(x) = T_{op} + (T_c - T_{op})\exp[\alpha_{mz}(x_{mz} - x)] \qquad (6.42b)$$

b) At $x = x_{mz}$, $dT_1/dx = dT_2/dx$, thus:

$$\left.\frac{dT_1}{dx}\right|_{x_{mz}} = -\frac{\rho_m J_m^2}{k_m}x_{mz} \qquad (S13.4)$$

$$\left.\frac{dT_2}{dx}\right|_{x_{mz}} = (T_c - T_{op})\alpha_{mz} = -(T_c - T_{op})\sqrt{\frac{wh_q}{k_m A_m}} \qquad (S13.5)$$

Thus,

$$-\frac{\rho_m J_m^2 x_{mz}}{k_m} = -(T_c - T_{op})\sqrt{\frac{wh_q}{k_m A_m}} \implies x_{mz} = \frac{(T_c - T_{op})k_m}{\rho_m J_m^2}\sqrt{\frac{wh_q}{k_m A_m}}$$

We have $A_m = wD$ and $J_m Dw = [J_{c_o}]_s dw$ or $J_m = [J_{c_o}]_s d/D$. Thus:

$$x_{mz} = \left(\frac{D}{d}\right)^2 \frac{(T_c - T_{op})}{\rho_m [J_{c_o}]_s^2}\sqrt{\frac{h_q k_m}{D}} \qquad (6.43)$$

c) Wilson argues that ΔE_h should be given by the energy needed to create $T_1(x)$ (Eq. 6.42a) within the MPZ [6.24]. Accordingly,

$$\Delta E_h = 2(D + d)w \int_o^{x_{mz}} \left[\int_{T_{op}}^{T_1(x)} C_{cd}(T)\, dT\right] dx \qquad (S13.6)$$

where C_{cd} is the conductor's volumetric heat capacity.

Solution to Problem 6.13

Another plausible scenario, as discussed in Problem 6.8, is that it is only necessary to raise the conductor's temperature to T_{cs} throughout the MPZ, because once this temperature is reached, Joule heating becomes active and the final temperature distribution is automatically established.

Let us first compute $2x_{mz}$ (Eq. 6.43) for the parameters given:

$$2x_{mz} = 2\left(\frac{D}{d}\right)^2 \frac{(T_c - T_{op})}{\rho_m[J_{co}]_s^2} \sqrt{\frac{h_q k_m}{D}} \tag{S13.7}$$

$$= 2\left(\frac{50\times10^{-6}\,\mathrm{m}}{8\times10^{-6}\,\mathrm{m}}\right)^2 \frac{(12\,\mathrm{K} - 10\,\mathrm{K})}{(2\times10^{-10}\,\Omega\,\mathrm{m})(2\times10^9\,\mathrm{A/m^2})^2}$$

$$\times \sqrt{\frac{(250\,\mathrm{W/m^2\,K})(1200\,\mathrm{W/m\,K})}{50\times10^{-6}\,\mathrm{m}}}$$

$$= 0.015\,\mathrm{m}\ (15\,\mathrm{mm})$$

The energy margin, ΔE_h, will be given by the change in enthalpy between $10\,\mathrm{K}$ and $12\,\mathrm{K}$ for the entire volume whose length is $2x_{mz}$. Thus:

$$\Delta E_h = 2x_{mz}(D+d)w \int_{10\,\mathrm{K}}^{12\,\mathrm{K}} C_{cd}(T)\,dT \tag{S13.8}$$

where $C_{cd}(T)$ is the heat capacity of the conductor, which may be roughly given by that of copper. For copper, we have:

$$\int_{10\,\mathrm{K}}^{12\,\mathrm{K}} C_{cu}(T)\,dT = h_{cu}(12\,\mathrm{K}) - h_{cu}(10\,\mathrm{K}) \tag{S13.9}$$

where $h_{cu}(12\,\mathrm{K})$ is the enthalpy $[\mathrm{J/m^3}]$ at $12\,\mathrm{K}$ and $h_{cu}(10\,\mathrm{K})$ is the enthalpy at $10\,\mathrm{K}$. With $h(12\,\mathrm{K}) = 42\,\mathrm{kJ/m^3}$ and $h(10\,\mathrm{K}) = 21\,\mathrm{kJ/m^3}$, Eq. $S13.8$ becomes:

$$\Delta e_h = (0.015\,\mathrm{m})(58\times10^{-6})(6.5\times10^{-3}\,\mathrm{m})(21\,\mathrm{kJ/m^3}) \tag{S13.10}$$

$$\simeq 120\,\mu\mathrm{J}$$

d)   Once the MPZ region is driven to T_c with an energy input of $\sim100\,\mu\mathrm{J}$, Joule heating will take over to further heat up the tape to create the temperature distribution given by $T_1(x)$. T_{max} occurs naturally at $x = 0$ and is given by:

$$T_{max} = T_1(x=0) = T_c + \frac{\rho_m J_m^2}{2k_m}x_{mz}^2 = T_c + \frac{D(T_c - T_{op})^2 h_q}{2d^2\rho_m[J_{co}]_s^2} \tag{S13.11}$$

Thus:

$$T_{max} = (12\,\mathrm{K}) + \frac{(50\times10^{-6}\,\mathrm{m})(12\,\mathrm{K} - 10\,\mathrm{K})^2(250\,\mathrm{W/m^2\,K})}{2(8\times10^{-6}\,\mathrm{m})^2(2\times10^{-10}\,\mathrm{m})(2\times10^9\,\mathrm{A/m^2})^2} \tag{S13.12}$$

$$= 12\,\mathrm{K} + 0.49\,\mathrm{K} \simeq 12.5\,\mathrm{K}$$

Problem 6.14: Stability of HTS magnets

This problem discusses the stability of high-T_c superconductors. In Fig. 1.2 of Chapter 1 it is indicated that the stability issue for HTS magnets should be easier than that for LTS magnets. Here we shall study the stability of HTS magnets in more detail and derive some quantitative results.

a) For an HTS magnet cooled in a bath of cryogen boiling at T_s and satisfying the Stekly criterion, show that the improvement in $[J_{c_o}]_{cd}$ from an LTS magnet cooled in a bath of liquid helium boiling at 4.2 K and satisfying the Stekly criterion is noticeable for liquid neon but is negligible for liquid nitrogen. Specifically, show:

$$\frac{[J_{c_o}]_{cd}|_{\text{HTS27}}}{[J_{c_o}]_{cd}|_{\text{LTS4.2}}} \equiv [\xi]^{27}_{[J_{c_o}]_{cd}} \sim 3 \qquad (6.44a)$$

$$\frac{[J_{c_o}]_{cd}|_{\text{HTS77}}}{[J_{c_o}]_{cd}|_{\text{LTS4.2}}} \equiv [\xi]^{77}_{[J_{c_o}]_{cd}} \sim 1 \qquad (6.44b)$$

where $[J_{c_o}]_{cd}|_{\text{HTS27}}$ is the current density (over conductor cross sectional area) for an HTS operating in a bath of liquid neon and satisfying the Stekly criterion, *i.e.* Eq. 6.12b; $[J_{c_o}]_{cd}|_{\text{LTS4.2}}$ is the overall current density for an LTS operating in a bath of liquid helium and satisfying the Stekly criterion; $[J_{c_o}]_{cd}|_{\text{HTS77}}$ is similarly defined. Assume the LTS magnet is wound with a superconductor-copper (RRR=100) composite and the HTS magnet is wound with a superconductor-silver (99.99% purity) composite. The approximate boiling heat transfer parameters given in Table 4.2 are adequate for the purpose of deriving Eq. 6.44. You may also use electrical resistivity data given in Fig. A4.1 (Appendix IV).

b) Now consider two adiabatic magnets—an LTS magnet operating at 4 K and an HTS magnet operating either at 27 K or 77 K—and show that the HTS magnet either at 27 K or 77 K has an MPZ energy margin ΔE_h much greater than that for the LTS magnet. Specifically show that:

$$\frac{\Delta E_h|_{\text{HTS27}}}{\Delta E_h|_{\text{LTS4.2}}} \equiv [\xi]^{27}_{\Delta E_h} \sim 10^6 \qquad (6.45a)$$

$$\frac{\Delta E_h|_{\text{HTS77}}}{\Delta E_h|_{\text{LTS4.2}}} \equiv [\xi]^{77}_{\Delta E_h} \sim 10^6 \qquad (6.45b)$$

$\Delta E_h|_{\text{HTS27}}$ and $\Delta E_h|_{\text{HTS77}}$ are MPZ energy margins for the HTS magnet operating, respectively, at 27 K and 77 K; similarly $\Delta E_h|_{\text{LTS4.2}}$ is the MPZ energy margin for the LTS magnet operating at 4 K.

In deriving Eqs. 6.44 and 6.45, as indicated in both equations, use a simple notation $[\xi]^{27}_P$ defined by:

$$[\xi]^{27}_P \equiv \frac{P|_{\text{HTS27}}}{P|_{\text{LTS4.2}}} \qquad (6.46)$$

where P can be any parameter, *e.g.* ρ_m, at 4.2 K or 27 K.

c) Discuss results of a) and b).

Solution to Problem 6.14

Table 6.7: Parameter Ratios for 27-K and 77-K Operations

$\xi^{27}_{\rho_m}$	~ 1	$\xi^{77}_{\rho_m}$	~ 10
$\xi^{27}_{q_{fm}}$	~ 10	$\xi^{77}_{q_{fm}}$	~ 10
$\xi^{27}_{k_m}$	~ 5	$\xi^{77}_{k_m}$	~ 1
$\xi^{27}_{k_i}$	~ 2	$\xi^{77}_{k_i}$	~ 5
$\xi^{27}_{C_m(T_{op})}$	~ 3000	$\xi^{77}_{C_m(T_{op})}$	~ 6000
$\xi^{27}_{\Delta T_{op}}$	~ 5	$\xi^{77}_{\Delta T_{op}}$	~ 10

Table 6.7 presents appropriate parameter ratios for operation of an LTS magnet at 4.2 K to that of an HTS magnet either at 27 K or 77 K. Ratio values are computed from data given in Table 4.2 (Chapter 4), Figs. A3.1 and A3.2 (Appendix III), and Fig. A4.1 (Appendix IV).

a) From Eq. 6.12b of Problem 6.4 we have an expression for $[J_{c_o}]_{cd}$ for cryostable operation valid for $A_m \gg A_s$:

$$[J_{c_o}]_{cd} \simeq \sqrt{\frac{f_p P_{cd} q_{fm}}{\rho_m A_m}} \qquad (A_m \gg A_s) \qquad (6.12b)$$

In the above equation, f_p is the fraction of conductor perimeter P_{cd} directly exposed to cryogen, q_{fm} is the minimum film boiling heat transfer flux, ρ_m is the matrix metal resistivity, and A_m is the matrix cross section.

Assuming LTS and HTS conductors have the same shape and dimensions, we compute the ratio of J_{cd} for the two magnets:

$$[\xi]_{[J_{c_o}]_{cd}} = \sqrt{\frac{1}{[\xi]_{\rho_m}}[\xi]_{q_{fm}}} \qquad (S14.1)$$

Inserting appropriate values given in Table 6.7 into Eq. $S14.1$, we obtain:

$$[\xi]^{27}_{[J_{c_o}]_{cd}} \sim \sqrt{\left(\frac{1}{1}\right)\left(\frac{10}{1}\right)} \sim 3 \qquad (6.44a)$$

$$[\xi]^{77}_{[J_{c_o}]_{cd}} \sim \sqrt{\left(\frac{1}{10}\right)\left(\frac{10}{1}\right)} \sim 1 \qquad (6.44b)$$

b) As we have studied in Problem 6.8, ΔE_h is the product of $\mathcal{V}_{mz}$ (MPZ volume) and energy density between T_{op} and T_c. $\mathcal{V}_{mz}$ is given by:

$$\mathcal{V}_{mz} \propto R_{mz1} R_{mz2}^2 \qquad (S14.2a)$$

$$\propto \sqrt{\frac{3 k_m (T_c - T_{op})}{\rho_m J_m^2}} \times \frac{3 k_i (T_c - T_{op})}{\rho_m J_m^2} \qquad (S14.2b)$$

Solution to Problem 6.14

By designating $\Delta T_{op} = T_c - T_{op}$, we shall first take the ratio of $\mathcal{V}_{mz}$ for the two magnets:

$$[\xi]_{\mathcal{V}_{mz}} = \sqrt{[\xi]_{k_m}} \times [\xi]_{k_i} \times \left(\frac{1}{[\xi]_{\rho_m}}\right)^{1.5} \times ([\xi]_{\Delta T_{op}})^{1.5} \qquad (S14.3)$$

In Eq. $S14.3$, J_m is assumed to be the same for both magnets.

The energy margin density Δe_h is given by:

$$\Delta e_h = \int_{T_{op}}^{T_c} C_m \, dT \sim C_m(T_{op}) \Delta T_{op} \qquad (S14.4)$$

Combining Eqs. $S14.3$ and $S14.4$, we obtain:

$$[\xi]_{\Delta E_h} = [\xi]_{\mathcal{V}_{mz}} \times [\xi]_{\Delta e_h}$$

$$= \sqrt{[\xi]_{k_m}} \times [\xi]_{k_i} \times \left(\frac{1}{[\xi]_{\rho_m}}\right)^{1.5} \times ([\xi]_{\Delta T_{op}})^{2.5} \times [\xi]_{C_m(T_{op})} \qquad (S14.5)$$

Inserting appropriate values from Table 6.7 into Eq. $S14.5$, we obtain:

$$[\xi]^{27}_{\Delta E_h} \simeq \sqrt{5} \times 2 \times \left(\frac{1}{1}\right)^{1.5} \times (5)^{2.5} \times 3000 \simeq 10^6 \qquad (6.45a)$$

$$[\xi]^{77}_{\Delta E_h} \simeq \sqrt{1} \times 5 \times \left(\frac{1}{10}\right)^{1.5} \times (10)^{2.5} \times 6000 \simeq 10^6 \qquad (6.45b)$$

c) Equation 6.44 indicates that an anticipated improvement in $[J_{c_o}]_{cd}$ of HTS over LTS for *cryostable magnets* is marginal: a factor of ~ 3 when an HTS magnet is operated in liquid neon and hardly any improvement when it is operated in liquid nitrogen. Operation in liquid hydrogen will be about the same with that in liquid neon. An implication is that both LTS and HTS cryostable magnets will have about the same size and weight.

In Problem 6.8 we found $\Delta E_h \sim 10\,\mu J$ for an LTS magnet. A factor of $\sim 10^6$ improvement in MPZ energy makes ΔE_h for the HTS magnet in the range $\sim 10\,J$, whether it is operated at 27 K or 77 K. Mechanical properties of magnets are essentially temperature-independent, making mechanical integrity of magnets also temperature-independent, as briefly discussed in Chapter 1. This implies that the size of mechanically-induced disturbances—the most troublesome disturbances afflicting adiabatic LTS magnets, as will be discussed more fully in Chapter 7—will also be temperature-independent and a six-fold increase in ΔE_h thus makes adiabatic HTS magnets almost absolutely stable against mechanical disturbances.

We can draw an important conclusion by comparing the stability performances of LTS and HTS with different cooling methods. There is no significant improvement in magnet size or weight for cryostable magnets. Adiabatic HTS magnets, however, are phenomenally more stable against disturbances than their LTS counterparts. Hence, *all* HTS magnets should be operated adiabatically at $\sim 20\,K$ and above; even at 20 K, $[\xi]^{20}_{\Delta E_h}$ is $\sim 2 \times 10^5$, still a comfortably high value.

References

[6.1] L.J. Donadieu and D.J. Rose, "Conception and design of large volume superconducting solenoid," *High Magnetic Fields*, Eds. H. Kolm, B. Lax, F. Bitter, and R. Mills (MIT Press and John Wiley & Sons, New York, 1962), 358.

[6.2] A.R. Kantrowitz and Z.J.J. Stekly, "A new principle for the construction of stabilized superconducting coils," *Appl. Phys. Lett.* **6**, 56 (1965).

Also see Z.J.J. Stekly, R. Thome, and B. Strauss, "Principles of stability in cooled superconducting magnets," *J. Appl. Phys.* **40**, 2238 (1969).

[6.3] Martin N. Wilson, *Superconducting Magnets* (Clarendon Press, Oxford, 1983).

[6.4] H.R. Hart, Jr., "Magnetic instabilities and solenoid performance: Applications of the critical state model," *Proc. 1968 Summer Study on Superconducting Devices and Accelerators*, (Brookhaven National Laboratory, Upton, NY, 1969), 571.

[6.5] T. Ogasawara, "Conductor design issues for oxide superconductors. Part 2: exemplification of stable conductors," *Cryogenics* **29**, 6 (1989).

[6.6] B.J. Maddock, G.B. James, and W.T. Norris, "Superconductive composites: heat transfer and steady state stabilization," *Cryogenics* **9**, 261 (1969).

[6.7] A.P. Martinelli and S.L. Wipf, "Investigation of cryogenic stability and reliability of operation of Nb_3Sn coils in helium gas environment," *Proc. Appl. Superconduc. Conf.* (IEEE Pub. 72CHO682-5-TABSC, 1977), 3311.

[6.8] M.N. Wilson, "Stabilization of superconductors for use in magnets," *IEEE Trans. Magn.* **MAG-13**, 440 (1977).

[6.9] Mitchell O. Hoenig and D. Bruce Montgomery, "Dense supercritical-helium cooled superconductors for large high field stabilized magnets," *IEEE Trans. Magn.* **MAG-11**, 569 (1975).

[6.10] Y. Iwasa, M.O. Hoenig, and D.B. Montgomery, "Cryostability of a small superconducting coil wound with cabled hollow conductor," *IEEE Trans. Magn.* **MAG-13**, 678 (1977).

[6.11] J.W. Lue, J.R. Miller, and L. Dresner, "Stability of cable-in-conduit superconductors," *J. Appl. Phys.* **51**, 772 (1980).

[6.12] L. Bottura, N. Mitchell, and J.V. Minervini, "Design criteria for stability in cable-in-conduit conductors," *Cryogenics* **31**, 510 (1991).

[6.13] See, for example, M.A. Hilal, C. Marinucci, P. Weymuth, G. Vecey, "Transient stability of forced flow cooled conductors," *Proc. 8th Sympo. Engr. Problems of Fusion Research* (IEEE Pub. NO. 79CH1441-5 NPS, 1979), 1774.

[6.14] See, for example, S.R. Shanfield, K. Agatsuma, A.G. Montgomery, and H.O. Hoenig, "Transient cooling in internally cooled, cabled superconductors (CI-CCS)," *IEEE Trans. Magn.* **MAG-17**, 2019 (1981).

[6.15] P.N. Haubenreich, J.N. Luton, and P.B. Thompson, "The role of the Large Coil Program in the development of superconducting magnets for fusion reactors," *IEEE Trans. Magn.* **MAG-15**, 520 (1979).

[6.16] H. Tsuji, E. Tada, K. Okuno, T. Ando, T. Hiyama, Y. Takahashi, M. Nishi, K. Yoshida, Y. Ohkawa, K. Koizumi, H. Nakajima, T. Kato, K. Kawano, T. Isono, M. Sugimoto, H. Yamamura, M. Satoh, M. Hasegawa, J. Yoshida, E. Kawagoe, Y. Kamiyauchi, M. Konno, H. Ishida, N. Itoh, M. Oshikiri, Y. Kurosawa, H. Nisugi, Y. Matsuzaki, H. Shirakata, and S. Shimamoto, "Evolution of the Demo Poloidal Coil program," *Proc. 11th Int'l Conf. Magnet Tech. (MT-11)* (Elsevier Applied Science, London, 1990), 806.

[6.17] See, for example, A.Vl. Gurevich and R.G. Mints, "Self-heating in normal metals and superconductors," *Rev. Mod. Phys.* **59**, 941 (1987).

T. Ito and H. Kubota, "On the classical theories of stability of pool-cooled superconductors," *Fusion Engineering and Design* **20**, 319 (1993).

[6.18] M.N. Wilson, "Heat transfer to boiling liquid helium in narrow vertical channels," *Proc. Int'l Inst. Refrig. Commission I* (Pergamon Press, Oxford, 1966), 109.

[6.19] J.R. Purcell, "Superconducting magnet for the ANL 12-ft hydrogen bubble chamber (Abstract)," *J. Appl. Phys.* **39**, 2622 (1968).

J.R. Purcell and Henri Desportes, "The NAL bubble chamber magnet," *Proc. 1972 Appl. Superconduc. Conf.* (IEEE Pub. No. 72CH0682-5-TABSC, 1972), 246.

[6.20] A.G. Prodell, "Eight-foot-diameter 30-kG superconducting magnet for the Brookhaven National Laboratory (Abstract)," *J. Appl. Phys.* **40**, 2109 (1969).

[6.21] A.D. Appleton and R.B. MacNab, "Performance of a large superconducting motor," *Proc. 3rd Int'l Conf. Cryo. Eng. Conf.* (Iliffe Science and Technology Publications, Surrey, 1970), 443.

[6.22] J. Wong, D.F. Fairbanks, R.N. Randall, and W.L. Larson, "Fully stabilized superconducting strip for the Argonne and Brookhaven bubble chambers," *J. Appl. Phys.* **39**, 2518 (1968).

[6.23] G. Bogner, C. Albrecht, R. Maier, and P. Parsch, "Experiments on copper- and aluminum-stabilized Nb-Ti superconductors in view of their application in large magnets," *Proc. 2nd Int'l Cryo. Eng. Conf.* (Iliffe Science and Technology Publication, Surrey, 1968), 175.

[6.24] M.N. Wilson and Y. Iwasa, "Stability of superconductors against localized disturbances of limited magnitude," *Cryogenics* **18**, 17 (1978).

[6.25] Michael J. Superczynski, "Heat pulses required to quench a potted superconducting magnet," *IEEE Trans. Magn.* **MAG-15**, 325 (1979).

[6.26] C.A. Scott, "Minimum heat pulse to quench a superconducting magnet," *Stability of superconductors*, (International Institute of Refrigeration, Commission A 1/2 Saclay, France, 1981), 189.

[6.27] Y. Iwasa, "Stability issues in high-performance superconducting magnets," *Cryogenics* **31**, 575 (1991).

[6.28] L. Bottura, "Stability, protection and ac loss of cable-in-conduit conductors – a designer's approach," *Fusion Eng. and Design* **20**, 351 (1993).

[6.29] M.M. Steeves, M. Takayasu, T.A. Painter, M.O. Hoenig, T. Kato, K. Okuno, H. Nakaji, and H. Tsuji, "Test results from the Nb3Sn US-Demonstration poloidal coil," *Adv. Cryo. Eng.* **37A**, 345 (1991).

[6.30] T.A. Painter, M.M. Steeves, M. Takayasu, C. Gung, M.O. Hoenig, H. Tsuji, T. Ando, T. Hiyama, Y. Takahashi, M. Nishi, K. Yoshida, K. Okuno, H. Nakajima, T. Kato, M. Sugimoto, T. Isono, K. Kawano, N. Koizumi, M. Oshikiri, H. Hanawa, H. Ouchi, M. Ono, H. Ishida, H. Hiue, J. Yoshida, Y. Kamiyauchi, T. Ouchi, F. Tajiri, Y. Kon, H. Shimizu, Y. Matsuzaki, S. Oomori, T. Tani, K. Oomori, T. Terakado, J. Yagyu, H. Oomori, *Test Data From The US-Demonstration Poloidal Coil Experiment*, (Plasma Fusion Center Report PFC/RR-92-1, MIT, Cambridge MA, January 1992).

[6.31] M. Takayasu, M.A. Ferri, C.Y. Gung, T.A. Painter, M.M. Steves and J.V. Minervini, "Measurements of ramp-rate limitation of cable-in-conduit conductors," *IEEE Trans. Applied Superconduc.* **3**, 456 (1993).

CHAPTER 7
AC, SPLICE, AND MECHANICAL LOSSES

7.1 Introduction

Although the perfect conductivity of superconductors is what has made super-conductivity an enduring fascination for scientists, engineers, and entrepreneurs, Type II superconductors suitable for magnets, as seen in Chapter 5, are magneti-cally hysteretic. That is, these superconductors are intrinsically dissipative when subjected to a time-varying magnetic field, transport current, or a combination of both. Furthermore, when a hard superconductor is processed into a compos-ite conductor in the form of multifilaments embedded in a normal metal matrix, other magnetic losses beside hysteresis come into play. These magnetic losses are commonly known as AC losses. In addition, when a multifilamentary conductor is wound into a magnet, the magnet is subjected to other dissipations. The sources of these dissipations include: 1) conductor splices (Joule heating); 2) Lorentz-force induced conductor (and even winding) motion, which results in frictional heating; and 3) Lorentz-force induced cracking in the winding impregnants, which also results in dissipation. Superconducting magnets in fusion reactors will also be subjected to an additional source of heating: neutron radiation. This radiation heating, however, is not discussed in this book.

The dissipation power density expressed by g_d in Eq. 6.1 of Chapter 6 lumps all these mostly non-Joule heating dissipation densities. Its size, when compared with the Joule dissipation density (g_j in Eq. 6.1) is minuscule. Despite its relatively small magnitude, it is critical in adiabatic superconducting magnets because the steady-state dissipation base line for superconducting magnets is, *by definition*, zero or nearly so. Compared with this zero base line, the dissipation density base line for water-cooled Bitter magnets is as great as tens of GW/m³. Because of this nearly infinite difference in the dissipation density base line, any nonzero dissipation, irrespective of its size, can be devastating in superconducting magnets. In Bitter magnets, on the other hand, dissipations other than Joule heating are completely negligible; they are indeed neglected in designing these magnets.

As mentioned above, the largest form of dissipation in superconducting magnets is Joule heating (other than that of splices), which actually occurs only *after* the conductor becomes nonsuperconducting; its density can reach tens of MW/m³— still only about 1/1000th of a Bitter magnet's. Because Joule heating density in the normal-state composite superconductor is by far the largest dissipation that can appear in the winding, as we have studied in Chapter 6, successful solutions to deal with Joule heating have evolved over the years. Indeed, even before any other losses were identified, let alone understood in superconducting magnets, a quantitative criterion—cryostability—was already available to build superconducting magnets that operated stably.

This chapter studies the disturbance term g_d, specifically of magnetic (AC), elec-trical (splice), and mechanical (frictional, epoxy cracking) origin. Among these losses, the most intractable are mechanical. Although considerable progress in understanding and dealing with these losses was achieved in the 1980s, it is still difficult to accurately quantify their energies and locations in the winding. A unique diagnostic technique based on acoustic emission (AE) has been developed and successfully applied to monitor high-performance magnets that are afflicted by mechanical disturbances, which are primary causes of premature quenches for these magnets. A brief description of the AE-based diagnostic technique is presented in this section.

7.2 AC Losses

There are three distinguishable AC loss energy densities in multifilamentary composites: 1) hysteresis, e_{hy}; 2) coupling, e_{cp}; and 3) eddy-current, e_{ed}. These losses are briefly discussed here. Each loss is treated independently in computing AC losses in multifilamentary composites.

7.2.1 Hysteresis Loss

As studied in Chapter 5, hysteresis loss occurs within superconducting filaments because each filament is made of hard superconductor which can sustain a nonzero electric field. The nonzero electric field is sustainable within the filament because the Type II superconductor really consists of finely divided nonsuperconducting regions in a sea of superconductivity. Fortunately, for the scope of this book, the hysteresis energy loss density per cycle of field excitation, e_{hy}, is derivable in closed analytical form for our standard Bean slab. Although we shall defer the derivation to the Problem Section (Problems 7.1~7.4), useful formulas for e_{hy} applicable to four common field excitations—sinusoidal, exponential decay, triangular, and trapezoid—are summarized in Table 7.1 (Eqs. 7.1~7.6). As is evident from the table, each of the four field excitations is characterized by two parameters: amplitude (B_m) and time constant (τ_m).

7.2.2 Coupling Loss

Coupling loss is actually another form of Joule heating generated within the matrix metal by inter-filament currents that are induced under a time-varying magnetic field (or transport current) excitation. Because of the complex geometry of a multifilamentary conductor, with twisted filaments as one source of complications, a straightforward analytical approach based on the Bean slab model similar to that used to derive closed-form formulas for hysteresis loss cannot be applied. More rigorous derivations are beyond the scope of this book; the reader is referred to the works of Ries and Brechna [7.1], Soubeyrand and Turck [7.2], Ogasawara [7.3], Hlásnik [7.4], and Wilson [6.3].

The key parameter used to compute e_{cp}—coupling energy loss density per cycle of field excitation—is the coupling time constant, τ_{cp}. It defines a decay time constant of inter-filament (coupled) currents induced in the conductor. In reality, these currents decay in a much more complicated way, governed by a diffusion equation whose solution is in the form of an infinite series containing an infinite number of time constants that for higher terms diminish rapidly. Experimentally, only the dominant term can be determined and it is used in most of the "phenomenological" approaches cited above. τ_{cp} is given by:

$$\tau_{cp} = \frac{\mu_o \ell_p^2}{8\pi^2 \rho_{ef}} \tag{7.7}$$

where ℓ_p is the twist pitch length of filaments and ρ_{ef} is the effective matrix resistivity for inter-filament currents. Using the Bean slab model, we shall derive an expression nearly identical to Eq. 7.7 later in the Problem Section.

Because the longer the coupled currents last, the greater the energy dissipated, the greater τ_{cp} is, the greater will be e_{cp}. As Wilson points out [6.3], e_{cp} may be viewed as a *fraction* of the total magnetic field (B_m) energy density stored in the composite, $B_m^2/2\mu_o$, and as $\tau_{cp} \to 0$, $e_{cp} \to 0$.

Table 7.1 also gives formulas for e_{cp} (Eqs. 7.8~7.11) for the same four common "slow" field excitations. Although none of Eqs. 7.8~7.11 (all from Wilson's text book [6.3]) is derived in the Problem Section, some of them will be used to solve practical problems.

7.2.3 Eddy-Current Loss

The eddy-current loss energy density, e_{ed}, has already been discussed in Chapter 2; its formulas, for the same field excitations, are also summarized in Table 7.1 (Eqs. 7.12~14); they, too, correspond to slow (quasi-static) excitations.

7.2.4 AC Power Densities

Although we have so far presented AC dissipation in the form of *energy* densities, under certain conditions it is more appropriate to express AC dissipation in the form of *power* densities. By examining the formulas summarized in Table 7.1, we may conclude that dependences of p_{hy}, p_{cp}, and p_{ed}, respectively, hysteresis, coupling, and eddy-current power loss densities, on time rate of change of external field, $\dot{B}_e$, are given by:

$$p_{hy} \propto \dot{B}_e \tag{7.15}$$

$$p_{cp} \propto \dot{B}_e^2 \tag{7.16}$$

$$p_{ed} \propto \dot{B}_e^2 \tag{7.17}$$

These equations suggest that during a long charging sequence, such as when the superconducting magnet of Hybrid III is energized at a relatively slow rate ($\sim$5 mT/s), the hysteresis loss, rather than the coupling and eddy-current losses, dominates AC dissipation in the magnet. When the magnet is discharged at a fast rate ($\sim$500 mT/s) on the other hand, which happens when the magnet is "dumped" in an emergency, the coupling and eddy-current losses, both proportional to $\dot{B}_e^2$, dominate. (Problem 7.7 studies AC losses for these two cases.)

Which is more important: energy loss or power loss? It depends on the cooling provided to the magnet. Obviously, it is desirable to maintain isothermal operation even in the presence of these losses and this can be achieved only if the cooling rate matches the dissipation rate.

Table 7.1: AC Energy Loss Density Formulas for Selected Field Excitations

Excitation Field, $B_e(t)$	
a) Sinusoidal $$B_e(t) = B_m \sin(2\pi t/\tau_m)$$	b) Exponential decay $$B_e(t) = B_m \exp(-t/\tau_m)$$
c) Triangular B_m, 0, $2\tau_m$	d) Trapezoid B_m, 0, τ_m, τ_m

Hysteresis Loss Energy Densities Per Cycle: e_{hy} [J/m³]

Case 1: $B_m \le B_p = \mu_0 J_c a$

a) $\dfrac{2B_m^3}{3\mu_0 B_p}$ (7.1)	b) $\dfrac{B_m^3}{24\mu_0 B_p}$ (7.2)	c, d) $\dfrac{B_m^3}{12\mu_0 B_p}$ (7.3)

Case 2: $B_m \gg B_p = \mu_0 J_c a$

a) $\dfrac{2B_p B_m}{\mu_0}$ (7.4)	b) $\dfrac{B_p B_m}{2\mu_0}$ (7.5)	c, d) $\dfrac{B_p B_m}{\mu_0}$ (7.6)

Coupling Loss Energy Densities Per Cycle: e_{cp} [J/m³]

a) $\left(\dfrac{B_m^2}{2\mu_0}\right)\dfrac{8\pi^2 \tau_m \tau_{cp}}{4\pi^2 \tau_{cp}^2 + \tau_m^2}$ (7.8)	b) $\left(\dfrac{B_m^2}{2\mu_0}\right)\dfrac{2\tau_{cp}}{\tau_m + \tau_{cp}}$ (7.9)
c) $\left(\dfrac{B_m^2}{2\mu_0}\right)\dfrac{8\tau_{cp}}{\tau_m}$ (7.10)	d) $\left(\dfrac{B_m^2}{2\mu_0}\right)\dfrac{8\tau_{cp}}{\tau_m}\left\{1 - \dfrac{\tau_{cp}}{\tau_m}\left[1 - \exp(-\tau_m/\tau_{cp})\right]\right\}$ (7.11)

Eddy-Current Loss Energy Densities Per Cycle: e_{ed} [J/m³]

a) $\dfrac{4\pi^2 \beta^2}{24\rho_m \tau_m}$ where $\beta^2 = [(aB_{m\perp a})^2 + (bB_{m\perp b})^2]$ (7.12)	
b) $\dfrac{\beta^2}{24\rho_m \tau_m}$ (7.13)	c, d) $\dfrac{\beta^2}{12\rho_m \tau_m}$ (7.14)

$B_{m\perp a}$: B_e component $\perp$ to conductor's narrow surface of thickness a.
$B_{m\perp b}$: B_e component $\perp$ to conductor's broad surface of width b.

7.2.5 Effective Matrix Resistivity

We shall briefly discuss the resistivity, ρ_{ef}, appearing in Eq. 7.7. It represents the matrix's effective resistivity for the flow of current *perpendicular* to the axis of filamentary conductors. Two models have been proposed by Carr for ρ_{ef} [7.5]:

$$\rho_{ef_1} = \frac{1 - \lambda_f}{1 + \lambda_f}\rho_m \tag{7.18a}$$

$$\rho_{ef_2} = \frac{1 + \lambda_f}{1 - \lambda_f}\rho_m \tag{7.18b}$$

where λ_f is the volumetric fraction of the superconducting filaments in the composite superconductors and ρ_m is the matrix resistivity.

The above expressions are based on the two limiting current distributions shown in Fig. 7.1. Figure 7.1a models the case in which the contact resistance at the filament surface is *zero* and current is drawn into the filaments as indicated schematically. With this flow distribution, the "apparent" cross section and distance for the passage of current are, respectively, increased and decreased, thereby effectively reducing the matrix resistivity and hence Eq. 7.18a. Similarly, Fig. 7.1b depicts the case for an *infinite* contact resistance that expels current from the filaments and hence Eq. 7.18b. Neither expression has been demonstrated rigorously, analytically or experimentally, to be correct. In practice, Eqs. 7.18a and 7.18b apply to composite superconductors, respectively, of Nb$_3$Sn and Nb-Ti.

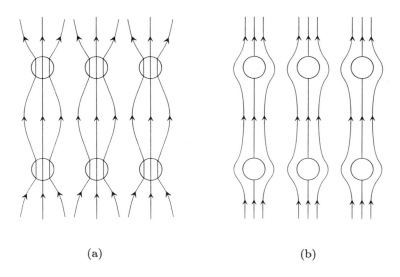

(a)   (b)

Fig. 7.1   Current distributions in a multifilamentary composite conductor *perpendicular* to the conductor axis. (a) Contact resistance between filament and matrix is zero—Eq. 7.18a. (b) Contact resistance is infinite—Eq. 7.18b.

7.3 Splice Resistance

Splice resistance is important in magnet design because in most magnets conductors must be spliced. Splicing must sometimes be superconducting or nearly so— less than $1\,p\Omega$ per splice for persistent-mode magnets such as those used in MRI and NMR devices—but generally it is resistive in nonpersistent-mode magnets. Splicing is needed primarily because it is economical to use graded conductors in a coil or in a magnet consisting of many coils; material handling of limited conductor length is also easier for big magnets.

In a given coil, the magnetic field in the inner region of the winding is considerably greater than that in the outer region. A conductor designed for the inner region of the winding will therefore contain superconducting filaments that are more than sufficient for use in the outer region of the winding. That is, it is generally advisable to wind the coil with graded conductors; for practical reasons only two grades of conductors are generally used in a given coil form. For a magnet consisting of a set of coils, each wound in a separate coil form, the conductors are naturally graded and each coil must be spliced to its nearest neighbors.

A resistive splice becomes a design issue only when: 1) it must be confined within a restricted space or conform to a specific configuration; 2) it is located deep inside the winding where there is a limited or zero amount of replenishable coolant; 3) it must withstand large forces; 4) there are so many of them that a net dissipation can affect the system's refrigeration capacity; or 5) it is not in direct contact with coolant as is the case in cryocooler-cooled magnets.

7.3.1 Lap Splice

A "shake hands" lap splice, shown in Fig. 7.2, is one of the most widely used splice designs; it is also quite suitable even for use within the winding. Its dissipation can be set to any desirable level by adjusting the overlap length, ℓ_{sl}.

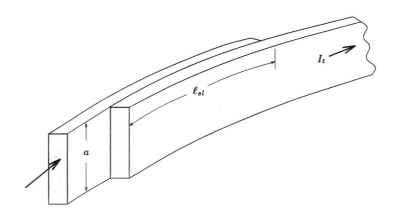

Fig. 7.2 Sketch of a typical "shake-hands" lap splice.

7.3.2 Contact Resistance

Joule dissipation, G_{sl}, in a lap splice such as the one shown in Fig. 7.2 is given by:

$$G_{sl} = R_{sl}I_t^2 \tag{7.19}$$

R_{sl} is the joint resistance and I_t is transport current. R_{sl} is given by:

$$R_{sl} = \frac{R_{ct}}{A_{ct}} = \frac{R_{ct}}{a\ell_{sl}} \tag{7.20}$$

where R_{ct} is the contact resistance in units of $[\Omega\,m^2]$ and a is the conductor width and ℓ_{sl} is the overlap length (Fig. 7.2). As remarked above, G_{sl} can be made arbitrarily small by making ℓ_{sl} sufficiently long.

Because the solder used to join two broad surfaces of the conductors is often an alloy of tin and lead, which means its conductivity does not improve with decreasing temperature as do the conductivities of pure metals, most of the contribution to contact resistance R_{ct} comes from the layer of this solder. Theoretically, $R_{ct} \simeq \rho_{sl}\delta_{sl}$, where ρ_{sl} is the solder resistivity and δ_{sl} is the solder layer thickness. (This expression neglects the "real" contact resistance at each solder-matrix metal interface.) In practice these kinds of theoretical equations are useful only as a general guide to create "good splices," meaning joints that have sufficient mechanical strength and a permissible level of splice resistance.

Table 7.2 presents values of R_{ct} for selected tin-lead solders. R_{ct} is magnetic field-dependent, increasing linearly with B. In some data, a nonlinearity is observed between zero and 1 T. These data are presented only as a general guide. In projects involving large magnets and where splice resistance is an important design issue, it is worthwhile to rely on actual measurements.

Table 7.2: Solder Contact Resistances at 4.2 K

Solder	R_{ct} [pΩ m^2]			
	0 T	1 T	2 T	9 T
Sn60-Pb40 [7.6]	3.0*	3.3*	3.6*	5.5*
	1.1*	1.6*	2.0*	5.3*
Sn50-Pb50 [7.7]	0.8	1.5*	1.7*	2.9*
	1.8	3.3*	3.7*	6.8*
Sn50-Pb50 [7.8]	$\rho_{sl} = 5.90(1+0.0081B)$ nΩ m (B in [T])			
Sn60-Pb40 [7.8]	$\rho_{sl} = 5.40(1+0.0089B)$			
USW† [7.9]	0.45 ($B = 0$ T)			
Sn60-Pb40‡ [7.10]	$R_{sl}(B)/R_{sl}(0) = 1 + 0.57B$ (B in [T])			

* Linear with B in this range.
† Between two scarfed copper surfaces ultrasonically welded.
‡ Splice resistance between two CIC conductors.

7.4 Mechanical Disturbances

The subject of mechanical disturbances was not a pressing issue in the early period (before 1975) when most magnets were built according to the Stekly stability criterion: magnets that are well-cooled and therefore have low overall current densities. Only when it became necessary to make magnets operate at high overall current densities ($\sim$100 A/mm^2 and above)—a real necessity in dipole and quadrupole magnets of high-energy physics particle accelerators—did mechanical disturbances become a critically important design issue. One obvious way to make a magnet "high performance" is to minimize the space occupied by coolant or even replace it altogether with field generating conductor or load-bearing material.

High-performance magnets are thus by necessity essentially adiabatic, and the power density equation (Eq. 6.1) for adiabatic magnets has the cooling term (g_q) set to zero. Under this condition, the only way to maintain $\dot{e}_h$, which is proportional to dT/dt of the conductor, at or near zero is to require that both g_j (Joule heating) and g_d (disturbance heating) be zero. If $\dot{e}_h$ can be maintained at or near zero, the Joule heating term will remain zero. This requires that the disturbance term must also remain essentially zero. This means AC losses, splice losses, mechanical losses, and any other losses must all be kept zero or nearly so. Thus, in adiabatic magnets no splices are placed within the winding. For this reason also, the magnets are generally swept at a "comfortably slow" rate, particularly as the operating current is approached, to keep the heat generation rate by AC and mechanical losses less than the limited heat removable rate available through thermal diffusion (the $\dot{e}_h$ and g_k terms).

7.4.1 Premature Quenches and Training

The absence of cooling makes adiabatic magnets prone to quench prematurely, sometimes at currents well below their intended operating currents. Through the use of AE technique described below, it has been demonstrated that virtually every incident of premature quench in adiabatic magnets is induced by a mechanical event, primarily either conductor motion or epoxy fracture. Fortunately, these mechanical events usually obey what is known in acoustic emission as the "Kaiser effect." It describes mechanical behavior observed during a sequence of cyclic loading in which mechanical disturbances such as conductor motion and epoxy fracture appear only when the loading responsible for events exceeds the maximum level achieved in the previous loading sequence. Thus an adiabatic magnet suffering premature quenches generally "trains" and improves its performance progressively to the point where it finally reaches its intended operating current. Obviously, the goal in designing an adiabatic magnet is to make it reach the operating current on the first try; quite recently such a remarkable feat was achieved with an adiabatic magnet for use in a 750-MHz (17.6 T) NMR system [7.11].

7.4.2 Conductor Motion and Epoxy-Resin Impregnation

Even if the space occupied by coolant is minimized and the winding no longer satisfies the Stekly criterion, it is still sufficiently loose for the conductor to move against a frictional force under the action of Lorentz forces. We may estimate the extent of the frictional displacement needed to adiabatically drive the unit conductor volume normal in a typical operating condition and show that this displacement is indeed within a possible range of motion, even in "tightly packed" windings. A conductor at $r = 0.2\,\mathrm{m}$, subjected to an induction in the z-direction (B_z) of 5 T and carrying a current density (over conductor cross section) in the θ-direction of $J_\theta = 200{\times}10^6\,\mathrm{A/m^2}$, experiences an r-directed Lorentz force density $f_{L_r} = J_\theta B_z$ of $2{\times}10^8\,\mathrm{N/m^3}$. Suppose that the conductor slides against the frictional force opposing this f_{L_r} by a distance Δr_f. Then a frictional energy density, e_f, generated by this motion over the unit conductor volume may be given by:

$$e_f = \mu_f f_{L_r} \Delta r_f \qquad (7.21)$$

where μ_f is the frictional coefficient. Inserting values of $\mu_f = 0.3$ and $e_f = 1300\,\mathrm{J/m^3}$, which is equivalent to the copper's enthalpy difference between 4.2 K and 5.2 K, and $f_{L_r} = 2\times10^8\,\mathrm{N/m^3}$, and solving Eq. 7.21 for Δr_f, we find:

$$\Delta r_f = \frac{e_f}{\mu_f f_{L_r}} = \frac{(1300\,\mathrm{J/m^3})}{(0.3)(2\times10^8\,\mathrm{N/m^3})} \simeq 20\times10^{-6}\,\mathrm{m} = 20\,\mu\mathrm{m}$$

A distance of $\sim20\,\mu\mathrm{m}$ is within the range of possibility in tightly packed windings. Indeed quench-inducing slips as small as $\sim10\,\mu\mathrm{m}$ ("microslips") were observed in an experiment [7.12]. One effective way to eliminate these microslips is to impregnate the winding with epoxy resin that leaves no void space for conductor motion; impregnation transforms the entire winding into one structural unit.

7.4.3 Impregnated Windings

Although conductor motion may be absent in the impregnated windings, two problems still remain. First, by the action of Lorentz forces, the entire winding body—in solenoidal magnets—tries to become barrel-shaped. Unless the winding is firmly anchored to the coil form to prevent this barrel-shaped deformation, interface motion occurs between the winding and the coil form; this motion generates heating, which in turn may result in a premature quench. It is possible to decouple the conductor from such frictional heating by means of a low thermal conductive sheet bonded to the inner surface of the winding [7.13]. Second, if the winding is firmly held to the coil form, high stresses are developed in the winding and the impregnant may fracture, resulting in another source of thermal disturbance.

In impregnated windings, there are two possible approaches to prevent epoxy fracture quench: 1) minimize the amount of energy induced by an epoxy fracture; and 2) eliminate fracture incidents altogether. Although there have been attempts to quantify fracture-induced energies at cryogenic temperatures [7.14, 7.15], our present understanding of the fracture mechanisms is not sufficiently advanced to permit this approach to be useful at the moment.

Great progress has been achieved in the second remedy for eliminating the fracture incidents. The techniques developed include controlling conductor tension during the winding process [7.16] and allowing the winding section to "float" in the coil form [7.16, 7.17; Maeda has pushed the floating winding concept to an extreme limit and achieved successful performance with "coilformless" solenoids [7.18].

7.4.4 Dry Windings

We have just seen that even in a tightly packed winding, if it is not impregnated ("dry") with organic materials such as epoxy resins, frictional sliding of the conductor is possible and can lead to a premature quench. Under certain conditions, however, it is possible to have dry winding magnets operate stably.

Considerable effort has been devoted in cryotribology work to find "tribological" approaches for preventing quenching in dry magnets [7.12, 7.19~7.23]. There are three possible approaches: 1) rely on cooling provided by the void-space helium; 2) minimize the frictional heating intensity; and 3) eliminate incidents of motion.

Because of an extremely limited space available for the coolant within the winding pack, the coolant is believed to have only a modest effect in balancing out motion-induced energies. Although its benefit has not been well quantified, there have been successful "quasi-adiabatic" magnets, among them, Hybrid III SCM [6.27].

A key element of the second approach to minimizing the frictional heating intensity is to achieve stable sliding by using insulating materials that, when paired with a copper-stabilized superconductor, exhibit a positive slope in their friction *vs* velocity plots. Based in part on the results of recent measurements on epoxy resins [7.22], we may conclude that winding materials having a sufficiently positive friction-velocity characteristic to provide much benefit at cryogenic temperatures are unlikely to exist. Most of the friction-velocity stabilization techniques that have been examined thus far rely on temperature-dependent creep mechanisms. In these materials, the strain rate sensitivity of the flow stress is responsible for the velocity dependence of the friction coefficient. In the absence of thermal activation at cryogenic temperatures these creep mechanisms become inoperative [7.22, 7.23].

In the third approach for eliminating incidents of motion (similar to the approach of eliminating fracture used in impregnated winding), Takao and Tsukamoto have shown that motion-induced energy can be reduced by increasing the winding stiffness [7.24]. It may also be possible to control the winding stiffness and configuration to eliminate incidents of conductor motion, at least over the critical volume of the winding, by matching ratios of the tangential to normal Lorentz force components to the prevailing static friction coefficient [7.25, 7.26]. It has been suggested both extremely high and low force ratios may be used, each ratio strategically located in the winding [7.26]. An extreme extension of this approach, of course, is epoxy impregnation. Under this design philosophy, the friction coefficient should be maximized. The ultimate goal in dry magnets is to achieve comparable structural integrity as in epoxy-impregnated windings.

7.5 Acoustic Emission Technique

Acoustic emission (AE) monitoring of superconducting magnets is the technique that has proven most successful in elucidating mechanical events occurring within the winding of high-performance magnets. Begun in the late 1970s [7.27~7.29], it was established in the 1980s that the two principal mechanical events afflicting high-performance magnets—conductor motion and epoxy fracture—can be detected by the AE technique [7.30~7.39].

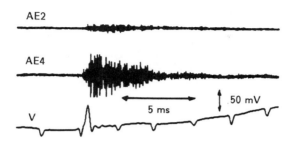

Fig. 7.3   Acoustic emission and voltage signals induced by a conductor motion event in a dipole magnet [7.31].

The most effective use of AE for superconducting magnets is the AE/voltage technique first reported by Brechna and Turowski in 1978 [7.29]. Because a sudden conductor motion event generates an AE signal and at the same time induces a voltage spike across the magnet, the simultaneous detection of AE and voltage signals at the time of a quench shows that it is induced by a conductor motion event. (A sudden shift in the position of a short length of the conductor in the presence of a magnetic field generates, through Faraday's law, a voltage pulse across the terminals of the winding.) The technique can also be used to demonstrate that a quench accompanied by an AE signal but not by a voltage spike is induced not by a conductor motion event but by an epoxy fracture event [7.30]. Examples of premature quenches induced by these two distinctive mechanical disturbances are shown below.

Figure 7.3 shows an early example of a successful application of the AE/voltage monitoring technique. The top two traces are AE signals and the bottom trace is the coil voltage in a 1.5-m long dipole [7.31]. Sensor AE2 was located at one end of the dipole and sensor AE4 at the other. The voltage pulse that precedes a rising resistive voltage was conductor-motion-induced. Sensor AE4 first records an AE signal, which after attenuation and a delay of ∼0.3 ms—the time delay discernable in an expanded scale not shown here—is recorded by sensor AE2. Since the AE wave propagates at speeds of 2∼5 km/s, a delay of ∼0.3 ms equals a travel distance of 0.6∼1.5 m; the event thus occurred near the AE4 sensor. This example demonstrated the technique's usefulness not only for quench source identification but also for source triangulation.

Figure 7.4 shows another example of a simultaneous capture of AE/voltage signals, recorded in a Nb_3Sn coil prematurely quenching at 520 A, the quench induced by a conductor motion event [7.38]. Ogitsu and others demonstrated that it is unnecessary to detect AE signals to infer mechanical disturbances [7.40]. By canceling out the charging voltages, they successfully extracted motion-induced voltage spikes from a quadrupole magnet.

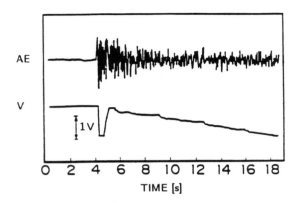

Fig. 7.4 Signals induced by a conductor motion event that prematurely quenched an Nb$_3$Sn coil. [7.38].

Another important source of quench in adiabatic magnets is epoxy fracture, which converts stress energy into heat. Because it is a mechanical event, epoxy fracture generates AE signals. The most significant feature, however, is that the event does not induce voltage spikes because it involves no relative motion of a current-carrying element in a magnetic field. Figure 7.5 shows an oscillogram of traces recorded at an epoxy fracture-triggered quench event in a test coil [7.35]. The top two traces are epoxy fracture-induced AE signals and the bottom trace is the quench voltage. Figure 7.6 shows another example of an epoxy fracture-induced quench recorded in a superconducting racetrack magnet [7.41].

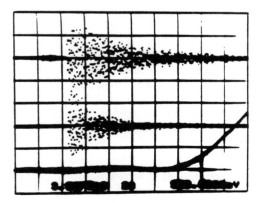

Fig. 7.5 Epoxy fracture-induced quench signals, AE (top two traces) and voltage (bottom trace) [7.35]. Time scale: 4 ms/div.

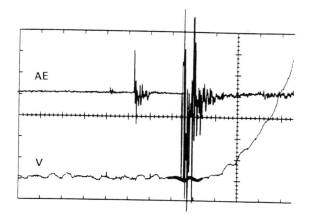

Fig. 7.6 AE (top) and coil voltage (bottom) traces recorded at
an epoxy fracture-induced quench in a racetrack magnet [7.41].
The voltage trace near the fracture point is enhanced to indicate
the absence of a pulse. Time scale: 5 ms/div.

"Time, time, what is time?
Swiss manufacture it.
French hoard it.
Italians squander it.
Americans say it is money.
Hindus say it does not exist ..." —O'Hara

"We can know the time, we can know a time. We can never know Time." —Ada

Problem 7.1:   Hysteresis loss—basic derivation

1. Without transport current

In this problem we first demonstrate, using the Bean slab, that in the *absence of transport current* the slab's hysteresis loss is given by the integral of its magnetization curve. As we briefly reviewed in Chapter 2, dissipation within a volume may be treated as the flow of Poynting vector $\vec{S}$ (Eq. 2.20, p.15). In the integral form, not including the electric energy term, Eq. 2.20 may be expressed by:

$$\int \left[-\oint_S \vec{E}\times\vec{H}\cdot d\vec{A} \right] dt = \int_V \left[\int \vec{E}\cdot\vec{J}\,dt + \frac{\mu_o H^2}{2} + \mu_o \int \vec{H}\cdot d\vec{M} \right] dV \qquad (7.22)$$

For a complete cycle, Eq. 7.22 becomes:

$$\oint \left[-\oint_S \vec{E}\times\vec{H}\cdot d\vec{A} \right] dt = \int_V \left[\oint \vec{E}\cdot\vec{J}\,dt + \mu_o \oint \vec{H}\cdot d\vec{M} \right] dV \qquad (7.23)$$

The nonzero right-hand side of Eq. 7.23 represents dissipation and is known as hysteresis loss. In purely magnetic and low-conductive materials such as ferrites, $\vec{J}=0$ and the hysteresis loss is purely magnetic. As evident from Eq. 7.23, the hysteresis loss in superconductors has an extra term coming from $\vec{E}\cdot\vec{J}$, which, in this case, is equal to $\vec{E}\cdot\vec{J}_c$.

Using a Bean slab of width $2a$, we can show that the hysteresis loss density, Δe_{hy}, may be given by the area under the magnetization curve. Namely,

$$\Delta e_{hy} \equiv \frac{1}{a}\int_0^a \left[\oint \vec{E}\cdot\vec{J}_c\,dt + \mu_o \oint \vec{H}\cdot d\vec{M} \right] dx \qquad (7.24a)$$

$$= -\mu_o \oint M\,dH_e \qquad (7.24b)$$

a) We shall now examine the case in which the external field H_e is raised by ΔH_e from H_e for a "virgin" Bean slab: $H_e < H_p = J_c a$, the critical field. Show that the incremental dissipation energy density, Δe_{hy}, is given by:

$$\Delta e_{hy} = \mu_o \left(\frac{H_e^2}{H_p} - H_e \right) \Delta H_e \qquad (7.25)$$

Derive Eq. 7.25 *two* ways, using: 1) Eq. 7.24a and 2) Eq. 7.24b.

b) Repeat a) for the slab in the critical state, *i.e.* $H_e > H_p$. Show that Δe_{hy} is given by:

$$\Delta e_{hy} = \mu_o \frac{H_p}{2}\Delta H_e \qquad (7.26)$$

Again, derive Eq. 7.26 two ways, using: 1) Eq. 7.24a and 2) Eq. 7.24b.

c) For the slab in the critical state, derive Eq. 7.26 by computing the Poynting energy flow (left-hand side of Eq. 7.23).

Solution to Problem 7.1

a) We shall solve this problem by considering
only the positive half $(0 \le x \le a)$ of the slab.
First, let us find an expression for $E_z(x)$ when
the field inside the slab between $x* = a - H_e/J_c$
and a, is raised by ΔH_e, as shown in Fig. 7.7.
$E_z(x)$ due to this field change, valid for $x* \le x \le a$, may be given by:

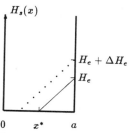

Fig. 7.7   Field profiles.

$$E_z(x)\,dt = \mu_o \Delta H_e \left[x - \left(a - \frac{H_e}{J_c} \right) \right] \quad (S1.1)$$

We thus can compute the first term in the right-hand side of Eq. 7.24a:

$$\int_0^a \left(\int \vec{E} \cdot \vec{J_c} \, dt \right) dx = J_c \mu_o \Delta H_e \int_{x*}^a \left[x - \left(a - \frac{H_e}{J_c} \right) \right] dx$$

$$= J_c \mu_o \Delta H_e \left[\frac{x^2}{2} - \left(a - \frac{H_e}{J_c} \right) x \right]_{x*=a-\frac{H_e}{J_c}}^a$$

$$= -a \mu_o \frac{H_e^2}{2H_p} \Delta H_e \qquad (S1.2)$$

Next, we shall compute the second term on the right-hand side of Eq. 7.24a. $-M$
for a partially critical-state slab is given by Eq. 5.4 (p. 164):

$$-M = H_e - \frac{H_e^2}{2H_p} \qquad (5.4)$$

Differentiating M, we have:

$$-dM = \left(1 - \frac{H_e}{H_p} \right) dH_e \qquad (S1.3)$$

The second term of the right-hand side of Eq. 7.24a is thus given by:

$$\int_0^a \left(\int \mu_o H_e \, dM \right) dx = - \left[\mu_o \left(H_e - \frac{H_e^2}{H_p} \right) \Delta H_e \right] a \qquad (S1.4)$$

Combining Eqs. $S1.2$ and $S1.4$, and dividing the sum by a, we can obtain an
expression for the right-hand side of Eq. 7.24a:

$$\frac{1}{a} \int_0^a \left(\oint \vec{E} \cdot \vec{J_c} \, dt + \mu_o \oint \vec{H} \cdot d\vec{M} \right) dx = -\mu_o \frac{H_e^2}{2H_p} \Delta H_e - \mu_o \left(H_e - \frac{H_e^2}{H_p} \right) \Delta H_e$$

$$= \mu_o \left(\frac{H_e^2}{2H_p} - H_e \right) \Delta H_e \qquad (S1.5)$$

Combining Eq. 5.4 and the right-hand side of Eq. $S1.5$, we also have:

$$\frac{1}{a} \int_0^a \left(\oint \vec{E} \cdot \vec{J_c} \, dt + \mu_o \oint \vec{H} \cdot d\vec{M} \right) dx = -\mu_o M \Delta H_e \qquad (S1.6)$$

Thus the hysteresis loss density, Δe_{hy}, may be computed either by Eqs. 7.24a or
by Eq. 7.24b.

Solution to Problem 7.1

b) When the slab is in the critical state, an increase in external field by ΔH_e creates electric field $E_z(x)$, given by:

$$E_z(x)\, dt = \mu_0 \Delta H_e x \qquad (S1.7)$$

We thus have:

$$\int_0^a \left(\int \vec{E}\cdot\vec{J_c}\, dt \right) dx = J_c\mu_0\Delta H_e \int_0^a x\, dx = \frac{a^2}{2}\mu_0 J_c\Delta H_e = a\mu_0\frac{H_p}{2}\Delta H_e \qquad (S1.8)$$

Because $M = H_p/2$ in the critical state and it is constant, $dM = 0$, we have:

$$\frac{1}{a}\int_0^a \left(\oint \vec{E}\cdot\vec{J_c}\, dt + \mu_0 \oint \vec{H}\cdot d\vec{M} \right) dx = \frac{1}{a}\int_0^a \left(\oint \vec{E}\cdot\vec{J_c}\, dt \right) dx$$

$$= \mu_0\frac{H_P}{2}\Delta H_e \qquad (S1.9)$$

Because $-M = H_p/2$ in the critical state, Eq. $S1.9$ can also be expressed by:

$$\frac{1}{a}\int_0^a \left(\oint \vec{E}\cdot\vec{J_c}\, dt + \mu_0 \oint \vec{H}\cdot d\vec{M} \right) dx = \frac{1}{a}\int_0^a \left(\oint \vec{E}\cdot\vec{J_c}\, dt \right) dx$$

$$= -\mu_0 M\Delta H_e \qquad (S1.10)$$

Again, we have shown that it is thus possible to compute Eq. 7.24 either way.

c) In this approach, we shall compute the Poynting energy flowing into the slab at $x = a$ (again, only the positive half of the slab is considered) per unit volume and the magnetic energy change per unit volume; the difference between these is the dissipation per unit volume. The Poynting vector $\vec{S}$ at $x = a$, for this slab geometry, is given by:

$$\vec{S} = \vec{E}(a)\times\vec{H_e} = -E_z(a)H_e\, \vec{i_x} \qquad (S1.11)$$

Note that as expected, the Poynting vector is pointing towards the slab. Because $E_z(a) = a\mu_0(dH_e/dt)$, we obtain the Poynting energy density:

$$\Delta e_s = s_e\, dt = \frac{1}{a}|S_x|\, dt = \mu_0 H_e\Delta H_e \qquad (S1.12)$$

The change in magnetic energy density, Δe_m, is given by:

$$\Delta e_m = \frac{\mu_0}{2a}\int_0^a [H_{s2}^2(x) - H_{s1}^2(x)]\, dx \qquad (S1.13)$$

where $H_{s2}(x)$ is the field distribution after an increase in field at $x = a$ of ΔH_e and $H_{s1}(x)$ is the original field distribution. In the critical stage, we have $H_{s2}(x) = H_{s1}(x) + \Delta H_e$. We thus have:

$$\Delta e_m = \mu_0 \left(H_e - \frac{H_p}{2} \right) \Delta H_e \qquad (S1.14)$$

Combining Eqs. $S1.12$ and $S1.14$, we have:

$$\Delta e_{hy} = \Delta e_s - \Delta e_m = \mu_0\frac{H_p}{2}\Delta H_e \qquad (7.26)$$

Problem 7.2: Hysteresis loss—basic derivation

2. With transport current

In this problem we shall derive the formula for hysteresis loss in the fully critical-state Bean slab in the presence of transport current. The derivation also demonstrates that in the presence of transport current the hysteresis loss *cannot* be determined from the area under the $-M$ *vs* H_e curve as is possible in the absence of transport current.

Figure 7.8 presents field distributions for a Bean slab of width $2a$ carrying a normalized transport current of i ($\equiv I_t/I_c$) and exposed to external fields of H_e (solid lines) and $H_e + \Delta H_e$ (dotted lines). As is the case in Problem 5.2, the order of excitation is transport current followed by external field.

a) Using the Poynting energy approach, similar to Problem 7.1 c), show that an expression for hysteresis energy density, Δe_{hy}, for a Bean slab carrying a normalized transport current i as external field is increased from H_e ($H_e > H_p$) to $H_e + \Delta H_e$ is given by:

$$\Delta e_{hy} = \mu_0 \frac{H_p}{2}(1 + i^2)\Delta H_e \tag{7.27}$$

b) Show that Eq. 7.27 cannot be given by the integral $-\int M \cdot dH_e$. That is, in the presence of transport current, show:

$$\mu_0 \frac{H_p}{2}(1 + i^2)\Delta H_e \neq -\int M \, dH_e = \mu_0 \frac{H_p}{2}(1 - i^2)\Delta H_e \tag{7.28}$$

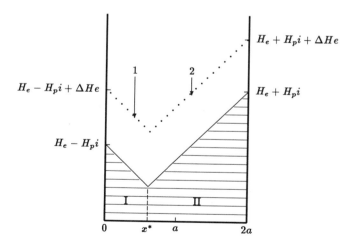

Fig. 7.8 Field distributions in a Bean slab carrying a normalized transport current of i. The dotted lines show the distribution after the field is increased by ΔH_e.

Solution to Problem 7.2

a) We shall first compute Poynting energy inputs at two surfaces, at $x = 0$ and $x = 2a$.

At $x = 0$, we have electric field $E_z(0)$, given by:

$$E_z(0) = -\mu_o a(1 - i)\frac{dH_e}{dt} \qquad (S2.1)$$

$\vec{E}(0)$ points in the $-z$-direction. Because the field at $x = 0$, $H_e - H_p i$, points in the $+y$-direction, we note that $\vec{S}(0) = \vec{E}(0) \times \vec{H}_e$ points in the $+x$-direction, or into the slab. The Poynting energy flux [W/m²] at $x = 0$, $\mathcal{E}_s(0)$, is given by:

$$
\begin{aligned}
\mathcal{E}_s(0) &= \int |S_x(0)|\, dt \\
&= a\mu_o(1 - i)(H_e - H_p i)\Delta H_e \qquad (S2.2)
\end{aligned}
$$

At $x = 2a$, we have electric field $E_z(2a)$, given by:

$$E_z(2a) = \mu_o a(1 + i)\frac{dH_e}{dt} \qquad (S2.3)$$

$\vec{E}(2a)$ points in the $+z$-direction. Because the field at $x = 2a$, $H_e + H_p i$, points in the $+y$-direction, we note that $\vec{S}(2a) = \vec{E}(2a) \times \vec{H}_e$ points in the $-x$-direction, or into the slab. The Poynting energy flux at $x = 2a$, $\mathcal{E}_s(2a)$, is given by:

$$
\begin{aligned}
\mathcal{E}_s(2a) &= \int |S_x(2a)|\, dt \\
&= a\mu_o(1 + i)(H_e + H_p i)\Delta H_e \qquad (S2.4)
\end{aligned}
$$

The total Poynting input energy density, e_s, is given by:

$$
\begin{aligned}
e_s &= \frac{1}{2a}[\mathcal{E}_s(0) + \mathcal{E}_s(2a)] \\
&= \mu_o(H_e + H_p i^2)\Delta H_e \qquad (S2.5)
\end{aligned}
$$

Next, we compute the difference in magnetic energy stored in the slab in the new state $(H_e + \Delta H_e)$ and in the original state (H_e). The computation is divided into two regions, $\Delta\mathcal{E}_{m1}$ for Region I $(0 \le x \le x*)$ and $\Delta\mathcal{E}_{m2}$ for Region II $(x* \le x \le 2a)$, where $x* = (1 - i)a$. Note that, as expected, $x* = a$ when $i = 0$.

$$
\begin{aligned}
\Delta\mathcal{E}_{m1} &= \frac{\mu_o}{2a}\int_0^{(1-i)a} [H_1'^2(x) - H_1^2(x)]\, dx \\
&= a\mu_o\left[H_e(1 - i) - \frac{H_p}{2}(1 - i^2)\right]\Delta H_e \qquad (S2.6a)
\end{aligned}
$$

Solution to Problem 7.2

$$\Delta \mathcal{E}_{m2} = \frac{\mu_o}{2a} \int_{(1-i)a}^{2a} [H_2'^2(x) - H_2^2(x)] \, dx$$

$$= a\mu_o \left[H_e(1+i) - \frac{H_p}{2}(1-i^2) \right] \Delta H_e \qquad (S2.6b)$$

The change in magnetic energy density stored in the slab, Δe_m, is given by:

$$\Delta e_m = \frac{1}{2a}(\Delta \mathcal{E}_{m1} + \Delta \mathcal{E}_{m2})$$

$$= \mu_o \left[H_e - \frac{H_p}{2}(1-i^2) \right] \Delta H_e \qquad (S2.7)$$

The hysteresis energy density Δe_{hy} is given by the difference between the Poynting energy density and the magnetic energy density. Thus:

$$\Delta e_{hy} = \mu_o \left\{ (H_e + H_p i^2) - \left[H_e - \frac{H_p}{2}(1-i^2) \right] \right\} \Delta H_e$$

$$= \mu_o \frac{H_p}{2}(1+i^2)\Delta H_e \qquad (7.27)$$

b) As studied in Problem 5.2, magnetization in the presence of transport current in the fully critical state is given by Eq. 5.18a, presented once again here:

$$-M(i) = \frac{H_p}{2}(1-i^2) \qquad (5.18a)$$

Thus the integration of the magnetization curve is given by:

$$-\mu_o \int M \cdot dH_e = \mu_o \frac{H_p}{2}(1-i^2)\Delta H_e \qquad (S2.8)$$

We therefore have demonstrated:

$$\mu_o \frac{H_p}{2}(1+i^2)\Delta H_e \neq -\int M \, dH_e = \mu_o \frac{H_p}{2}(1-i^2)\Delta H_e \qquad (7.28)$$

That is, in the presence of transport current, hysteresis loss *cannot* be given by the area under the $-M$ vs H_e plot. In fact, the area under the $-M$ vs H_e plot may be much smaller than the actual hysteresis loss.

Problem 7.3: Hysteresis loss (no transport current)

1. "Small" amplitude cyclic field

Figure 7.9 shows magnetization traces for a virgin Bean slab (width $2a$) subjected to an external field H_e. As indicated by arrows in the figure, H_e initially increases from 0 to a peak field of H_m (trace A), where $H_m < H_p = J_c a$, and decreases back to 0 (trace B), continues to $-H_m$ (trace C), and increases from $-H_m$ back to 0 (trace D). It cycles back to H_m (trace E), completing a full cycle. From symmetry, traces B and D are mirror images of each other and likewise traces C and E are also mirror images of each other. For the triangular and trapezoid excitations of Table 7.1 that swing between 0 and $B_m = \mu_o H_m$, the up-swing follows trace F indicated in the figure. Expressions for traces A, B, D, and F are given by:

$$\text{Trace A:}\qquad -M(H_e) = H_e - \frac{H_e^2}{2H_p} \qquad\qquad (5.4)$$

$$\text{Trace B:}\qquad -M(H_e) = H_e + \frac{H_e^2 - 2H_m H_e - H_m^2}{4H_p} \qquad (7.29a)$$

$$\text{Trace E:}\qquad -M(H_e) = H_e - \frac{H_e^2 + 2H_m H_e - H_m^2}{4H_p} \qquad (7.29b)$$

$$\text{Trace F:}\qquad -M(H_e) = H_e - \frac{H_e^2 + H_m^2}{4H_p} \qquad\qquad (7.29c)$$

a) Derive Eq. 7.1 given in Table 7.1 for hysteresis energy density, e_{hy}, under a cyclic (B→C→D→E) B-field excitation, between B_m and $-B_m$.

b) Derive Eq. 7.2 for an exponential decay from B_m to 0.

c) Derive Eq. 7.3 for the triangular and trapezoid excitations of Table 7.1.

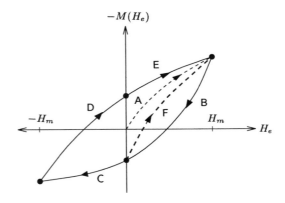

Fig. 7.9 Magnetization traces for "small" amplitude
field cycles with no transport current.

Solution to Problem 7.3

a) Because the slab carries no transport current, e_{hy} for a complete field cycle is given by:

$$e_{hy} = \mu_0 \oint_{-H_m}^{H_m} -M(H_e)\, dH_e = 2\mu_0 \int_0^{H_m} -M(H_e)\, dH_e \qquad (S3.1)$$

where $-M(H_e)$ is given by the difference between Eqs. 7.29b and 7.29a. We thus have:

$$e_{hy} = 2\mu_0 \int_0^{H_m} \left[\left(H_e - \frac{H_e^2 + 2H_m H_e - H_m^2}{4H_p} \right) \right. $$
$$\left. - \left(H_e + \frac{H_e^2 - 2H_m H_e - H_m^2}{4H_p} \right) \right] dH_e \qquad (S3.2)$$

$$= 2\mu_0 \int_0^{H_m} \left(-\frac{H_e^2}{2H_p} + \frac{H_m^2}{2H_p} \right) dH_e = 2\mu_0 \left[-\frac{H_e^3}{6H_p} + \frac{H_m^2 H_e}{2H_p} \right]_0^{H_m}$$

$$= 2\mu_0 \left(-\frac{H_e^3}{6H_p} + \frac{H_m^3}{2H_p} \right) = \frac{2\mu_0 H_m^3}{3H_p} \qquad (S3.3)$$

$$e_{hy} = \frac{2B_m^3}{3\mu_0 B_p} \qquad (7.1)$$

b) If the slab is exposed to H_m from the virgin state and then the field decays exponentially, e_{hy} is nearly half the area enclosed by traces F and B. Thus:

$$e_{hy} = \frac{\mu_0}{2} \int_0^{H_m} \left[\left(H_e - \frac{H_e^2 + H_m^2}{4H_p} \right) \right.$$
$$\left. - \left(H_e + \frac{H_e^2 - 2H_m H_e - H_m^2}{4H_p} \right) \right] dH_e \qquad (S3.4)$$

$$e_{hy} = \frac{B_m^3}{24\mu_0 B_p} \qquad (7.2)$$

c) When the slab is exposed to a periodic excitation field varying between 0 and H_m, then e_{hy} is given by the area enclosed by traces F and B. Thus:

$$e_{hy} = \mu_0 \int_0^{H_m} \left[\left(H_e - \frac{H_e^2 + H_m^2}{4H_p} \right) \right.$$
$$\left. - \left(H_e + \frac{H_e^2 - 2H_m H_e - H_m^2}{4H_p} \right) \right] dH_e \qquad (S3.5)$$

$$e_{hy} = \frac{B_m^3}{12\mu_0 B_p} \qquad (7.3)$$

An important point to note is that in each case e_{hy} is proportional to the 3rd power of B_m. Note also that because $H_p = J_c a$, e_{hy} for small field excursions depends inversely on $J_c a$.

Problem 7.4: Hysteresis loss (no transport current)
2. "Large" amplitude cyclic field

This problem is similar to Problem 7.3 except that H_m now exceeds H_p so that the slab is driven to the critical state during most of the field excursion. Specifically, we treat the case where the amplitude of the cyclic external field H_m is at least twice H_p, i.e. $H_m > 2H_p$. Note that $\mu_o J_c a$ is typically about ~0.1 T so that in most field excitations, this condition is met.

Figure 7.10 presents the magnetization for the slab subjected to a cyclic field of amplitude H_m, where $H_m > 2H_p$. As indicated in the figure, the complete field cycle consists of B, C, D, E, F, and G. (The initial trace A may safely be neglected for this case.) $-M(H_e)$ for trace B is given by:

$$\text{Trace B:} \qquad -M(H_e) = H_e - H_m + \frac{H_p}{2} + \frac{(H_m - H_e)^2}{4H_p} \qquad (7.30)$$

For traces C and D, $-M(H_e) = -H_p/2$, as indicated in Fig. 7.10. Trace E is a mirror image of trace B and similarly, traces F and G are mirror images, respectively, of traces C and D.

a) Show that an expression of e_{hy} derived from one complete field excursion of Fig. 7.10 is given by:

$$e_{hy} = 2\mu_o H_p H_m \left(1 - \frac{2H_p}{3H_m}\right) = \frac{2B_p B_m}{\mu_o} \left(1 - \frac{2B_p}{3B_m}\right) \qquad (7.31)$$

b) Use Eq. 7.31 to derive Eq. 7.4 for sinusoidal excitation with a field amplitude $H_m \gg H_p$. Clearly for triangular and trapezoid excitations shown in Table 7.1, e_{hy} (Eq. 7.6) is half that given by Eq. 7.4; for exponential decay, e_{hy} (Eq. 7.5) for $H_m \gg H_p$ is half that given by Eq. 7.6.

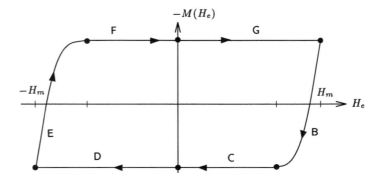

Fig. 7.10 Magnetization traces for "large" amplitude field cycles with no transport current.

Solution to Problem 7.4

a) Again, we start with:

$$e_{hy} = \mu_o \oint_{-H_m}^{H_m} -M(H_e)\,dH_e = 2\mu_o \int_0^{H_m} -M(H_e)\,dH_e \qquad (S4.1)$$

The integration must be divided into two regions, from $H_e = 0$ to H_{bc} and from H_{bc} to H_m, where H_{bc} is the field at which traces B and C meet. Thus:

$$H_{bc} - H_m + \frac{H_p}{2} + \frac{(H_m - H_{bc})^2}{4H_p} = -\frac{H_p}{2} \qquad (S4.2)$$

Solving Eq. $S4.2$ for H_{bc}, we obtain: $H_{bc} = H_m - 2H_p$. Equation $S4.1$ may now be integrated:

$$e_{hy} = 2\mu_o \int_0^{H_{bc}} \left(\frac{H_p}{2} + \frac{H_p}{2} \right) dH_e$$

$$+ 2\mu_o \int_{H_{bc}}^{H_m} \left\{ \frac{H_p}{2} - \left[H_e - H_m + \frac{H_p}{2} + \frac{(H_m - H_e)^2}{4H_p} \right] \right\} dH_e \qquad (S4.3)$$

$$= 2\mu_o \left[H_p H_{bc} + \frac{H_m^2}{2} + \frac{H_{bc}^2}{2} - H_m H_{bc} - \frac{(H_m - H_{bc})^3}{12H_p} \right] \qquad (S4.4)$$

Inserting $H_{bc} = H_m - 2H_p$ into Eq. $S4.4$, we obtain:

$$e_{hy} = 2\mu_o \left[H_p(H_m - 2H_p) + \frac{H_m^2}{2} + \frac{(H_m - 2H_p)^2}{2} - H_m(H_m - 2H_p) - \tfrac{2}{3}H_p^2 \right]$$

$$= 2\mu_o \left(H_p H_m - \tfrac{2}{3}H_p^2 \right)$$

$$= 2\mu_o H_p H_m \left(1 - \frac{2H_p}{3H_m} \right) \qquad (S4.5)$$

$$e_{hy} = \frac{2B_p B_m}{\mu_o} \left(1 - \frac{2B_p}{3B_m} \right) \qquad (7.31)$$

b) For $B_m \gg B_p$ Eq. 7.31 approximates to:

$$e_{hy} = \frac{2B_p B_m}{\mu_o} \qquad (7.4)$$

Both the B_m^3 and B_m dependences of e_{hy} (Eqs. 7.3 and 7.4) have been verified experimentally [7.42]. Because triangular and trapezoid excitations cover half the $M(H_e)$ area covered by the sinusoidal excitation, their e_{hy} (Eq. 7.6) is half that given by Eq. 7.4. For the exponential decay, the $M(H_e)$ covered is half that covered by the triangular and trapezoid excitations, giving rise to e_{hy} (Eq. 7.5) that is 1/4 of Eq. 7.4.

An important point to note for all the cases considered here is that e_{hy} increases with a, requiring "submicron" filaments [7.43] if superconductors are to be used in 50-Hz or 60-Hz electrical devices.

Problem 7.5: Coupling time constant

In this problem we shall derive the coupling time constant, τ_{cp}, an indispensable parameter in computing the coupling loss. Figure 7.11 presents a two-filament model of a multifilamentary composite conductor subjected to a time-varying magnetic induction $\dot{B}$ directed in the z-direction. As indicated in the figure, ℓ_p represents a twist pitch length.

Starting with Eq. 6.15 (p. 219), an expression for the magnetic diffusion time constant τ_{mg}, use a "plausible" argument to show that an expression for the coupling time constant for a Bean slab of width $2a$, τ_{cp}, is given by:

$$\tau_{cp} = \frac{\mu_o \ell_p^2}{4\pi^2 \rho_{ef}} \tag{7.32}$$

Here ρ_e in Eq. 6.15, representing the matrix electrical resistivity, is replaced by ρ_{ef}, either of the effective matrix electrical resistivities given by Eq. 7.18.

Solution to Problem 7.5

From Fig. 7.11, it is appropriate to model a conductor of length $\ell_p/2$ as one Bean slab of thickness $2a$, because over this length induced currents are similar to those in the slab. From Eq. 6.15, we have:

$$\tau_{mg} = \frac{\mu_o}{\rho_e} \left(\frac{2a}{\pi}\right)^2 \tag{6.15}$$

Substituting $2a = \ell_p/2$ and $\rho_e = \rho_{ef}$ into Eq. 6.15, we obtain:

$$\tau_{cp} = \frac{\mu_o \ell_p^2}{4\pi^2 \rho_{ef}} \tag{7.32}$$

Note that τ_{cp} derived above is off by a factor of 2 from that given by Eq. 7.7.

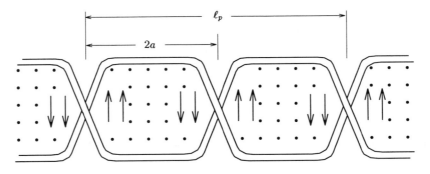

Fig. 7.11   Two-filament model of a multifilamentary composite wire.

Problem 7.6: Hysteresis loss of an Nb_3Sn strand

In the US-DPC program, a pair of ~1-m diameter coils was tested at JAERI. The US coils, wound with a CIC conductor containing many Nb_3Sn multifilamentary strands, was placed in JAERI's superconducting pulsed magnet system and subjected to an AC field excitation.

During design of the coil, the magnetization of a test sample consisting of many short lengths of strands carrying no transport current was measured to estimate AC losses of the conductor under a cyclic excitation between 0 and 3 T. Figure 7.12 gives the magnetization curve of the test sample [7.44]. The ordinate $(-M)$ unit has been adjusted from raw data to correspond to magnetization for *one* filament of diameter d_f in the strands.

a) Show that the hysteresis loss density per cycle generated within one filament as computed directly from the $-M$ vs. B_e plot shown in Fig. 7.12 is ~85 kJ/m³. You may assume that $d_f = 2a$, where $2a$ is the Bean slab width. Further, you may treat the *entire* filament as a hard superconductor obeying Bean's model.

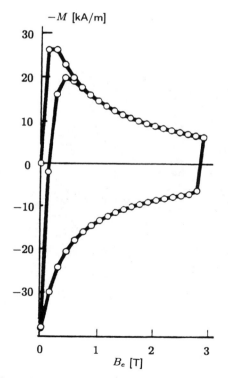

Fig. 7.12   Magnetization behavior deduced for *one* filament of the strands [7.44].

Problem 7.6: Hysteresis loss of an Nb₃Sn strand

b) Based on an I_c measurement of the strands, the critical current density, J_c [A/m²], of each filament as a function of B_e [T] is found to be [7.44]:

$$J_c(B_e) = 6.1 \times 10^9 e^{-0.22B_e}$$

What is the *effective* filament diameter d_{f_e} consistent with the magnetization data given in **a)**, the measured $J_c(B_e)$ given above, and the AC loss theory based on Bean's model? Note that here you cannot assume J_c to be field independent as is usually assumed in the simple analysis. You may assume $\mu_o H_p \ll 1$ T for the entire field range.

c) Theoretically, each filament in the strand is of perfectly circular cross section with a diameter (d_f) of 6 μm. Is your effective diameter d_{f_e} obtained in **b)** greater or smaller than d_f? Explain your discrepancy, if any, using information contained in the microphotograph of the strand cross section shown in Fig. 7.13. Each square-like cross section shows the cross section of one filament.

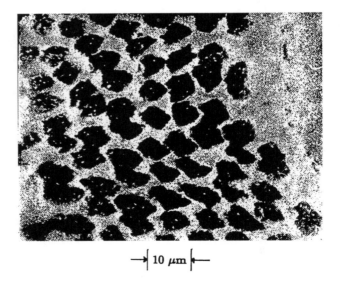

→| 10 μm |←

Fig. 7.13   Microphotograph of a section of one strand, showing cross sections of many individual filaments [7.44].

Solution to Problem 7.6

a) The hysteresis loss density $[\text{J/m}^3]$ is given by the total area enclosed under the $-M$ vs B_e curve, as B_e [T] is swept $0 \to 2.8 \to 0$. Thus:

$$e_{hy} = -\oint M \, dB_e \tag{S6.1}$$

$$\simeq (-\tilde{M}\!\uparrow - \tilde{M}\!\downarrow)B_m \tag{S6.2}$$

where $-\tilde{M}\!\uparrow$ and $\tilde{M}\!\downarrow$ are, respectively, the average magnetization over the magnetic induction range from 0 to B_m and that over the range from B_m to 0. Substituting into Eq. S6.2 $-\tilde{M}\!\uparrow \simeq 15\,\text{kA/m}$, $\tilde{M}\!\downarrow \simeq -15\,\text{kA/m}$, and $B_m = 2.8\,\text{T}$, we have:

$$e_{hy} \simeq (30\,\text{kA/m})(2.8\,\text{T}) = 84\,\text{kJ/m}^3$$

b) We shall assume $\mu_o H_p \ll 1$ T, and thus $-M = H_p/2 = J_c a/2$. By substituting $2a = d_{f_e}$, we have: $-M = J_c d_{f_e}/4$. e_{hy} is thus given by:

$$e_{hy} = -\oint \mu_o M \, dH_e$$

$$= 2 \int_0^{2.82\,\text{T}} \frac{J_c d_{f_e}}{4} \, dB_e \tag{S6.3}$$

Substituting the given $J_c(B_e)$ function into Eq. S6.3, we have:

$$e_{hy} = 84 \times 10^3 \text{ J/m}^3$$
$$= \frac{d_{f_e}(6.1 \times 10^9 \text{ A/m}^2)}{2} \int_0^{2.82\,\text{T}} e^{-0.22B_e} \, dB_e$$
$$= \frac{d_{f_e}(6.1 \times 10^9 \text{ A/m}^2)}{(2)(0.22)} \left(1 - e^{-0.62}\right) \text{(T)}$$
$$= \frac{d_{f_e}(6.1 \times 10^9 \text{ A/m}^2)(0.46)}{0.44} \text{(T)}$$
$$= (d_{f_e})6.4 \times 10^9 \text{ J/m}^3 \tag{S6.4}$$

Solving for d_{f_e} from Eq. S6.4, we obtain $d_{f_e} = 13\ \mu\text{m}$.

c) Because $d_{f_e} = 13\ \mu\text{m}$, $d_{f_e} \simeq 2d_f$. As seen from the microphotograph of Fig. 7.13, the most likely reason for $d_{f_e} \simeq 2d_f$ is that some filaments are sintered to their neighbors, making d_{f_e} more like twice the actual filament size, or $d_{f_e} \sim 2d_f$.

Problem 7.7: AC losses in Hybrid III SCM

This problem deals with AC losses generated in the Hybrid III SCM during its operation. Hybrid III's typical operational sequence is described below:

Step 1: SCM is charged from 0 to 800 A in a period of 1200 s. This charging rate corresponds to a field sweep rate at the innermost winding radius at the magnet midplane of 4 mT/s. During this sequence, a significant dissipation apparently takes place, resulting in a net rise in bath temperature of ~0.1 K, from 1.70 to 1.80 K.

Step 2: SCM is charged from 800 to 1800 A in a period of 900 s. No measurable increase in bath temperature is observed.

Step 3: In the final leg of the charging sequence, SCM goes from 1800 to 2100 A in a period of 600 s. Again, with no apparent increase in bath temperature. The SCM is now generating 12.3 T at the magnet center.

Step 4: With SCM held at 2100 A, the insert is energized and discharged at a constant rate to an induction, typically between 0 and 22.7 T. Again, during this charging-discharging sequence, no measurable increase in bath temperature is observed.

Panic: In the event of an insert malfunction, the insert is "tripped," forcing its field to decay from 22.7 to 0 T in a time period of ~0.3 s. Because of large AC dissipations expected in the SCM under this emergency condition, the SCM is automatically "dumped," resulting in a decay of its current from 2100 to 0 A with an effective time constant of ~10 s.

As noted above, AC losses are important only during Step 1. Because of a rapid decrease in the insert's fringing field during *Panic*, the SCM, particularly the Nb-Ti coil, is driven normal, forcing the SCM to dump.

Table 7.3: Pertinent Conductor Parameters

Parameter		Nb_3Sn Coil*	Nb-Ti Coil*
Overall width, a	[mm]	9.50	9.20
Overall thickness, b	[mm]	4.50	2.60
Filament diameter, d_f	[μm]	50	75
Twist pitch length, ℓ_p	[mm]	100	100
Filament #, N_f		1000	2500
Total conductor length, ℓ_{cd}	[m]	1700	8100
J_c @ 0~5 T, 1.8 K	[GA/m²]	5	3
ρ_m @ 0~8 T, 1.8 K	[nΩ m]	0.5	0.5

* For the purpose of this problem, each conductor is assumed to be wound with only one grade of conductor.

Problem 7.7: AC losses in Hybrid III SCM

a) Justify a temperature increase of $\sim$0.1 K (from 1.70 to 1.80 K at 1 atm) as the SCM is charged from 0 to 800 A, at which point it is generating 4.8 T. In computing AC losses generated by the SCM's own field, assume the field decreases linearly from 5 T at $r = 0.22$ m (the innermost radius of the Nb$_3$Sn coil; see Table 3.1, p. 59) to $\sim$0 T at $r = 0.45$ m (the outermost radius of the Nb-Ti coil) over the *entire length* of each coil. Note that the magnet vessel contains 250 liters of superfluid liquid helium.

b) Show that the narrow cooling channels provided in each double pancake are sufficient to transport AC losses generated in the double pancake to annular spaces at the i.d. and o.d. Note that there are 32 double pancakes in the Nb-Ti coil, each double pancake having cooling channels of 1-mm height and occupying $\sim$40% of the pancake surface area (Fig. 6.13, p. 235). The pancake i.d. and o.d. are, respectively, 658 mm and 907 mm.

c) During *Panic* each coil is subjected to a rapid decrease in the insert's fringing field. For the Nb-Ti conductor at the innermost turn at the magnet midpoint ($r = 329$ mm, $z = 0$), $\Delta|B_e|$ is estimated to be $\sim$1 T, taking place in a time period of 0.3 s, or $\Delta|\dot{B}_e| \sim 3$ T/s. Show that the average temperature of a unit combined volume (conductor and liquid helium adjacent to it) located at the innermost radius at the midpoint will exceed T_λ.

"Burst Disk" and Diffuser for Hybrid III Cryostat

As discussed in Problems 3.14~3.16, the major fault condition in hybrid magnets is triggered by an insert burnout; the Hybrid III SCM is designed to be discharged quickly in the event of an insert burnout— the event described above as *Panic*. (Problems 8.1 and 8.2 discuss this fast discharge mode in more detail.)

One critical consequence of this fast discharge is a rapid rise in the cryostat pressure. The Hybrid III cryostat is, therefore, equipped with a "burst disk" to keep the pressure increase in the cryostat below 1 atm; a 40-μm thick aluminum foil disk, with an active diameter of 70 mm, is placed in vacuum fittings (Fig. 7.14). When a pressure increase of 1 atm is reached in the cryostat, the foil ruptures, relieving the cryostat pressure. As indicated in the figure, a diffuser is placed at the burst disk exit to minimize the exit pressure loss of the vapor released from the cryostat.

Fig. 7.14   Burst disk (with diffuser) arrangement for Hybrid III cryostat.

Solution to Problem 7.7

a) The major source of dissipation during Step 1 is hysteresis loss. For the Nb_3Sn coil, we have:

$$E_{hy1} = V_{f1}\frac{B_{p1}\tilde{B}_{m1}}{2\mu_o} \tag{S7.1}$$

where V_{f1} is the total volume of the filaments in the coil. The factor 2 is needed here because we are computing E_{hy1} only during a field excursion from 0 to $\tilde{B}_{m1}$, which is an average of the maximum field the coil is subjected to. V_{f1} is given by:

$$V_{f1} = N_{f1}\ell_{cd1}\left(\frac{\pi d_{f1}^2}{4}\right) \tag{S7.2}$$

where N_{f1} is the total number of filaments in the conductor; ℓ_{cd1} is the total conductor length; and d_{f1} is the filament diameter. Combining Eqs. $S7.1$ and $S7.2$ and noting that $B_{p1} = \mu_o J_{c1}d_{f1}/2$, we obtain:

$$E_{hy1} = N_{f1}\ell_{cd1}\left(\frac{\pi d_{f1}^2}{4}\right)\left(\frac{J_{c1}d_{f1}\tilde{B}_{m1}}{4}\right) \tag{S7.3}$$

Inserting appropriate values (Table 7.3) into Eq. $S7.3$ and with $\tilde{B}_{m1} = 4.3$ T (an average of 5.0 T at $r = 216$ mm and 3.6 T at $r = 328$ mm), we obtain:

$$
\begin{aligned}
E_{hy1} &= (1000)(1700\,\text{m})\frac{\pi(50\times10^{-6}\,\text{m})^2}{4} \\
&\quad \times \frac{(5\times10^9\,\text{A/m}^2)(50\times10^{-6}\,\text{m})(4.3\,\text{T})}{4} \\
&= (3.3\times10^{-3}\,\text{m}^3)(270\times10^3\,\text{J/m}^3) \simeq 900\,\text{J}
\end{aligned}
\tag{S7.4}
$$

Similarly, we can compute E_{hy2}, the hysteresis energy generated in the Nb-Ti coil:

$$E_{hy2} = V_{f2}\frac{B_{p2}\tilde{B}_{m2}}{2\mu_o} \tag{S7.5}$$

Inserting appropriate values into Eq. $S7.5$, we obtain:

$$
\begin{aligned}
E_{hy2} &= (2500)(8100\,\text{m})\frac{\pi(75\times10^{-6}\,\text{m})^2}{4} \\
&\quad \times \frac{(3\times10^9\,\text{A/m}^2)(75\times10^{-6}\,\text{m})(1.8\,\text{T})}{4} \\
&= (89.4\times10^{-3}\,\text{m}^3)(101\times10^3\,\text{J/m}^3) \simeq 9000\,\text{J}
\end{aligned}
\tag{S7.6}
$$

The total hysteresis loss released into the liquid is thus ~ 10000 J.

Solution to Problem 7.7

The mass of 250 liters of liquid helium at 1.70 K is 37 kg. The enthalpy of 1-atm helium at 1.70 K is 1280 J/kg and that at 1.80 K is 1530 J/kg (Table A2.1, Appendix II), or a net change in enthalpy of 250 J/kg. With a total mass of 37 kg, an increase in bath temperature from 1.70 to 1.80 K requires a net energy input of 9250 J, which is close to the total hysteresis dissipation computed above.

Note that during this field sweep, the system's refrigeration rate, matched to a quiescent load of ~10 W, may increase by a marginal amount as the fluid is heated; the increase may, however, be neglected.

b) As computed above, the total hysteresis energy dissipation in the Nb-Ti coil is 9000 J, taking place over a period of 1200 s, or an overall hysteresis dissipation rate of ~8 W. For each double pancake, the dissipation rate would be ~0.2 W. Under the most conservative condition, the total channel cross section for each double pancake would be ~4 cm^2 (40% of the circumference corresponding to the innermost diameter of 658 mm times a channel height of 0.5 mm—note that the 1-mm high channel is shared by two pancakes), or a heat flux of 0.05 W/cm^2. Since heat can flow radially both inward and outward, an appropriate value for the channel length is one quarter the difference between o.d. and i.d., or ~6 cm.

Because dissipation is taking place over the entire channel length, Eq. 4.3 (p. 118) is applicable. Thus:

$$X(T_b) = \frac{q_c^{3.4}}{4.4} L \qquad (4.3)$$

From Fig. 4.4, we have: $X(T_b = 1.8\,K) = 350$. With $L = 6$ cm and solving Eq. 4.3 for q_c, we obtain: $q_c = 5.1$ W/cm^2, which is clearly greater than the minimum required value of 0.05 W/cm^2. That is, the channels are sufficient to remove the hysteresis dissipation during a charge up from 0 to 800 A. This conclusion has been validated by actual runs.

c) In a rapidly changing field, the most important losses are coupling (e_{cp}) and eddy (e_{ed}). We shall consider here only e_{cp} because it alone is sufficient to drive the conductor-helium unit volume to T_λ.

We shall first compute τ_{cp}, the coupling time constant, for the Nb-Ti conductor.

$$\tau_{cp} = \frac{\mu_o \ell_p^2}{8\pi^2 \rho_{ef}} \qquad (7.7)$$

For Nb-Ti composite, ρ_{ef} given by Eq. 7.18b is generally used. In terms of $\gamma_{c/s}$, the copper-to-superconductor ratio, ρ_{ef} is given by:

$$\rho_{ef} = \frac{1 + \lambda_f}{1 - \lambda_f} \rho_m$$

$$= \frac{\gamma_{c/s} + 2}{\gamma_{c/s}} \rho_m \qquad (S7.7)$$

Solution to Problem 7.7

With $\gamma_{c/s} \sim 3$ (deducible from the conductor parameters given in Table 7.3) and inserting appropriate values into Eq. 7.7, we obtain:$\tau_{cp} = 0.2\,\mathrm{s}$, which is comparable with $\tau_m = 0.3\,\mathrm{s}$ for the insert discharge.

We shall apply Eq. 7.11 of Table 7.1, inserting a factor of $1/2$ (for discharge only), we obtain:

$$e_{cp} = \tfrac{1}{2}\left(\frac{B_m^2}{2\mu_o}\right)\frac{8\tau_{cp}}{\tau_m}\left\{1 - \frac{\tau_{cp}}{\tau_m}[1 - \exp(-\tau_m/\tau_{cp})]\right\} \qquad (S7.8)$$

$$= \tfrac{1}{2}\frac{(1\,\mathrm{T})^2}{2(4\pi\times10^{-7}\,\mathrm{H/m})}\frac{8(0.2\,\mathrm{s})}{0.3\,\mathrm{s}}\left\{1 - \frac{0.2\,\mathrm{s}}{0.3\,\mathrm{s}}\left[1 - \exp\left(-\frac{0.3\,\mathrm{s}}{0.2\,\mathrm{s}}\right)\right]\right\}$$

$$= \frac{10^7}{3\pi}[1 - \tfrac{2}{3}(1 - 0.223)] = 0.51\times10^6\,\mathrm{J/m^3}$$

Consider a unit length (1 cm) of conductor. Because the conductor's cross section is $(0.92\,\mathrm{cm})\times(0.26\,\mathrm{cm}) = 0.24\,\mathrm{cm^2}$, it has volume $\mathcal{V}_{cd}$ of $0.24\,\mathrm{cm^3}$. Over this conductor length, helium occupies 0.4 cm length (40% filling) and 0.5 mm channel depth (1-mm deep channel is shared by conductors of the top and bottom pancakes) over the conductor width of 2.6 mm. Thus for a unit conductor length, helium occupies a volume $\mathcal{V}_{he}$ of $5.2\times10^{-3}\,\mathrm{cm^3}$. The total dissipation energy over unit conductor length, E_{cp}, will thus be given by:

$$E_{cp} = e_{cp}\mathcal{V}_{cd} \qquad (S7.9)$$

$$= (0.51\,\mathrm{J/cm^3})\times(0.24\,\mathrm{cm^3})$$

$$= 0.12\,\mathrm{J}$$

The total thermal energy, ΔE_{th} needed to raise the unit conductor (and accompanying liquid helium) from 1.8 K to T_λ is given by:

$$E_{th} = [h_{cu}(T_\lambda) - h_{cu}(1.8\,\mathrm{K})]\mathcal{V}_{cd} + [h_{he}(T_\lambda) - h_{he}(1.8\,\mathrm{K})]\mathcal{V}_{he} \qquad (S7.10)$$

Inserting $\Delta h_{cu} \simeq 0.1\,\mathrm{mJ/cm^3}$ and $\Delta h_{he}{\sim}290\,\mathrm{mJ/cm^3}$ into Eq. S7.10, we have:

$$E_{th} = (0.1\,\mathrm{mJ/cm^3})(0.24\,\mathrm{cm^3}) + (290\,\mathrm{mJ/cm^3})(5.2\times10^{-3}\,\mathrm{cm^3}) \qquad (S7.11)$$

$$\simeq 0.02\,\mathrm{J}$$

Because $E_{cp} \gg E_{th}$, the entire helium volume surrounding the unit conductor volume be heated well above T_λ, making it impossible for the conductor to recover.

"Oh figures! You can make figures do whatever you want." —Ned Land

Problem 7.8: AC losses in the US-DPC Coil

This problem deals with a US-DPC Experiment on AC loss measurements conducted at JAERI (Naka, Japan) in late 1990 [6.29]. The US-DPC consists of three double-pancake (DP) coils, A, B, and C, each wound with a CIC conductor containing Nb$_3$Sn multifilamentary strands. The three DP coils are stacked and connected in series, with DP-B sandwiched between DP-A and DP-C. The 3-double pancake US-DPC Coil assembly is in turn sandwiched between two coils, DPC-U1 and DPC-U2 [7.45] that provide a background field to the US-DPC coil. Figure 7.15 shows the overall assembly of the experiment. Figure 7.16 shows a simplified flow diagram for supercritical helium forced through the double pancakes.

The pertinent parameters of DP-B and the CIC conductor used to wind the pancakes are summarized in Table 7.4. Values are approximate and some parameters are simplified for this problem. Table 7.5 presents DP-B field data, generated by three DP coils, at radial locations across the DP-B's midplane at a transport current of 25 kA. For the purpose of this problem, you may assume that $|B|$ at each radial location is constant in the axial direction. If needed, you may use a linear interpolation for $|B|$ between two radial locations.

Figure 7.17 shows the waveform of a transport current pulse applied to the US-DPC. The transport current rises from 0 to 25 kA in a ramp-up time τ_m of 1.75 s, remains at 25 kA for 3.0 s, and decreases from 25 to 0 kA in a ramp-down time τ_m of 1.75 s. In this particular run, the total steady-state helium mass flow rate through DP-B before, during, and after the pulsing was constant at 13 g/s. The helium at the *inlet* of the DP-B remained at a constant temperature and at 6.1 atm at all times. Figure 7.18 shows the corresponding measured temperature trace, $T_{ex}(t)$, at the *exit* of the DP-B. Note that there is a *long* time delay and that the exit temperature for $0 \leq t < 20$ s is the steady-state temperature before the application of the current pulse. The steady-state temperature is restored for $t > 250$ s. The measured steady-state pressure drop between the inlet and exit was ~1 atm. You may make the following additional assumptions:

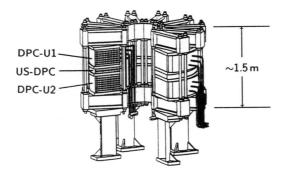

DPC-U1
US-DPC
DPC-U2

~1.5 m

Fig. 7.15   Drawing of the US-DPC Coil placed in the DPC
Test Assembly. Based on a drawing appearing in [7.45].

Problem 7.8: AC losses in the US-DPC Coil

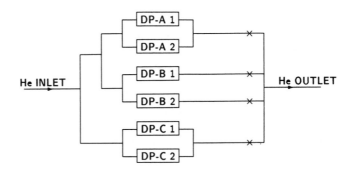

Fig. 7.16   Schematic diagram showing helium flow
in three double pancakes of the US-DPC Coil [6.30].

- The current-pulse-induced AC losses were generated in the strands only and were totally released to the helium flowing through the DP-B. This energy transfer was completed well before $t \sim 225\,\mathrm{s}$.

- The inlet and exit helium pressures remained constant at all times.

- No heating was generated in the DP-B by transient turbulence induced in the conductor's helium flow resulting from the current ramping. Steady-state frictional flow losses through the DP-B are, however, present.

Table 7.4: Parameters of DP-B and CIC Conductor

Double-Pancake B			CIC Conductor		
Winding i.d., $2a_1$	[mm]	1000	# strands		225
Winding o.d., $2a_2$	[mm]	1820	Strand diameter d_{st}	[mm]	0.78
Winding height, $2b$	[mm]	50	Total $A_{cu} + A_{nc}$	[mm^2]	107
Total # turns in DP-B		33	Total conductor length	[m]	150

Table 7.5: DP-B Midplane Field Data @ 25 kA

r [m]	$|B|$ [T]	r [m]	$|B|$ [T]
0.5	5.0	0.8	0.5
0.6	3.0	0.9	2.0
0.7	1.5	—	—

Problem 7.8: AC losses in the US-DPC Coil

Figure 7.19 presents measured total AC losses per cycle [mJ/cm³], at 4.2 K, *vs*
B_m [T] data for a *single* Nb₃Sn strand used in the DP-B conductor. The strand
was subjected to external field pulses (inset); the strand carried no transport
current. Data, originally based on those of Takayasu [7.46], are for cycle times
of 24, 12, and 6 s; only those for $\tau_m = 6$ s are presented in the figure. You may
assume the total AC losses are composed only of hysteresis (e_{hy}) and coupling
(e_{cp}). A valid theoretical e_{cp} line for this strand at $\tau_m = 6$ s is also plotted in
Fig. 7.19. With B_m given in tesla: $e_{cp} = 0.53B_m^2$ mJ/cm³. Also note that for this
strand, $\tau_{cp} \ll \tau_m$, where τ_{cp} is the coupling time constant.

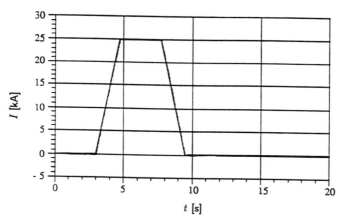

Fig. 7.17   Transport current pulse applied to the US-DPC [6.30].

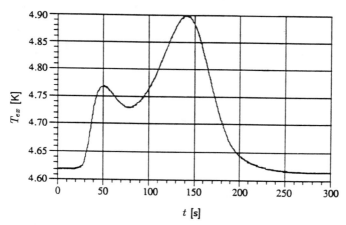

Fig. 7.18   Temperature waveform measured at the exit of DP-B
for the transport current pulse shown in Fig. 7.17 [6.30].

Problem 7.8: AC losses in the US-DPC Coil

a) Explain the general shape of the temperature trace $T_{ex}(t)$ shown in Fig. 7.18.

b) Using the exit temperature trace, $T_{ex}(t)$ presented in Fig. 7.18, determine the total AC losses generated in the DP-B, E_{ac} [J], by the transport current pulse. You should derive, by clearly specifying the thermodynamics control volume, appropriate equations required for loss computation. To simplify numerical computation, you may use an average C_p of 4.0 J/g K for helium in the temperature and pressure ranges of interest.

c) Using the data of Fig. 7.19, compute E_{ac} generated in the DP-B under the current pulse of Fig. 7.17 and demonstrate that it is within ~10% of E_{ac} determined in **b)**. Assume that there are no strand-to-strand coupling losses in the DP-B conductor. Also assume that the conductor's critical current at 5 T is 100 kA. *Hint:* To compute E_{ac} from Fig. 7.19 data with the accuracy required for this problem, it is necessary to divide the DP-B radially into 5 segments and use an appropriate value of B_m (given in Table 7.5) for each segment to determine its e_{hy} and e_{cp}.

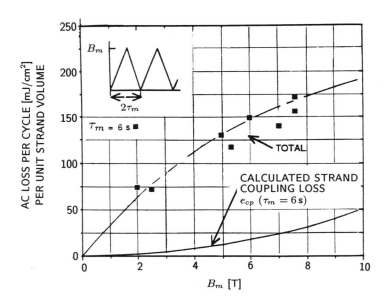

Fig. 7.19 Total AC losses per cycle for unit strand volume [mJ/cm^3], at 4.2 K, *vs* B_m [T] data of a *single* DP-B conductor strand subjected to external field pulses (inset); the strand carried no transport current. The continuous solid curve gives the coupling loss density per cycle for this strand at $\tau_m =6$ s: $e_{cp} =0.53B_m^2$ mJ/cm^2, with B_m in tesla. Based on data presented in [7.46].

Solution to Problem 7.8

a) The first temperature peak with $\Delta T = 4.77\,\text{K} - 4.62\,\text{K} = 0.15\,\text{K}$ occurring at $t = 50\,\text{s}$ is a result of the current ramp-up that induced the first dose of AC losses in the coil. The second peak with $\Delta T = 4.90\,\text{K} - 4.73\,\text{K} = 0.17\,\text{K}$ occurring at $t = 135\,\text{s}$ is a result of the current ramp-down that induced the second dose of AC losses in DP-B. Because each ramp current has the same $|dB/dt|$, each should have generated the same heating in the DP-B or the same ΔT. The measured ΔT from each ramp agrees reasonably well.

b) The control volume encloses the DP-B; helium enters into and leaves from the control volume, respectively, at the inlet and exit of the DP-B. Within this control volume, the following power equation for *helium* is valid during the current ramping:

$$\frac{dE_{cv}}{dt} = \frac{dE_{ac}}{dt} - \dot{W}_{he} + \dot{m}[h_{in}(t) - h_{ex}(t)] \qquad (S8.1)$$

where E_{cv} is the internal energy of the fluid; $\dot{W}_{he}$ is the work (actually time rate of change), except the $p\,dV$ work, performed *by* the fluid external to the control volume; $h_{in}(t)$ and $h_{ex}(t)$ are, respectively, the specific helium enthalpies at the inlet and exit. For this particular case, $h_{in}(t)$ and $h_{ex}(t)$ may be expressed as:

$$h_{in}(t) = (h_{in})_{ss} \qquad (S8.2a)$$
$$h_{ex}(t) = (h_{ex})_{ss} + \xi_{ex}(t) \qquad (S8.2b)$$

where $(h_{in})_{ss}$ and $(h_{ex})_{ss}$ are steady-state enthalpies at the inlet and exit; and $\xi_{ex}(t)$ is the "transient" enthalpy at the exit, resulting from the current pulse. Integrating Eq. S8.1 with time between $t = 0$ and $t = 250\,\text{s}$, and realizing that $E_{cv}(0) = E_{cv}(250\,\text{s})$, we have:

$$0 = E_{ac} - \int_0^{250\,\text{s}} \dot{W}_{he}\,dt$$
$$+ \dot{m}\Big[(h_{in})_{ss} - (h_{ex})_{ss}\Big](250\,\text{s}) - \dot{m}\int_0^{250\,\text{s}} \xi_{ex}(t)\,dt \qquad (S8.3)$$

Because $\dot{W}_{he}$, if any, does not change during the current pulsing, it remains constant at all times and the following equation is valid at *all times*, including during the 250-s period of interest:

$$\int_0^{250\,\text{s}} \dot{W}_{he}\,dt = \dot{m}\Big[(h_{in})_{ss} - (h_{ex})_{ss}\Big](250\,\text{s}) \qquad (S8.4)$$

In this particular case, because the fluid has no work interaction (except the $p\,dV$ work) external to the control volume, $\dot{W}_{he} = 0$ and $(h_{in})_{ss} = (h_{ex})_{ss}$, which is identical to the condition for J-T (isenthalpic) expansion.

Solution to Problem 7.8

[For the range of temperature and pressure for this case, the J-T expansion coefficient, $\mu_{JT} = (\partial T/\partial p)_h$, is *negative*, making $(T_{ex})_{ss} > (T_{in})_{ss}$ because $(p_{in})_{ss} > (p_{ex})_{ss}$. However, *measured* values were $(T_{ex})_{ss} = 4.62\,\text{K}$ and $(T_{in})_{ss} = 4.73\,\text{K}$. The only reasonable explanation for this discrepancy is inaccuracy in temperature measurement involving different sensors. That is, *differential* measurement with one sensor, *e.g.* between inlet and exit, is generally accurate; absolute agreement involving different sensors is generally much more difficult. Because of this confusing discrepancy in temperatures, T_{in} was not given in the problem statement.]

Combining Eqs. $S8.3$ and $S8.4$, and recognizing that $\xi_{ex}(t)$ is given by $C_p \Delta T_{ex}(t)$, with $\Delta T_{ex}(t) = T_{ex}(t) - T_{ex}(0) = T_{ex}(t) - 4.62\ \text{K}$, we have:

$$E_{ac} \simeq \dot{m} C_p \int_0^{250\,\text{s}} \Delta T_{ex}(t)\, dt \qquad (S8.5)$$

The temperature integral of Fig. 7.18 has a value of about $26\,\text{K s}$. With $\dot{m} = 13\,\text{g/s}$ and $C_p = 4.0\,\text{J/g K}$, we obtain $E_{ac} \simeq 1350\,\text{J}$.

c)   As may be inferred from the field data given in Table 7.5, the field is not uniform over DP-B. Because both e_{hy} and e_{cp} are field-dependent, it is necessary to take into account the field nonuniformity in computing these loss densities. Following the suggestion given in the problem statement, we divide the DP-B into five radial segments as in Table 7.5. We follow the procedure given below:

- Compute e_{hy} in each radial segment; take an average of r-weighted e_{hy}, $\tilde{e}_{hy}$; and multiply $\tilde{e}_{hy}$ with total strand volume, V_{st}, in the DP-B to obtain E_{hy}, the total hysteresis loss.

- Compute e_{cp} in each segment; average r-weighted e_{cp} and obtain $\tilde{e}_{cp}$; and then multiply $\tilde{e}_{cp}$ with V_{st} to finally obtain E_{cp}.

- The total AC dissipation E_{ac} in the DP-B is given by: $E_{ac} = E_{hy} + E_{cp}$.

The total strand volume V_{st} (Table 7.4): $V_{st} = (150 \times 10^2\ \text{cm})(107 \times 10^{-2}\ \text{cm}^2)$ or $V_{st} = 16050\ \text{cm}^3$.

Hysteresis: Because the transport current increases from 0 to $25\,\text{kA}$ linearly with time and $25\,\text{kA}$ is $1/4$ of I_c ($100\,\text{kA}$), the multiplying factor $(1+i^2)$ has a peak value of 1.0625 at $25\,\text{kA}$; its average is ~ 1.02, which is neglected in computation of $e_{hy}(I_t)$.

As e_{hy} is independent of dB/dt, $e_{hy}(B_m)$ at each of the five radial locations may be estimated by taking the difference between the total losses (the line connecting data points) and $e_c = 0.53 B_m^2\ \text{mJ/cm}^3$. The results are given in Table 7.6.

With $\sum r = 3.5$ m and $\sum re_{hy} = 221\ \text{m mJ/cm}^3$, we have $\tilde{e}_{hy}$ of $63\,\text{mJ/cm}^3$.

The total hysteresis loss: $E_{hy} = \tilde{e}_{hy} V_{st} = 1011\ \text{J}$.

Solution to Problem 7.8

Table 7.6: Hysteresis Loss Density Computation

r [m]	$|B|$ [T]	e_{total}* [mJ/cm³]	e_{cp}† [mJ/cm³]	e_{hy} [mJ/cm³]	re_{hy} [m mJ/cm³]
0.5	5.0	136	13	123	62
0.6	3.0	93	5	88	53
0.7	1.5	50	1	49	34
0.8	0.5	18	0	18	14
0.9	2.0	66	2	64	57

* From Fig. 7.19 data points.
† Based on $e_{cp} = 0.53B_m^2$.

Coupling: Because e_{cp} depends on τ_m, the solid line given in Fig. 7.19 cannot be used directly. Its value, corresponding to $\tau_m = 6$ s, however, can be scaled for our case ($\tau_m = 1.75$ s). Although p_{cp}, coupling loss *power* density, scales as $(B_m/\tau_m)^2$, e_{cp} scales as B_m^2/τ_m because $e_{cp} \propto \tau_m p_{cp}$. Table 7.7 presents the appropriate values needed to compute e_{cp} with $\tau_m = 1.75$ s.

With $\sum r = 3.5$ m and $\sum re_{cp} = 41$ m mJ/cm³, we have $\tilde{e}_{cp}$ of 12 mJ/cm³.

The total coupling loss: $E_{cp} = \tilde{e}_{cp}V_{st} = 193$ J.

Total AC loss: Combining E_{hy} and E_{cp}, we have: $E_{ac} = E_{hy} + E_{cp} = 1204$ J, which is ~150 J less than that determined in b) based on the $T_{ex}(t)$ trace.

The measured strand AC losses of the DP-B conductor are greater by ~150 J than those based on the *single* strand data (Fig. 7.19). This discrepancy, however, is not due to strand-to-strand coupling losses in the DP-B because they were independently measured to be negligible. Because the single-strand AC loss measurements were performed before the strands were configured into the CIC conductor, the discrepancy might be due to a slight difference between the test strand and actual strands contained in the DP-B.

Table 7.7: Coupling Loss Density Computation

r [m]	$|B|$ [T]	e_{cp}* [mJ/cm³]	e_{cp}† [mJ/cm³]	re_{cp} [m mJ/cm³]
0.5	5.0	13	45	23
0.6	3.0	5	17	10
0.7	1.5	1	3	2
0.8	5.0	~0	~0	~0
0.9	2.0	2	7	6

* From Fig. 7.19 plot.
† Scaled by 6/1.75.

Problem 7.9: Splice dissipation in Hybrid III Nb-Ti coil

This problem uses the Hybrid III Nb-Ti coil to illustrate estimation of splice dissipation. Splice dissipation is important in magnets operated in a bath of 1.8-K superfluid helium such as Hybrid III SCM because the cost of refrigeration is higher at 1.8 K than at 4.2 K.

The Hybrid III Nb-Ti coil consists of 32 double pancakes, each double pancake is wound with two grades of 9.2-mm wide Nb-Ti composite strip. In each single pancake, a "shake-hands" splice (Fig. 7.2, p. 266) between the high-field (HF) grade conductor and low-field (LF) grade conductor occurs at $r = 378$ mm covering a 90° arc; in each double pancake there are thus two such splices. In addition, there are two more splices in each double pancake, at $r = 455$ mm over a 90° arc, connecting the pancakes within the double pancake, which in turn is connected to the next double pancake. Altogether the Hybrid III Nb-Ti coil has a total of 64 splices at $r = 378$ mm and 64 splices at $r = 455$ mm.

Compute the total splice dissipation rate in the Hybrid III Nb-Ti coil at 2100 A. Each splice was soldered with Sn50-Pb50 solder. Assume that the splices at $r = 378$ mm are at 3 T and those at $r = 455$ mm are at 1 T.

Mechanical Properties of Tin-Lead Solders

Table 7.8 presents selected properties of tin-lead solders [7.8]. Based on a factor of 3~4 improvement in tensile strength (σ_U) in these alloys from room temperature to cryogenic temperatures, it is reasonable to expect a similar improvement in shear strength (σ_{sh}) over the same temperature range; in shake-hand lap joints, shear strength is a more important property than tensile strength.

Table 7.8: Mechanical Properties of Sn-Pb Solders

Composition[%-%]	T_{mb} [C°]*	T_{ml} [C°]*	σ_U [MPa]†	σ_{sh} [MPa]†
Sn100-Pb0	232	232	12	18
Sn60-Pb40	183	190	44	39
@ 77 K	—	—	124 (90)‡	—
@ 4.2 K	—	—	173 (149)	—
Sn50-Pb50	183	215	45	40
@ 77 K	—	—	114 (89)	—
@ 4.2 K	—	—	171 (132)	—
Sn40-Pb60	183	238	44	39
@ 77 K	—	—	105 (79)	—
@ 4.2 K	—	—	158 (112)	—
Sn0-Pb100	327	327	12	12

* T_{mb}: temperature at which melting begins; T_{ml}: molten temperature.

† Unless otherwise specified, at room temperature.

‡ Yield strength.

Solution to Problem 7.9

To make a conservative estimate, we shall use R_{ct} data having the higher values of R_{ct} for Sn50-Pb50 (Table 7.2, p. 267): R_{ct} of $3.3 \times 10^{-12}\,\Omega\,m^2$ at 1 T and $4.1 \times 10^{-12}\,\Omega\,m^2$ at 3 T.

Splice resistance at 378 mm: By applying Eq. 7.20, we compute resistance R_{hl} at $r = 378\,mm$:

$$R_{hl} = \frac{R_{ct}}{a\ell_{hl}} \qquad (S9.1)$$

where a is the conductor width and ℓ_{hl} is the splice overlap length. ℓ_{hl} is given by: $\ell_{hl} = \pi r_{hl}/2$, where r_{hl} is the winding radius at which the splice takes place. With $R_{ct} = 4.1 \times 10^{-12}\,\Omega\,m^2$, $a = 9.2 \times 10^{-3}\,m$, $\pi/2 = 1.57$ (90° arc), and $r_{hl} = 0.378\,m$, we have:

$$R_{hl} = \frac{(4.1 \times 10^{-12}\,\Omega\,m^2)}{(9.2 \times 10^{-3}\,m)(1.57 \times 0.378\,m)} = 0.75\,n\Omega \qquad (S9.2)$$

Splice resistance at 455 mm: Similarly, resistance R_{pp} at $r = 455\,mm$ is given by:

$$R_{pp} = \frac{R_{ct}}{a\ell_{pp}} \qquad (S9.3)$$

where a is the conductor width and ℓ_{pp} is the splice overlap length. ℓ_{pp} is given by: $\ell_{pp} = \pi r_{pp}/2$, where r_{pp} is the winding radius at which the pancake-pancake splice takes place. With $R_{ct} = 3.3 \times 10^{-12}\,\Omega\,m^2$, $a = 9.2 \times 10^{-3}\,m$, $\pi/2 = 1.57$, and $r_{pp} = 0.455\,m$, we have:

$$R_{pp} = \frac{(3.3 \times 10^{-12}\,\Omega\,m^2)}{(9.2 \times 10^{-3}\,m)(1.57 \times 0.455\,m)} = 0.50\,n\Omega \qquad (S9.4)$$

Total splice resistance: Total splice resistance, R_{sl}, is given by:

$$R_{sl} = 64R_{hl} + 64R_{pp} = 48\,n\Omega + 32\,n\Omega = 80\,n\Omega \qquad (S9.5)$$

Total dissipation: Total dissipation at $I_{op} = 2100\,A$, P_{sl}, is given by:

$$P_{sl} = R_{sl}I_{op}^2 = (80 \times 10^{-9}\,\Omega)(2.1 \times 10^3\,A)^2 = 0.35\,W \qquad (S9.6)$$

That is, total dissipation at 2100 A is a fraction of a watt. This estimate is consistent with data obtained from Hybrid III runs. When the system is allowed to reach the lowest possible temperature with no transport current, the bath temperature reaches 1.65 K; the ultimate temperature is unchanged even with a current of 2100 A, indicating that the amount of extra dissipation by splices is indeed negligible compared with an estimated quiescent refrigeration load of ~10 W.

Problem 7.10: A splice for CIC conductors

Here we discuss a "shake-hands" splice for CIC conductors developed at FBNML [7.10]. It is used to join CIC conductors in a Nb-Ti coil (Coil C), the outermost coil of the three superconducting coils (Coils A, B, and C), for the NHMFL's 45-T hybrid magnet [3.13]. The Nb-Ti coil consists of 29 double pancakes, each wound with a CIC conductor. (The drawing of a double pancake shown in Fig. 3.6 is based on this double pancake.) The conductor contains 135 Nb-Ti multifilamentary strands encased in a steel conduit [7.47]. (Table 8.5 in Chapter 8 gives a more complete list of the conductor parameters.) Each of the 29 splice-resistances in the Nb-Ti coil must not exceed $0.33\,n\Omega$ in a field of $3.5\,T$; this limits the total dissipation for the 29 splices to $1\,W$ at the design operating current of $10\,kA$. Note that because it is difficult to splice CIC conductors within a pancake winding, each double pancake in this coil is wound with one grade of CIC conductor.

Figure 7.20 shows the cross section of the splice where two "peeled" (from conduit) bundles of cabled strands are tightly placed within a copper channel, separated by a septum which extends over the entire overlap length, and filled with solder (Sn60-Pb40, Table 7.2). As with all shake-hands splices, the overlap length is used to control the splice resistance. Because the strands in each bundle are transposed with a twist pitch length shorter than the septum length, each strand contacts the septum; the distance between strands on one side to those on the other side is thus defined by the septum thickness—the shortest distance between the bundles. A stainless steel enclosure (not shown in Fig. 7.20), a part of the CIC helium circuit, houses each splice.

A top view of a 64-cm long scale model of the splice is shown in Figure 7.21. The splice occupies nearly the entire loop; current enters the loop from one end through one bundle and leaves the loop from the other end through the other bundle. Current commutation takes place within the splice.

Figure 7.22 shows zero-field voltage *vs* current data obtained for this model splice with the splice immersed in a bath of liquid helium at $4.2\,K$ [7.10].

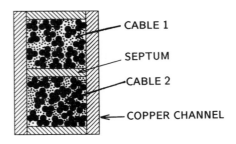

Fig. 7.20   Schematic drawing of the cross section of Coil C's shake-hands splice for CIC conductors, each containing 135 strands [7.10]. The actual void space filled with solder (dotted) is much less than appears in this drawing.

Problem 7.10: A splice for CIC conductors

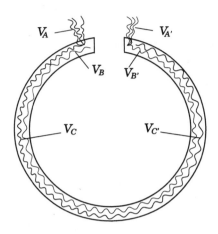

Fig. 7.21 Scale model splice, 64 cm long. For clarity, only one strand from each bundle is shown within the splice; also the two strands are configured side-by-side rather than top-to-bottom as is actually the case shown in Fig. 7.20.

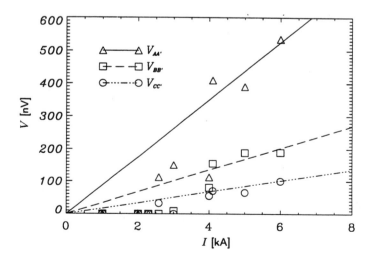

Fig. 7.22 Zero field voltage vs current data for the splice immersed in a bath of liquid helium at 4.2 K [7.10].

Problem 7.10: A splice for CIC conductors

  a) Explain the shape of each voltage plot in Fig. 7.22.

  b) What is the overall splice resistance for this scale model?

  c) Make a voltage *vs* distance plot at 6 kA. Assume the splice to be electrically symmetric about its midpoint. Does voltage increase linearly with distance from entrance (tap A in Fig. 7.21) to exit (tap A') ? If it does not, give a new way of interpreting the data.

  d) Estimate the septum resistance (R_{sp}) and reinterpret the data. The septum is 3 mm thick, 10 mm wide, and 60 cm long. Assume $\rho_{cu} = 1.0 \times 10^{-10}\,\Omega\,\mathrm{m}$ (RRR$\simeq$173), a value for OFHC (oxygen-free-high-conductivity) copper at 4.2 K and zero field.

  e) Estimate R_{sl} in a background field of 3.5 T, the exposure field for these splices in actual operation. Also estimate a total Joule dissipation due to 29 of these splices at 10 kA. Is it less than a maximum limit of 1 W? Use the Kohler plot (Fig. A4.2, Appendix V) to account for magnetoresistive effect.

Stability of a CIC Splice in a Time-Varying Magnetic Field

As exemplified by Coil C's splice, whose cross section is shown in Fig. 7.20, a CIC splice is essentially a solid mass of conductive metal. Under a time-varying magnetic field, a large amount of AC loss is thus generated in the CIC splice. To keep the splice superconducting at all times, it must be well cooled.

The most critical event in hybrid systems is an insert burnout. The burnout causes a sudden drop in the insert's fringing field to which a splice is exposed, heating the splice. We may make the most conservative estimate of this heating density, Δe_{ac}, by equating it to the net change in magnetic energy stored in the splice. Thus:

$$\Delta e_{ac} = \frac{2\vec{B}_{ff} \cdot \Delta \vec{B}_{ff} - |\Delta B_{ff}|^2}{2\mu_o} \tag{7.33}$$

where $\vec{B}_{ff}$ is the total fringing magnetic induction at the splice before the burnout and $\Delta \vec{B}_{ff}$ is a decrease in that magnetic induction due to insert burnout. Note that for Coil C, $\vec{B}_{ff}$ is a vector sum of the fringing magnetic inductions generated by the insert and Coils A, B, and C. (Coils A and B are two inner Nb_3Sn coils.)

To keep the splice superconducting during an insert burnout, the total heating generated in the splice must be absorbed by the liquid helium layer (δ_{he}) that surrounds the splice. This assumption of a thin liquid layer soaking up energy is generally valid for subcooled, 1.8-K liquid. For this splice having rectangular cross section, a_{sl} wide and b_{sl} high, stability requires that:

$$a_{sl}b_{sl}\Delta e_{ac} \leq 2(a_{sl} + b_{sl})\delta_{he}\Delta h_{he} \tag{7.34}$$

For Coil C's splice located at the magnet midplane, $|\vec{B}_{ff}|\sim 3.5\,\mathrm{T}$, $|\Delta \vec{B}_{ff}|\sim 0.1\,\mathrm{T}$, $a_{sl}\sim 16\,\mathrm{mm}$, $b_{sl}\sim 33\,\mathrm{mm}$, and $\Delta h_{he} = 0.29 \times 10^6\,\mathrm{J/m^3}$, Eq. 7.34 gives: $\delta_{he} \geq 5\,\mathrm{mm}$.

Solution to Problem 7.10

a) Each V *vs* I plot shows that V is essentially zero for currents up to 2~3 kA, eventually merging to a straight line asymptote (Fig. 7.22). The only reasonable explanation is to assume that the solder is superconducting up to 2~3 kA—note that the splice was in zero background field. Either self field or current densities are sufficient to destroy solder's superconductivity above these currents.

b) From the slope of the solid straight line (Fig. 7.22), we have: $R_{sl} = 87\,\text{p}\Omega$ as the overall splice resistance at 4.2 K and in zero field.

c) Using the same data, we can construct a V *vs* distance plot (Fig. 7.23), from entrance A (V_A) to exit A' ($V_{A'}$). The plot shows a steep voltage rise at both end regions. This "end" effect is not well understood. If it is due to current commutation from the filaments to the matrix within each strand end, then this commutation resistance should be evident at currents even below 2~3 kA.

The voltage between B (V_B) and B' ($V_{B'}$) increases linearly with distance. We may divide splice resistance (R_{sl}) into two components, one corresponding to the end effect (R_{ee}) and the other corresponding to the main region (R_{jt}): $R_{sl} = R_{ee} + R_{jt}$. R_{ee} is independent of a splice's overlap length (ℓ_{sl}), while R_{jt} varies inversely with ℓ_{sl}. For this scale model, we have $R_{ee} = 54\,\text{p}\Omega$ and $R_{jt} = 33\,\text{p}\Omega$.

d) Using the appropriate values for the septum, we have: $R_{st} = 50\,\text{p}\Omega$. Since R_{jt} is 33 pΩ, less than the septum's 50 pΩ, the septum is apparently "shunted" by the other three sides, whose combined resistance is ~100 pΩ.

e) At 3.5 T and 4.2 K, $\Delta\rho_{cu}(B)/\rho_{cu}(0)$ for this copper (RRR = 173), from a Kohler plot, is ~2, or the splice resistance is 3 times the value at zero field: $R_{sl}(3.5\,\text{T}) = 261\,\text{p}\Omega$. The total Joule dissipation at 10 kA for 29 splices thus becomes ~0.8 W, still within the maximum limit.

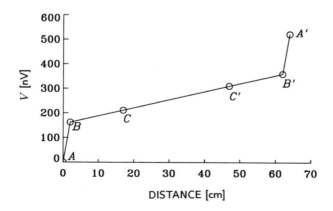

Fig. 7.23   Voltage *vs* distance plot at 6 kA.

Problem 7.11: Loss due to "index" number

This problem deals with nonzero resistance in superconductors due to "index" number, n. This index becomes an important conductor specification parameter for "persistent-mode" magnets, because these magnets must maintain field decay rates below a certain level, typically 0.01 ppm (part per million) over 1-hr period. Field decay rates are directly proportional to the total resistance of the "shorted loop" of such a magnet. (Figure 8.11 of Problem 8.9 presents a basic circuit for persistent-mode magnets.) As we shall see, the index can contribute significantly to the total resistance.

It has been experimentally determined that in "real" conductors, electric field E *vs* transport current density J characteristics may be given by:

$$E = E_c \left(\frac{J}{J_c} \right)^n \tag{7.35}$$

Here, E_c is the critical electric field that defines critical current density J_c. In practice, E_c ranges from 0.1 to 1 μV/cm. n is known as the index number and for an "ideal" superconductor $n = \infty$. Equation 7.35 implies that for "real" superconductors, *i.e.* those with $n < \infty$, the electric field is not zero even for current densities below J_c. It should be noted that Eq. 7.35 is based on measurement at or above J_c; below J_c, E decreases to a level too small to be measured easily in practice. [The question is if Eq. 7.35 is valid even at J quite well below ($<80\%$) J_c where most persistent magnets are operated.]

It is believed that the finiteness of the index number is caused by a nonuniformity ("sausaging effect") in the diameter of the filaments [7.48, 7.49]. Figure 7.24 shows three E *vs* J plots corresponding to three superconductors having the same J_c, one with $n = \infty$ and the others with $n < \infty$, $n_1 > n_2$. (The curves are not to scale, especially below J_c, so that the nonzero aspect of E in this region can be magnified.) Note that at $J_{op} < J_c$, $E = 0$ for conductor with $n = \infty$ and $0 < E_1 < E_2$ for conductors with n_1 and n_2 ($< n_1$).

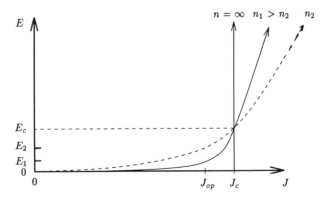

Fig. 7.24 Superconducting transition at three different index numbers.

Problem 7.11: Loss due to "index" number

Consider a persistent-mode superconducting magnet wound with a conductor whose E vs J curve at the magnet's maximum field region corresponds to $n < \infty$ of Fig. 7.24. Because field strength decreases away from the maximum field region, located in the vicinity of the magnet midplane at the innermost winding radius, the conductor's J_c also increases and J_{op}/J_c decreases away from the maximum field region. Therefore, it is necessary to consider the index voltage only in the maximum field region. Let ℓ_{mx} be the total conductor length in this region.

a) Show that the current decay rate, $dI_{op}(t)/dt$, because of index-induced resistive voltage, is given by:

$$\frac{dI_{op}(t)}{dt} = -\frac{E_c}{L_m}\left(\frac{I_{op}}{I_c}\right)^n \ell_{mx} \tag{7.36}$$

where L_m is the magnet self inductance.

b) Suppose the field decay rate must be less than a critical level, $\Delta H/(H_o \tau_p)$; show that the I_{op}/I_c must satisfy the following equation:

$$\frac{I_{op}}{I_c} \leq \left[\frac{L_m I_{op}}{E_c \ell_{mx}}\left(\frac{\Delta H}{H_o \tau_p}\right)\right]^{1/n} \tag{7.37}$$

c) For the following set of magnet parameters, solve I_{op}/I_c for $n = 10, 20, 25, 50$. $I_{op} = 300\,\text{A}$; $L_m = 100\,\text{H}$; $E_c = 0.1\,\mu\text{V/cm}$; $\ell_{mx} = 10^3\,\text{m}$; and $\Delta H/(H_o\tau_p) = 10^{-8}/\text{hr}$.

Experimental Determination of Index Number

The technique most widely used to determine index number n of a superconductor is through the conductor's V vs I plot, a typical shape of which is shown in Fig. 7.25. If $E(J)$ is assumed to be given by Eq. 7.35, then it follows immediately that n may be determined by measured values of V_1, V_2, I_1, and I_2 (Fig. 7.25):

$$n = \frac{\ln(E_2/E_1)}{\ln(I_2/I_1)} \tag{7.38}$$

Typically, $E_1 = 0.1\,\mu\text{V/cm}$ and $E_2 = 1.0\,\mu\text{V/cm}$.

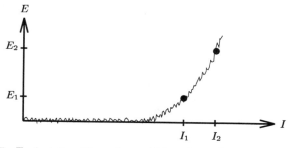

Fig. 7.25 Typical E vs I trace from which to determine an index number.

Solution to Problem 7.11

a) By integrating E field over a conductor length ℓ_{mx}, we obtain a total index-induced voltage, V_n, at I_{op} given by:

$$V_n = E\ell_{mx} = E_c \left(\frac{I_{op}}{I_c}\right)^n \ell_{mx} \tag{S11.1}$$

This resistive voltage causes a current decay, given by:

$$-L_m \frac{dI_{op}}{dt} = V_n$$

$$\frac{dI_{op}}{dt} = -\frac{V_n}{L_m} = -\frac{E_c}{L_m}\left(\frac{I_{op}}{I_c}\right)^n \ell_{mx} \tag{7.36}$$

b) Because I_{op} and H_o are directly related, we have:

$$\frac{dH}{dt} \equiv -\left(\frac{\Delta H}{\tau_p}\right) \propto -\frac{E_c}{L_m}\left(\frac{I_{op}}{I_c}\right)^n \ell_{mx}$$

$$\left(\frac{\Delta H}{H_o\tau_p}\right) = \frac{E_c}{L_m I_{op}}\left(\frac{I_{op}}{I_c}\right)^n \ell_{mx} \tag{S11.2}$$

Solving for I_{op}/I_c from Eq. $S11.2$, we obtain:

$$\frac{I_{op}}{I_c} \leq \left[\frac{L_m I_{op}}{E_c \ell_{mx}}\left(\frac{\Delta H}{H_o\tau_p}\right)\right]^{1/n} \tag{7.37}$$

c) By inserting $\Delta H/(H_o\tau_p) = 10^{-8}$ /hr $= 2.78\times10^{-12}$ /s and other parameter values into Eq. 7.37, we have:

$$\frac{I_{op}}{I_c} = \left[\frac{(300\,\text{A})(100\,\text{H})}{(10^{-7}\,\text{V/cm})(10^5\,\text{cm})}(2.78\times10^{-12}\,\text{/s})\right]^{1/n} \tag{S11.3}$$

From Eq. $S11.3$, we have: $I_{op}/I_c = 0.31$ for $n = 10$; $I_{op}/I_c = 0.56$ for $n = 20$; $I_{op}/I_c = 0.68$ for $n = 30$; and $I_{op}/I_c = 0.79$ for $n = 50$.

These values indicate that for a conductor of $n = 10$, I_{op} must be kept below 31% of I_c—rather an inefficient use of the conductor; for $n = 50$, I_{op} may be increased to 79% of I_c. In most cases, $n > 25$ is adequate to make I_{op}/I_c at least about 70% and $n \sim 30$ is not an impossibly high value of n for Nb-Ti multifilamentary conductors presently available commercially.

Furthermore, results of a recent experimental study [7.50] show that n, determined from a V vs I plot at and *above* I_c, instead of remaining constant actually *increases* for currents *below* I_c. In one test conductor, n is found to increase from $\sim$30 at I_c to 42 at $I/I_c = 0.86$, 123 at $I/I_c = 0.81$, and 145 at $I/I_c = 0.75$. The results thus indicate that as long as I_{op} is not chosen too close to I_c, even conductors with n values as low as $\sim$20 at I_c may be usable for persistent-mode magnets.

Problem 7.12: Frictional sliding

This problem covers the basics of frictional sliding. As discussed in the introductory section, high-performance superconducting magnets are susceptible to quench because of mechanical disturbances, specifically of frictional motion and epoxy-resin cracking. Figure 7.26 presents three friction coefficient (μ_f) vs sliding velocity (v) curves for two surfaces of different material undergoing relative motion; μ_f is defined as the ratio of the force acting normal to the direction of motion to that acting parallel to the direction of motion.

The solid curve, with both positive slope ($d\mu_f/dv > 0$) and negative slope regions, represents the most general case. The positive slope region results from the interfacial creep behavior, while the negative slope region arises from the breakdown of the creep mechanism. As the sliding speed increases past the maximum point (μ_{fpk} at v_{pk}), μ_f decreases because of the limited time available for interfacial contact [7.22, 7.23].

The dotted curve represents sliding characteristics whose maximum μ_f occurs so near the ordinate of zero velocity that only the negative slope region exists in practice. If the positive slope region is a manifestation of interfacial creep behavior, which diminishes with temperature, then this negative-slope-region-only behavior exhibited at cryogenic temperatures by many magnet winding materials, e.g. copper/nylon, copper/polyethylene, copper/Teflon [7.19, 7.20], is consistent. Note that all these pairs have a positive slope region at room temperature.

The dashed curve represents the behavior of creep-resistant materials and is quite prominently observed at 4.2 K with material pairs like copper/G-10, and copper/phenolic.

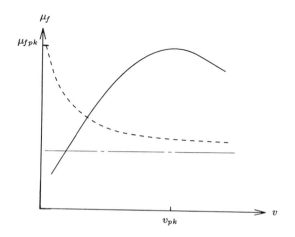

Fig. 7.26   Three friction coefficient vs sliding velocity curves.

Problem 12: Frictional sliding

Figure 7.27 shows a mass M on a smooth surface subjected to two forces, F_n acting in the direction normal to the surface and F_x acting parallel to the surface, in the x-direction. This model can be used to represent, in the simplest degree, conductor motion taking place inside the magnet winding, in which both F_n and F_x are components of the Lorentz force acting over a length of the conductor.

The friction coefficient between the mass and surface may be characterized by the solid curve of Fig. 7.26. For the positive and negative sloped regions, assume that $F_\mu(v)$ may be approximated with the following simple expressions:

Positive-sloped region: $\qquad F_\mu(v) = av \qquad (v < v_{pk}) \qquad\qquad (7.39a)$

Negative-sloped region: $\qquad F_\mu(v) = -bv \qquad (v > v_{pk}) \qquad\qquad (7.39b)$

where a and b are both positive constants and v_{pk} is the velocity corresponding to μ_{fpk}, the peak μ_f. In reality, a and b are constant only over a small velocity range. Generally, μ_f vs v plots shown in Fig. 7.26 have the v-axis given in a log scale.

a) Show that the sliding motion of the mass in the positive-sloped region is stable. Specifically, show that an expression of $v(t)$ in this region in response to a step force function, $F_x(t) = F_x < \mu_{fpk} F_n$ for $t > 0$, is given by:

$$v(t) = \frac{F_x}{a}\left(1 - e^{-at/M}\right) \qquad\qquad (7.40a)$$

Note that $v(0) = 0$.

b) Show that the sliding motion of the mass in the negative-sloped region is unstable. Specifically, show that an expression of $v(t)$ in response to a step force function, $F_x(t) = F_x > \mu_{fpk} F_n$ for $t > 0$, is given by:

$$v(t) = \frac{F_x}{b}\left(e^{bt/M} - 1\right) \qquad\qquad (7.40b)$$

Note that $v(0) = 0$.

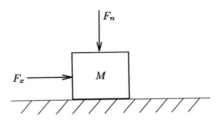

Fig. 7.27   Mass M subjected to lateral force F_x and sliding on a smooth surface.

Solution to Problem 7.12

The equation of motion is given by:

$$M\frac{dv}{dt} = F_x - F_\mu \qquad (S12.1)$$

where F_μ, the friction force, has the minus sign because the force acts in the direction opposite to that of F_x.

a) Combining Eq. $S12.1$ and $7.39a$ with $F_x < \mu_{fpk}F_n$, we have an expression for the equation of motion for M given by:

$$M\frac{dv}{dt} + av = F_x \qquad (S12.2)$$

Solving Eq. $S12.2$ for $v(t)$, we obtain:

$$v(t) = \frac{F_x}{a}\left(1 - e^{-at/M}\right) \qquad (7.40a)$$

where the constant of integration is taken to make $v = 0$ at $t = 0$. Clearly, the motion is stable with a steady-state velocity of F_x/a.

b) For the case when the μ_f vs v curve is negative-sloped ($F_x > \mu_{fpk}F_n$), we have:

$$M\frac{dv}{dt} - bv = F_x \qquad (S12.3)$$

We can solve Eq. $S12.3$ for $v(t)$, obtaining:

$$v(t) = \frac{F_x}{b}\left(e^{bt/M} - 1\right) \qquad (7.40b)$$

where the constant of integration is again taken to make $v = 0$ at $t = 0$. Here, unlike in the positive-sloped case, the motion is unstable, with v increasing exponentially with time. In real magnets, the velocity is limited by the physical arrangement of the windings.

This "stick-slip" type of motion occurring within the winding causes premature quenches in high-performance adiabatic magnets. Typically, each stick-slip event covers a distance of $\sim 10\,\mu$m and hence it is appropriate to call a stick-slip event "microslip" [7.12]. Even at this small displacement, a microslip can still cause a quench in high-performance magnets. Because, as remarked above, positive-sloped μ_f vs v behavior is based on an interfacial creep mechanism, which depends on thermal activation, it ceases at cryogenic temperatures, particularly at 4.2 K, making it virtually impossible to guarantee stable motion with materials most often used in the magnet winding.

Problem 7.13: Source location with AE signals

This problem illustrates the usefulness of AE signals in pinpointing the location of a mechanical event in superconducting magnets.

Figure 7.28 presents an oscillogram showing two traces of AE signals recorded from an epoxy-impregnated test coil equipped with two AE sensors. The top trace corresponds to the sensor attached to the top end flange of the test coil and the bottom trace corresponds to the sensor at the bottom end flange [7.51]. Because the test coil is 0.1 m long axially from the top flange to the bottom flange, we may take the axial distance separating the two sensors to be roughly 0.1 m. Although it is not evident from this oscillogram, the mechanical event responsible for the AE signals did trigger a quench.

Before the experiment, a support rod holding the test coil in a cryostat was tapped, sending a mechanical impulse down the support rod and through the test coil. From these measurements, the acoustic wave propagation velocity in the coil's axial direction was determined to be 2000 m/s. (A separate set of measurements determined the acoustic wave propagation velocity along the *conductor axis* to be ~5000 m/s. This velocity is very close to that given by $\sqrt{E_{cu}/\varrho_{cu}}$, where E_{cu} and ϱ_{cu} are, respectively, copper's Young's modulus and density. The propagation along the *coil's axis* corresponds to the turn-to-turn propagation, in which the wave must travel from one conductor to the next through a thin layer of epoxy.)

Based on the traces given in Fig. 7.28, show that the mechanical event responsible for the AE signals occurred very close to the top end of the test coil. How would the two traces change if the event occurred at the coil midplane?

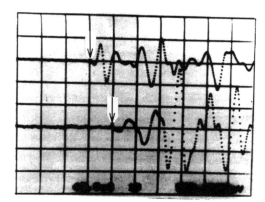

Fig. 7.28   Oscillogram showing two AE signals recorded with two AE sensors attached to an epoxy-impregnated test coil, 0.1 m long [7.51]. The top and bottom traces correspond, respectively, to the sensor attached to the top flange and the other sensor attached to the bottom flange. The arrow in each trace indicates the arrival instance of the signals. Time scale: 50 μs/div.

Solution to Problem 7.13

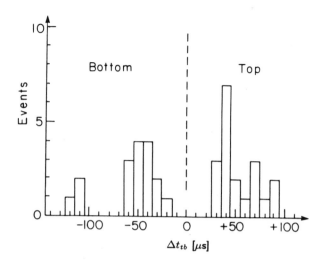

Fig. 7.29   Crack distribution histogram expressed in terms of signal arrival time difference between the top and bottom sensors for an epoxy-impregnated test coil with both ends rigidly clamped [7.51].

Since the wave propagation velocity in the axial (turn-to-turn) direction in the test coil is 2000 m/s, a transit time between the two end flanges, separated by a distance of 0.1 m, is 50 μs, which is a delay time between the top and bottom trace shown in Fig. 7.28. Thus, it is reasonable to assume that the mechanical event responsible for these signals originated near the top flange.

If the start of the bottom trace were ahead of the start of the top trace by 50 μs, then the disturbance would have originated near the bottom flange. Clearly, if the disturbance originated at the coil midplane, the start of each trace would coincide.

Figure 7.29 presents crack distribution histogram in terms of signal arrival time difference between the top and bottom sensors, Δt_{tb}, for the test coil [7.51]. The data show that quench-inducing mechanical events occur principally near coil ends, more or less equally at either top or bottom. In this series of measurements, both ends of the coil were rigidly clamped. Clamping creates large shear stresses near the coil ends [7.17], which in turn induce cracking events that are responsible for quenching. Note that the floating winding technique avoids this clamping.

"Clocks are slow on Sundays." —*Holly Golightly*

Acoustic Emission Sensor for Cryogenic Environment

As discussed in the introductory section of this chapter, AE monitoring has become an almost indispensable technique in elucidating mechanical events occurring within the winding of high-performance superconducting magnets. AE sensors commonly available commercially are, however, generally unsuitable, mainly because they are not designed to withstand the large temperature excursion that would occur when the superconducting magnet is cooled. To permit reliable AE monitoring of superconducting magnets, AE sensors specifically designed for use in cryogenic environment were developed in the early 1980s by Tsukamoto and others at FBNML. The sensors have proven to last almost indefinitely, withstanding many temperature excursion cycles.

Sensor Details: Figure 7.30 shows an exploded view of an FBNML AE sensor. The sensor's key component is a piezoelectric wafer (lead-zirconate-titanate)

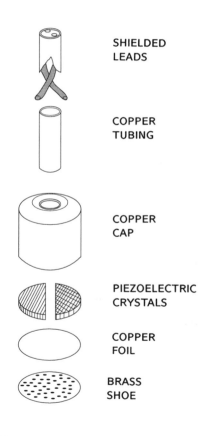

10 mm in diameter and 2 mm thick. As indicated in the figure, the disk is cut into two halves, obtainable from a supplier in paired (not random) halves. The two halves are soldered to a copper foil disk with their polarities reversed, making the AE a differential type. At the input end of a differential charge amplifier, signals from each half are subtracted to enhance AE signals and at the same time reduce noise signals. The disk assembly is mounted on a brass shoe and enclosed by a copper cap; the disk, the brass shoe, and the copper cap are soldered at their rims. The flexible copper foil allows the sensor elements to move relative to the brass shoe and copper cap during temperature cycling.

Sensor Mounting: The sensor is housed within a sturdy diecast aluminum box and spring-loaded (from both sides of the box) directly on a flat surface of the coil form. Spring loading is quite effective in ensuring a good acoustic coupling between the sensor and the magnet.   An organic compound often used to improve mechanical contact is ineffective at cryogenic temperatures—it simply becomes another hard layer and does not act as an adhesive.

SHIELDED LEADS

COPPER TUBING

COPPER CAP

PIEZOELECTRIC CRYSTALS

COPPER FOIL

BRASS SHOE

Fig. 7.30   Exploded view of AE sensor.

Problem 7.14: Conductor-motion-induced voltage pulse

This problem studies the relationship between a voltage pulse and conductor motion. In the 1980s voltage pulse signals induced by conductor motion were investigated in detail to understand the mechanism of motion-induced quench events.

Figure 7.31 illustrates a conductor motion model [7.52] in which length ℓ of a current-carrying (I_t) conductor exposed to a background field B_e is deflected like a beam under the action of Lorentz force F_L, given by:

$$F_L = \ell_{df} I_t B_e \qquad (7.41)$$

Deflection involving short lengths of conductor can happen within the winding particularly if the void space within the winding is not filled with epoxy resin.

From the theory of beams, an expression for the maximum deflection, Δy_{mx} (Fig. 7.31), is given by:

$$\Delta y_{mx} = \frac{F_L \ell_{df}^3}{384 E_{cd} I_{bb}} = \frac{I_t B_e \ell_{df}^4}{384 E_{cd} I_{bb}} \qquad (7.42)$$

where E_{cd} and I_{bb} are, respectively, the conductor's Young's modulus and moment of inertia. Equation 7.42 is valid for a beam fully clamped at both ends and subjected to a uniformly distributed force, the sum of which is given by Eq. 7.41. For a conductor, which is b thick and $a(> b)$ wide (high), we have: $I_{bb} = b^3 a/12$.

If this deflection occurs rapidly in the presence of a background field, B_e, which is uniform and directed as indicated in Fig. 7.31, the deflecting conductor will induce voltage V across it, satisfying Faraday's law:

$$\int V \, dt = A_{sh} B_e \qquad (7.43)$$

where A_{sh} is the shaded area indicated in the figure. The theory of beams gives the shaded area in terms of ℓ_{df} and Δy_{mx}:

$$A_{sh} = \tfrac{8}{15} \ell_{df} \Delta y_{mx} \qquad (7.44)$$

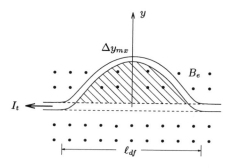

Fig. 7.31   Conductor deflection model based on theory of beams [7.52].

Problem 7.14: Conductor-motion-induced voltage pulse

a) Many voltage pulses are induced by conductor motion during a magnet charge-up. By accurately monitoring these pulses, we may deduce approximate sizes of ℓ_{df} for a given operating condition. Show that an expression relating ℓ_{df} and $\int V \, dt$, based on the deflecting beam model presented here, is given by:

$$\ell_{df} = \left(\frac{720 E_{cd} I_{bb}}{I_t B_e^2} \int V \, dt \right)^{1/5} \tag{7.45}$$

b) Figure 7.32 presents a set of voltage and AE signals recorded with a superconducting dipole magnet (for an MHD facility) as it was energized [7.31]. The pulse, $\sim$100 mV in amplitude lasting $\sim$0.5 ms shown in the trace, occurred at $I_t = 1300$ A, clearly induced by a conductor motion event, which also generated AE signals. This particular event did not lead to a quench. The conductor is a Nb-Ti composite superconductor having rectangular cross section. For the values of parameters given in Table 7.9, compute ℓ_{df}.

c) Compute the maximum deflection Δy_{mx} for the above parameters.

Table 7.9: Conductor Parameters

Conductor thickness, a	[mm]	3
Conductor width, b	[mm]	6
Young's modulus, E_{cd}	[GPa]	110
Transport current, I_t	[A]	1300
Background magnetic induction, B_e [T]		4

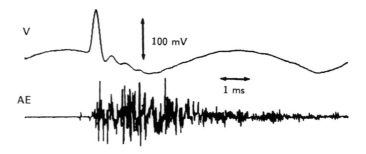

Fig. 7.32 Voltage and AE signals recorded during a conductor motion event in a superconducting dipole [7.31].

Solution to Problem 7.14

a) To determine the length of the moving conductor segment, we start by combining Eqs. 7.43 and 7.44 to obtain:

$$\int V\,dt = \tfrac{8}{15}\ell_{df}\Delta y_{mx}B_e \tag{S14.1}$$

Substituting Δy_{mx} given by Eq. 7.42 into Eq. S14.1, we obtain:

$$\int V\,dt = \frac{8\ell_{df}I_t B_e^2 \ell_{df}^4}{15 E_{cd}I_{bb}384}$$

$$= \frac{I_t B_e^2 \ell_{df}^5}{720 E_{cd}I_{bb}} \tag{S14.2}$$

Solving Eq. S14.2 for ℓ_{df}, we obtain:

$$\ell_{df}^5 = \frac{720 E_{cd}I_{cd}}{I_t B_e^2}\int V\,dt$$

$$\ell_{df} = \left(\frac{720 E_{cd}I_{bb}}{I_t B_e^2}\int V\,dt\right)^{1/5} \tag{7.45}$$

b) From the voltage pulse of Fig. 7.32, we can make a rough estimate of $\int V\,dt$. With a pulse amplitude of $\sim 100\,\mathrm{mV}$ lasting $\sim 0.5\,\mathrm{ms}$, the area under the triangular shaped pulse is $\sim 25\times 10^{-6}\,\mathrm{V\,s}$. By substituting appropriate values into Eq. 7.45, we have:

$$\ell_{df} = \left[\frac{720(150\times 10^9\,\mathrm{Pa})(3\times 10^{-3}\,\mathrm{m})^3(6\times 10^{-3}\,\mathrm{m})}{(1300\,\mathrm{A})(4\,\mathrm{T})^2(12)}(25\times 10^{-6}\,\mathrm{V\,s})\right]^{1/5} \tag{S14.3}$$

$$\simeq 7\,\mathrm{cm}$$

That is, based on this beam deflection model, a conductor length of $\sim 10\,\mathrm{cm}$ must deflect suddenly in order to produce a voltage of $\sim 100\,\mathrm{mV}$ lasting $\sim 0.5\,\mathrm{ms}$.

c) With $I_{bb} \simeq 14\times 10^{-12}\,\mathrm{m}^4$ and $\ell_{df} \simeq 7\times 10^{-2}\,\mathrm{m}$, and substituting other appropriate values into Eq. 7.42, we obtain:

$$\Delta y_{mx} \sim \frac{(1300\,\mathrm{A})(4\,\mathrm{T})(7\times 10^{-2}\,\mathrm{m})^4}{384(150\times 10^9\,\mathrm{Pa})(14\times 10^{-12}\,\mathrm{m}^4)} \tag{S14.4}$$

$$\sim 0.2\,\mathrm{mm}$$

A deflection of $\sim 0.2\,\mathrm{mm}$ over a conductor length of $\sim 10\,\mathrm{cm}$ within a dry winding is plausible for the conductor of this size.

Problem 7.15: Disturbances in HTS magnets

This problem studies the effects of operating temperature on disturbances as applied to HTS magnets.

   a) Discuss as quantitatively as possible how AC losses, splice losses, and mechanical disturbances vary with temperature, particularly over the operating temperature range of HTS magnets, 20~100 K.

   b) Give reasons why disturbances are not as important for HTS magnets as they are for LTS magnets.

Field Orientation Anisotropy in BiPbSrCaCuO (2223) Tapes

In polycrystalline HTS such as silver-sheathed BiPbSrCaCuO (2223) tapes, transport current flows principally along the crystal's a-b plane; whenever current crosses the grain boundaries it flows along the c-axis. As discussed briefly in Chapter 1, the superconductor's critical current density is determined by flux pinning sites that balance the Lorentz force density. Because HTS have anisotropic flux pinning sites, their critical currents are also anisotropic; the effect is pronounced at higher temperatures as seen from representative data shown in Fig. 7.33.

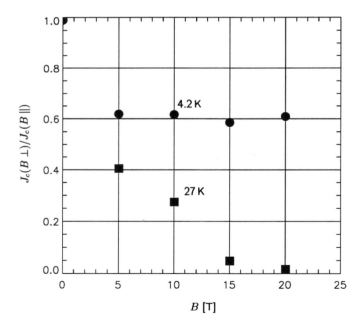

Fig. 7.33 $J_c(B\perp)/J_c(B\parallel)$ *vs* B plots for a silver-sheathed BiPbSrCaCuO (2223) tape at 4.2 K (circule) and 27 K (squares) [7.53]. $J_c(B\perp)$ (perpendicular field) data and $J_c(B\parallel)$ (parallel field) data are for two *nearly* identical tapes. $J_c(B\parallel)$ data are shown in Fig. A5.3; the two tapes have nearly identical $J_c(0)$ values at each temperature.

Solution to Problem 7.15

a) *AC hysteresis loss:* If HTS's J_c is about the same as LTS's in their respective temperature ranges then e_{hy} would remain the same for the same filament size.

AC coupling loss: e_{cp} would be affected to the extent τ_{cp} is affected, which will be through ℓ_p^2 and ρ_m: $e_{cp} \propto \tau_{cp} \propto \ell_p^2/\rho_m$. Hence for the *same* ℓ_p, coupling losses will be less for HTS magnets than LTS magnets perhaps by a factor of $\sim$10 if HTS's operating temperature is $\sim$80 K. To become a magnet-grade superconductor, HTS must eventually be of twisted multifilaments with twist pitch lengths comparable to those in LTS; limited success has been achieved recently toward this goal with silver-sheathed BSCCO tapes [7.54].

AC eddy-current loss: Because $e_{ed} \propto 1/\rho_m$, eddy-current losses will also be $\sim$1/10 for HTS than those for LTS, again if HTS magnets are operated near 80 K.

Splices: Because resistivities of solders are weakly dependent on temperature, the solder's contribution to R_{ct}, say at $\sim$80 K, should only be slightly higher than that at 4 K. However, because the resistivities of copper and silver are $\sim$10 times greater at 80 K than those at 4 K, the matrix metal's contribution to R_{ct} will not be negligible. Splice losses may thus increase by a factor as much as $\sim$10.

Mechanical: Because the effect of temperature on friction coefficients over the range 4$\sim$100 K is slight—if there is an increase, it is less than a factor of 2 [7.19, 7.20]—the amplitude of frictional heating, for example, will increase only slightly as the temperature is raised from 4 to 80 K.

b) We may make the following observations on the magnitude of disturbance-induced losses as operating temperature is increased from 4 to near 80 K:

- Increase by a factor of $\sim$10, *e.g.* splice;
- Unchanged, *e.g.* hysteresis loss and mechanical;
- Decrease by a factor of $\sim$10, *e.g.* coupling and eddy-current.

As seen in Problem 6.14, over this temperature span the energy margin in adiabatic magnets, on the other hand, increases by a factor of as much as $\sim$10^6. Therefore, disturbances are clearly not pressing design issues for HTS magnets as they are for LTS magnets. This assessment is valid for HTS magnets operating at a temperature above $\sim$20 K. (Note that because electrical resistivities of both copper and silver remain roughly constant over the temperature span 4 to $\sim$30 K, the losses by splice, coupling, and eddy-current—each dominated by normal-metal electrical resistivity—also remain constant over this temperature span.)

References

[7.1] G. Ries and H. Brechna, "AC losses in superconducting pulsed magnets," (KFK Report 1372, Gesellschaft fur Kernforschung M.B.H. Karlsruhe, 1972).

[7.2] J.P. Soubeyrand and B. Turck, "Losses in superconducting composite under high rate pulsed transverse field," *IEEE Trans. Magn.* **MAG-15**, 248 (1979).

[7.3] T. Ogasawara, Y. Takahashi, K. Kanbara, Y. Kubota, K. Yasohama, and K. Yasukochi, "Alternating field losses in superconducting wires carrying dc transport currents. Part 2: multifilamentary composite conductors," *Cryogenics* **21**, 97 (1981).

[7.4] I. Hlásnik, "Review of ac losses in superconductors," *IEEE Trans. Magn.* **MAG-17**, 2261 (1981).

[7.5] W.J. Carr, Jr., "Conductivity, permeability, and dielectric constant in a multifilament superconductor," *J. Appl. Phys.* **46**, 4043 (1975).

[7.6] J.F. Maguire (an internal memo, FBNML, unpublished 1979).

[7.7] A.M. Hatch, R.C. Beals, and Y. Iwasa (Avco Everett Research Laboratory report, unpublished 1975).

[7.8] R.W. Fast, W.W. Craddock, M. Kobayashi, and M.T. Mruzek, "Electrical and mechanical properties of lead/tin solders and splices for superconducting cables," *Cryogenics* **28**, 7 (1988).

[7.9] J.W. Hafstrom, D.H. Killpatrick, R.C. Niemann, J.R. Purcell, and H.R. Thresh, "Joining NbTi superconductors by ultrasonic welding," *IEEE Trans. Magn.* **MAG-13**, 94 (1977).

[7.10] Magnet Technology Division, FBNML (internal reports, unpublished 1993).

[7.11] Y. Kawate, R. Ogawa, and R. Hirose (personal communication, 1994).

[7.12] H. Maeda, O. Tsukamoto, and Y. Iwasa, "The mechanism of friction motion and its effect at 4.2 K in superconducting magnet winding models," *Cryogenics* **22**, 287 (1982).

[7.13] Y. Iwasa, J.F. Maguire, and J.E.C. Williams, "The effect on stability of frictional decoupling for a composite superconductor," *Proc. 8th Symp. on Engr. Problems of Fusion Research,* (IEEE Publication 79CH1441-5, 1979), 1407.

[7.14] Y. Yasaka and Y. Iwasa, "Stress-induced epoxy cracking energy release at 4.2 K in epoxy-coated superconducting wires," *Cryogenics* **24**, 423 (1984).

[7.15] S. Fuchino and Y. Iwasa, "A cryomechanics technique to measure dissipative energies of ∼10 nJ," *Exp. Mech.* **30**, 356 (1990).

[7.16] E.S. Bobrov and J.E.C. Williams, "Direct optimization of the winding process for superconducting solenoid magnets (linear programming approach)," *IEEE Trans. Magn.* **MAG-17**, 447 (1981).

[7.17] E.S. Bobrov, J.E.C. Williams, and Y. Iwasa, "Experimental and theoretical investigation of mechanical disturbances in epoxy-impregnated superconducting coils. 2. Shear stress-induced epoxy fracture as the principal source of premature quenches and training—theoretical analysis," *Cryogenics* **25**, 307 (1985).

[7.18] H. Maeda, M. Urata, H. Ogiwara, S. Miyake, N. Aoki, M. Sugimoto, and J. Tani, "Stabilization for wind and react Nb$_3$Sn high field insert coil," *Proc. 11th Intnl. Conf. Magnet Tech. (MT-11)* (Elsevier Applied Science, London, 1990), 1114.

[7.19] R.S. Kensley and Y. Iwasa, "Frictional properties of metal insulator surfaces at cryogenic temperatures," *Cryogenics* **20**, 25 (1980).

[7.20] R.S. Kensley, H. Maeda, and Y. Iwasa, "Transient slip behavior of metal/insulator pairs at 4.2 K," *Cryogenics* **21**, 479 (1981).

[7.21] A. Iwabuchi and T. Honda, "Temperature rise due to frictional sliding of SUS316 vs SUS316L and SUS316 vs polyimide at 4 K," *Proc. 11th Intl. Conf. Magnet Tech. (MT11)* (Elsevier Applied Science, London, 1990), 686.

[7.22] P.C. Michael, D. Aized, Y. Iwasa, and E. Rabinowicz, "Mechanical properties and static friction behavior of epoxy mixes at room temperature and at 77 K," *Cryogenics* **30**, 775 (1990).

[7.23] P.C. Michael, Y. Iwasa, and E. Rabinowicz, "Reassessment of cryotribology theory," *Wear* **174**, 163 (1994).

[7.24] T. Takao and O. Tsukamoto, "Stability against the frictional motion of conductor in superconducting windings," *IEEE Tran. Magn.* **27**, 2147 (1991).

[7.25] M. Urata and H. Maeda, "Relation between radial stress and quench current for tightly wound dry solenoids," *IEEE Trans. Mag.* **MAG-23**, 1596 (1987).

[7.26] P.C. Michael, E.S. Bobrov, Y. Iwasa, and M. Arata, Stabilization of dry-wound high-field NbTi solenoids, *IEEE Trans. Appl. Superconduc.* **3**, 316 (1993).

[7.27] H. Nomura, K. Takahisa, K. Koyama, and T. Sakai, "Acoustic emission from superconducting magnets," *Cryogenics* **17**, 471 (1977).

[7.28] Curt Schmidt and Gabriel Pasztor, "Superconductors under dynamic mechanical stress," *IEEE Tran. Magn.* **MAG-13**, 116 (1977).

[7.29] H. Brechna and P. Turowski, "Training and degradation phenomena in superconducting magnets," *Proc. 6th Intl. Conf. Magnet Tech. (MT-6)* (ALFA, Bratislava, Czechoslovakia, 1978), 597.

[7.30] O. Tsukamoto, J.F. Maguire, E.S. Bobrov, and Y. Iwasa, "Identification of quench origins in a superconductor with acoustic emission and voltage measurements," *Appl. Phys. Lett.* **39**, 172 (1981).

[7.31] O. Tsukamoto, M.F. Steinhoff, and Y. Iwasa, "Acoustic emission triangulation of mechanical disturbances in superconducting magnets," *Proc. 9th Symp. Engr. Problems of Fusion Research* (IEEE Pub. No. 81CH1715-2 NPS, 1981), 309.

[7.32] O. Tsukamoto and Y. Iwasa, "Sources of acoustic emission in superconducting magnets," *J. Appl. Phys.* **54**, 997 (1983).

[7.33] S. Caspi and W.V. Hassenzahl, "Source, origin and propagation of quenches measured in superconducting dipole magnets," *IEEE Trans. Magn* **MAG-19**, 692 (1983).

[7.34] H. Iwasaki, S. Nijishima, and T. Okada, "Application of acoustic emission method to the monitoring system of superconducting magnet," *Proc. 9th Intl. Conf. Magnet Tech. (MT-9)* (Swiss Institute for Nuclear Research, Villigen, 1985), 830 (1985).

[7.35] Y. Iwasa, E.S. Bobrov, O. Tsukamoto, T. Takaghi, and H. Fujita, "Experimental and theoretical investigation of mechanical disturbances in epoxy-impregnated superconducting coils. 3. Fracture-induced premature quenches," *Cryogenics* **25**, 317 (1985).

[7.36] K. Yoshida, M. Nishi, H. Tsuji, Y. Hattori, and S. Shimamoto, "Acoustic emission measurement on large coils at JAERI," *Adv. Cryogenic Eng.* **31**, 277 (1986).

[7.37] O.O. Ige, A.D. McInturff, and Y. Iwasa, "Acoustic emission monitoring results from a Fermi dipole," *Cryogenics* **26**, 131 (1986).

[7.38] H. Maeda, A. Sato, M. Koizumi, M. Urata, S. Murase, I. Takano, N. Aoki,

M. Ishihara, E. Suzuki, "Application of acoustic emission technique to a multi-filamentary 15.1 tesla superconducting magnet system," *Adv. Cryogenic Eng.* **31**, 293 (1986).

[7.39] J. Chikaba, F. Irie, K. Funaki, M. Takeo, and K. Yamafuji, "Instabilities due to mechanical strain energy in superconducting magnets," *IEEE Tran. Magn.* **MAG-23**, 1600 (1987).

[7.40] T. Ogitsu, K. Tsuchiya, and A. Devred, "Investigation of wire motion in superconducting magnets," *IEEE Trans. Magn.* **27**, 2132 (1991).

[7.41] K. Ikizawa, N. Takasu, Y. Murayama, K. Seo, S. Nishijima, K. Katagiri, and T. Okada, "Instability of superconducting racetrack magnets," *ibid.*, 2128 (1991).

[7.42] See, for example, P.F. Dahl, G.H. Morgan, and W.B. Sampson, "Loss measurements on twisted multifilamentary superconducting wires," *J. Appl. Phys.* **40**, 2083 (1969).

[7.43] See, for example, A. Février, "Latest news about superconducting A.C. machines," *IEEE Trans. Magn.* **24**, 787 (1988).

[7.44] R.B. Goldfarb and M. Takayasu (personal communication, 1989).

[7.45] H. Tsuji, M. Nishi, K. Yoshida, E. Tada, K. Kawano, K. Koizumi, H. Yamamura, M. Oshikiri, Y. Takahashi, T. Ando, and S. Shimamoto, "Thermal design and verification tests of the Nb-Ti demo poloidal coils (DPC-U1, U2)," *IEEE Trans. Magn.* **MAG-24**, 1303 (1988).

[7.46] M. Takayasu, C.Y. Gung, M.M. Steves, B. Oliver, D. Reisner and M.O. Hoenig, "Calorimetric measurement of ac loss in Nb$_3$Sn superconductors," *Proc. 11th Int'l Conf. Magnet Tech. (MT-11)* (Elsevier Applied Science, London, 1990), 1033.

[7.47] J.E.C. Williams M. Baker, E.S. Bobrov, Y. Iwasa, M.J. Leupold, V.J. Stejskal, R.J. Weggel, A. Zhukovsky, C.Y. Gung, J. Miller, T. Painter, S. Van Sciver, "The development of a niobium-titanium cable-in-conduit coil for a 45 T hybrid magnet," *IEEE Trans. Magn.* **30**, 1633 (1994).

[7.48] W.H. Warnes and D.C. Larbalestier, "Determination of the average critical current from measurements of the extended resistive transition," *IEEE Trans. Magn.* **MAG-23**, 1183 (1987).

[7.49] J.E.C. Williams, E.S. Bobrov, Y. Iwasa, W.F.B. Punchard, J. Wrenn, A. Zhukovsky, "NMR magnet technology at MIT," *IEEE Trans. Magn.* **28**, 627 (1992).

[7.50] Y. Iwasa and V.Y. Adzovie, "The index number (n) below 'critical' current in Nb-Ti superconductors," (to be published 1995).

[7.51] H. Fujita, T. Takaghi, and Y. Iwasa, "Experimental and theoretical investigation of mechanical disturbances in epoxy-impregnated superconducting coils. 4. Prequench cracks and frictional motion," *Cryogenics* **25**, 323 (1985).

[7.52] Y. Iwasa, "Conductor motion in the superconducting magnet—a review," *Stability of superconductors in helium I and helium II* (International Institute of Refrigeration, Paris France, 1981), 125.

[7.53] J. Fujikami, K. Sato, Y. Iwasa, M. Yunus, H. Lim, and J.B. Kim (preliminary data, FBNML, 1994).

[7.54] R. Schwall (personal communication, 1994).

CHAPTER 8
PROTECTION

8.1 Introductory Remarks

As remarked in Chapter 1, superconducting magnets may be divided into two classes. Class 1 magnets are large and generally cryostable. Magnets for fusion reactors belong to Class 1. Class 2 magnets are small and operate at high overall current densities under adiabatic conditions. Magnets for MRI and NMR systems are in Class 2. This chapter addresses the protection issue, focusing principally on Class 2 magnets; more specifically it addresses quench-induced conductor overheating, which can cause irreparable damage to magnets. As remarked in Chapter 1, protection against overheating increases in difficulty with operating temperature and therefore becomes a more important design issue for HTS magnets. At least for the near future most HTS magnets, chiefly because of their processing requirements, will be of Class 2.

Montgomery [8.1] and Thome [8.2] give surveys of "failures" in magnets and magnet systems. Of the 115 entries that Montgomery classified into six areas of failure (Table 8.1), the most frequent (29) are insulation related, followed by 25 entries related to mechanical support. The Table 8.1 data must be viewed with the understanding that the data base for this survey may include more Class 1 magnets than Class 2 magnets. (Although there have been more Class 2 magnets built, mostly for MRI, failures in these commercial magnets are seldom reported and are thus generally not included in this kind of survey.) This is probably the reason why the data tend to show more failures related to insulation (large discharge voltages, as discussed in Problem 8.1) and mechanical (large forces), both characteristics of Class 1 magnets, than those related to conductor (high overall current densities), characteristic of Class 2 magnets.

Before focusing on the problem of overheating, we discuss here briefly two important problem areas in superconducting magnets: 1) mechanical and 2) electrical.

8.1.1 Mechanical

There are three kinds of problems in superconducting magnets that are mechanical

Table 8.1: Classification of Failures [8.1]

Area of Failure	No. Entry
Insulation	29
Mechanical	25
System Performance	21
Conductor	17
External Systems	16
Coolant	7

in origin: 1) mechanical-disturbance induced quenches; 2) strain-induced conductor damage; and 3) structural failure. Subject 1 was treated in Chapter 7. Subjects 2 and 3 are only briefly described here. Strain damage is serious in strain-sensitive conductors such as Nb_3Sn and HTS. Because strain-dependent critical current density data for Nb-Ti and Nb_3Sn are available [3.33, 3.34], and because it is now possible to quantify the stresses to which conductors are subjected accurately incidents of strain damage that render the entire magnet useless are quite rare.

Lorentz forces and fatigue contribute to mechanical failures, particularly the forces that appear under fault conditions which are generally more difficult to anticipate. Fortunately, because of the collective experience gained in the magnet community over the past 30 years, blatant structural failures in superconducting magnets are now virtually nonexistent. A significant advance in structural analysis through the use of finite-element techniques has also reduced incidents of failure. Because there are very few active superconducting magnets that have been operating for 10 years or more, it is difficult to gauge their "fatigue proofness." The Hybrid II [3.5], in operation at FBNML since 1981, may qualify as one of these few magnets.

8.1.2 Electrical

The most serious electrical failure in magnets is arcing that either permanently damages a section of the winding or, if arcing is mild, simply short circuits two neighboring conductors. Either way, the magnet is no longer a reliable system component. An arc-inducing high voltage is another problem that may arise in the protection technique most widely used against overheating. (This is discussed in Problems 8.9~8.11, and discussion on this subject is deferred until then.)

The voltage taps used to monitor the magnet must not fail because they typically provide signals to a quench-detection system protecting the magnet. Despite careful and methodical procedures used to mount them to the magnet for *permanent* service, some do eventually become open- or short-circuited. The key word for mounting voltage taps to the magnet is ruggedness.

8.2 Protection for Class 2 Magnets

Within the winding of Class 2 superconducting magnets, there are three dangerous failures that can be induced during a magnet quench: 1) overheating (meltdown); 2) high-voltage arcing; and 3) overstressing. Of these accidents, overheating, particularly meltdown, must be avoided at all cost; the other two, because they are in some lucky instances repairable, are not as irreversible as the first one. Arcing and overstressing are addressed in Problems 8.9~8.11 where a widely used protection technique against overheating is discussed in detail.

The thermal process that may eventually lead to the meltdown of a section of the winding in a superconducting magnet is quite complex; protection against overheating is one of the magnet issues that has received a great deal of attention since the early days of superconducting magnet technology [8.3~8.5]. Interest in this subject is unabated, driven by consideration of two types of magnets that will be of paramount interest and importance in the coming years: 1) high-field (> 12 T) magnets for NMR spectroscopy and 2) HTS magnets.

A high-field NMR magnet, typically consisting of as many as a dozen nested solenoids, is generally operated in persistent mode. The persistent-mode operation is needed in NMR magnets to satisfy requirements of temporal field stability. To minimize the system's refrigeration load, the magnet is physically uncoupled from the external power source. Ensuring a suitable conversion of magnetic energy into thermal energy in the event of a quench in such an *isolated* Class 2 magnet is generally difficult and challenging.

8.2.1 Meltdown

The meltdown of a section of the winding by overheating is not uncommon. It results from conversion into thermal energy of the magnetic energy stored by the entire magnet over a *fraction*—as small as 1%—of the winding volume. To heat up copper, a good representative material of the winding, from 4 K to its melting temperature of 1356 K entirely by the adiabatic conversion of the magnetic energy stored only within its *own* volume, the intial magnetic induction, B_0, would have to be $\sim$115 T, as demonstrated below:

$$\frac{B_0^2}{2\mu_o} = h_{cu}(1356\,\text{K}) - h_{cu}(4\,\text{K}) \simeq 5.2 \times 10^9 \, \text{J/m}^3 \tag{8.1}$$

$$B_0 \simeq \sqrt{2(4\pi \times 10^{-7}\,\text{H/m})(5.2 \times 10^9 \, \text{J/m}^3)} \sim 115\,\text{T}$$

That even a modest magnet, *e.g.* a 4-T magnet, is known to have been permanently damaged by overheating attests that an unfavorable energy concentration can occur in *real* magnets. (If a *local* magnetic energy density is converted *locally* into heat, the maximum *hot-spot* temperature for a 4-T magnet would only be $\sim$40 K; even for a 25-T magnet it is still only $\sim$200 K.)

8.2.3 Magnet Quench

The prediction of the time-dependent temperature distribution within the winding of a quenching coil is difficult, particularly if the coil is part of a magnet comprised of many coils. It can be quantified only after the coil's time-dependent current and voltage are solved, through an analysis of normal zone growth. Thermal-diffusion-propelled normal zone growth proceeds one dimensionally along the conductor in unfilled windings that are cooled by liquid helium [8.3, 8.6~8.11], while it is generally three dimensional in epoxy-impregnated adiabatic windings [8.12~8.21]. It has been verified experimentally that in adiabatic windings the transverse (turn-to-turn) propagation velocity of a normal zone is proportional to the longitudinal (along conductor axis) propagation velocity [8.17]. The power density equation appropriate for adiabatic windings may be given by Eq. 6.1 (p. 203) with the disturbance (g_d) and cooling (g_q) terms set to zero: $g_d = g_q = 0$.

Quench propagation in an adiabatic winding involves two coupled processes: electrical and thermal. Simply stated, the propelling source of normal zone propagation is Joule heating, which is controlled by the winding's electric circuit. The resistances appearing in the electric circuit are in turn controlled by the winding's "thermal" circuit. In a magnet comprised of more than one winding, the complexity increases owing to inductive coupling among windings.

8.2.4 Self-Protecting Magnet

A superconducting magnet is said to be *self-protecting* if it can be protected against overheating by having normal zone spread out quickly over most of its winding volume. How fast this process takes place may be gauged by the normal-zone propagation (NZP) velocity. Self-protecting magnets generally have "high" NZP velocities. Ideally all isolated Class 2 magnets should be self-protecting.

8.3 Computer Simulation

Because of the coupled nature of quench processes in adiabatic magnets, particularly those comprised of more than one coil, quench analysis is best performed with the aid of a computer. From Wilson's early attempt in 1968 [8.13] to comprehensive works by Kadambi in 1986 [8.16] and Kuroda in 1989 [8.18], computer-aided quench simulation work is still in progress. Recently, it has been applied to the study of a quenching composite HTS tape [8.21].

Here we briefly describe the quench simulation codes for adiabatic, solenoidal windings that have evolved at FBNML, starting in 1985 with Williams' work [8.14]. The basic postulate of the FBNML codes is that the complex thermal diffusion process that controls normal-zone propagation within the winding may be expressed by a single parameter U_t, the H-, T-, and J-dependent transverse propagation velocity. The complex effects of the winding's thermal properties are incorporated into U_t and the codes are simplified immensely without sacrifice in accuracy. As discussed in Problem 8.8, U_t is related to longitudinal propagation velocity, U_ℓ, whose functional dependencies on H, T, and J can be derived from Eq. 6.1 (See Problem 8.7). U_t thus depends both on time and space within the winding.

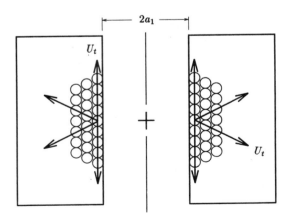

Fig. 8.1   Quenching in an adiabatic, solenoidal winding.

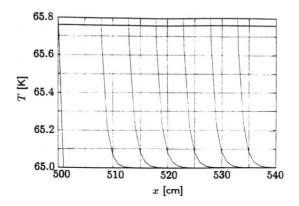

Fig. 8.2   Simulation of normal-zone propagation in a composite HTS tape [8.21].

Figure 8.1 pictorially shows quench propagation within an adiabatic solenoidal winding, which in this case is a close-packed hexagonal arrangement of round wire, impregnated with epoxy resin. Note that quenching in the figure is initiated at the innermost radius of the winding midplane. The turn-to-turn transit time by transverse propagation velocity (U_t) is generally shorter than the circumferential transit time by longitudinal velocity U_ℓ because of the following condition, valid in most windings:

$$\frac{d_{cd}}{U_t} \ll \frac{2\pi a_1}{U_\ell} \tag{8.2}$$

where a_1 is the innermost winding radius and d_{cd} is the conductor diameter.

Figure 8.2 presents the results of a numerical analysis of the nonlinear partial differential equation (Eq. 6.1) computed for a propagating normal zone along the axis of an adiabatic composite HTS tape [8.21]. The temperature profiles are computed at a time interval of 2 seconds; an equal distance between adjacent profiles shows that normal zone propagation velocity is constant. This result indeed confirms our basic postulate that the normal zone does propagate at a *constant* velocity when key parameters—field, temperature, and current density—are constant.

"ACCIDENT, n. An inevitable occurrence due to the action of immutable natural laws." —Ambrose Bierce

Problem 8.1: Active protection

This problem discusses an active protection technique widely used in large magnet systems. Originally proposed by Maddock and James in 1968 [8.22], its basic premise is to protect the magnet by dissipating most of the stored energy into a "dump" resistor connected across the magnet terminals. With most of the potentially dangerous energy dissipated elsewhere, the hot spot is heated only during the brief time that the magnet current is decaying to zero. As we shall see in this problem, the faster the current decay rate, the smaller the hot-spot temperature. To achieve this fast rate of current discharge, however, the magnet terminals have to withstand higher voltages. Thus, the magnet designer must choose a hot-spot temperature on one hand and a terminal discharge voltage on the other: two conflicting requirements, as is often the case. What all this leads to is another criterion for current density over the matrix metal cross section at operating current, $[J_{op}]_m$.

Figure 8.3 presents the basic circuit for this active protection technique, commonly known as "detect-and-dump." The magnet is represented by inductance L_m; the dump resistor, connected across the magnet terminals and usually located outside the cryostat, is represented by R_D. Switch S is opened when a nonrecovering normal zone, represented by $r(t)$, appears within the magnet. The stored magnet energy, E_m, at the magnet's operating current I_{op} is given by $L_m I_{op}^2/2$.

This active protection technique requires two sequentially executed actions: 1) detection of a small non-recovering normal zone; and 2) opening of switch S that forces the magnet to discharge through the dump resistor. The drawback of the technique is that both actions are subject to failure. Detection of this normal zone is not easy because of the presence of a large inductive voltage: usually the action is needed while the magnet is being charged up, rather than after it has been fully charged to I_{op} and is in the quiescent state. Problems 8.3 and 8.4 discuss quench detection techniques useful for this protection scheme.

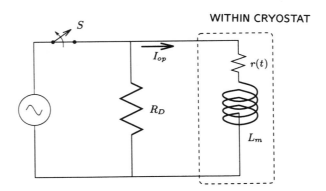

Fig. 8.3   Magnet circuit for the detect-and-dump active protection.

Problem 8.1: Active protection

a) With switch S open, a differential equation for magnet current $I(t)$ is given by:

$$L_m \frac{dI(t)}{dt} + [r(t) + R_D]I(t) = 0 \tag{8.3}$$

Show that for the case $r(t) \ll R_D$, the magnet current $I(t)$ after switch S is opened is given by:

$$I(t) = I_{op} \exp\left(-\frac{R_D t}{L_m}\right) \tag{8.4}$$

b) Starting with the power energy density given by Eq. 6.1 (p. 203) with $g_k = g_d = g_q = 0$, show that an expression for a "protection" criterion on operating current density in the conductor matrix, $[J_{op}]_m \equiv I_{op}/A_m$, is given by:

$$[J_{op}]_m \leq \sqrt{\left(\frac{A_{cd}}{A_m}\right) \frac{V_D I_{op} Z(T_i, T_f)}{E_m}} \tag{8.5}$$

where A_{cd} is the overall conductor cross sectional area. $V_D = R_D I_{op}$ is known as the "dump" voltage. $Z(T_i, T_f)$ is defined as:

$$Z(T_i, T_f) = \int_{T_i}^{T_f} \frac{C_{cd}(T)}{\rho_m(T)} dT \tag{8.6}$$

where $C_{cd}(T)$ is the conductor's heat capacity and $\rho_m(T)$ is the matrix metal resistivity, both functions of temperature. T_i and T_f are the initial and final temperatures of the hot spot—generally the region in the winding originally driven normal.

From Eq. 8.5 we can derive an expression for V_D:

$$V_D = \left(\frac{A_m}{A_{cd}}\right) \frac{[J_{op}]_m^2 E_m}{I_{op} Z(T_i, T_f)}$$

Equation 8.7 states that V_D increases linearly with E_m and quadratically with $[J_{op}]_m$ and decreases inversely with I_{op}.

c) Consider a Nb-Ti composite strip, $a = 10\,$mm and $b = 3\,$mm, with a volumetric copper-to-superconductor ratio $(\gamma_{c/s})$ of 4. With $f_p = 0.5$ (fraction of conductor perimeter exposed to liquid helium) and $q_{fm} = 0.36\,$W/cm^2, compute I_{op}, the operating current that satisfies the Stekly criterion. For this value of I_{op}, what must V_D be to make I_{op} satisfy Eq. 8.5 when $E_m = 10\,$MJ? Take $T_f = 100\,$K. Note that $A_{cd} = A_m + A_s$, where A_s is the superconductor cross sectional area. Also take RRR of the matrix copper to be 50.

d) Repeat c) for V_D when $E_m = 100\,$MJ.

Solution to Problem 8.1

a) For the case $r(t) \ll R_D$, Eq. 8.3 may be approximated by:

$$L_m \frac{dI(t)}{dt} + R_D I(t) = 0 \tag{S1.1}$$

which, for $I(0) = I_{op}$, has a solution given by:

$$I(t) = I_{op} \exp\left(-\frac{R_D t}{L_m}\right) \tag{8.4}$$

b) Equation 6.1, with $g_k = g_d = g_q = 0$, is given by:

$$A_{cd} C_{cd}(T) \frac{dT}{dt} = \frac{\rho_m(T) I^2(t)}{A_m}$$

$$C_{cd}(T) \frac{dT}{dt} = \rho_m(T) J_m^2(t) \frac{A_m}{A_{cd}} \tag{S1.2}$$

where $J_m(t) = I(t)/A_m$. From Eq. 8.4, we have:

$$J_m(t) = [J_{op}]_m \exp\left(-\frac{R_D t}{L_m}\right) \tag{S1.3}$$

Combining Eqs. S1.2 and S1.3 and placing temperature-dependent parameters on one side and time-dependent parameters on the other side of the equation, we have:

$$\frac{C_{cd}(T)}{\rho_m(T)} dT = \left(\frac{A_m}{A_{cd}}\right) J_m^2(t) dt \tag{S1.4a}$$

$$= \left(\frac{A_m}{A_{cd}}\right) [J_{op}]_m^2 \exp\left(-\frac{2R_D t}{L_m}\right) dt \tag{S1.4b}$$

Integrating Eq. S1.4b between initial temperature T_i and final temperature T_f for the hot spot in the left-hand side and between $t = 0$ and $t = \infty$ in the right-hand side and noting that T_f is the upper limit to the hot-spot temperature, we obtain:

$$\int_{T_i}^{T_f} \frac{C_{cd}(T)}{\rho_m(T)} dT \equiv Z(T_i, T_f) \geq \left(\frac{A_m}{A_{cd}}\right) [J_{op}]_m^2 \int_0^\infty \exp\left(-\frac{2R_D t}{L_m}\right) dt$$

$$= \left(\frac{A_m}{A_{cd}}\right) [J_{op}]_m^2 \frac{L_m}{2R_D} \tag{S1.5}$$

By noting that $V_D = R_D I_{op}$, and thus $R_D = V_D/I_{op}$, and that $E_m = L_m I_{op}^2/2$ and thus $L_m = 2E_m/I_{op}^2$, we may rewrite Eq. S1.5:

$$Z(T_i, T_f) \geq \left(\frac{A_m}{A_{cd}}\right) \frac{[J_{op}]_m^2 E_m}{V_D I_{op}} \tag{S1.6}$$

Solution to Problem 8.1

Solving for $[J_{op}]_m$ from Eq. $S1.6$, we have:

$$[J_{op}]_m \leq \sqrt{\left(\frac{A_{cd}}{A_m}\right)\frac{V_D I_{op} Z(T_f, T_i)}{E_m}} \qquad (8.5)$$

Figure 8.4 presents Z vs T plots for copper, aluminum, and silver for $T_i = 4\,\mathrm{K}$.

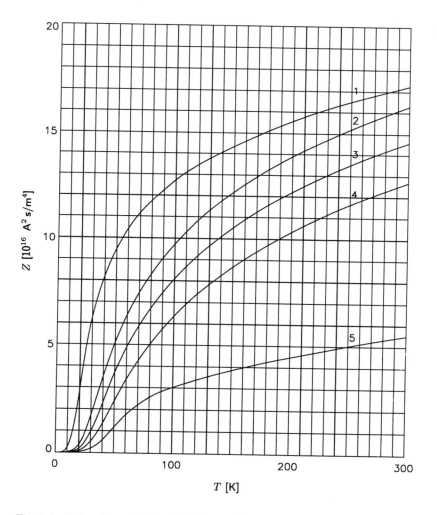

Fig. 8.4 Z functions: 1. Silver (99.99%); 2. Copper (RRR 200); 3. Copper (RRR 100); 4. Copper (RRR 50); 5. Aluminum (99.99%).

Solution to Problem 8.1

c) With $a = 10\,\text{mm}$, $b = 3\,\text{mm}$, and a volumetric copper-to-superconductor ratio $(\gamma_{c/s})$ of 4, we have: $P_{cd} = 26 \times 10^{-3}\,\text{m}$; $A_m = 24 \times 10^{-6}\,\text{m}^2$; and $A_s = 6 \times 10^{-6}\,\text{m}^2$. For the conductor satisfying the Stekly stability criterion, we have, modifying Eq.6.12a (p. 217) for $I_{op} < I_{c_o}$:

$$[J_{op}]_{cd} = \sqrt{\frac{f_p P_{cd} A_m q_{fm}}{\rho_m (A_s + A_m)^2}} \qquad (S1.7)$$

where $[J_{op}]_{cd} = I_{op}/(A_m + A_s)$. With $f_p = 0.5$, $q_{fm} = 0.36\,\text{W/cm}^2$ or $3.6\,\text{kW/m}^2$, and $\rho_m = 5 \times 10^{-10}\,\Omega\,\text{m}$, we obtain:

$$[J_{op}]_{cd} = \sqrt{\frac{(0.5)(26 \times 10^{-3}\,\text{m})(24 \times 10^{-6}\,\text{m}^2)(3.6 \times 10^3\,\text{W/m}^2)}{(5 \times 10^{-10}\,\Omega\,\text{m})(30 \times 10^{-6}\,\text{m}^2)^2}} \qquad (S1.8)$$

$$\simeq 5 \times 10^7\,\text{A/m}^2$$

With $I_{op} = [J_{op}]_{cd}(A_s + A_m)$, we have: $I_{op} = 1500\,\text{A}$ or $[J_{op}]_m = 6.25 \times 10^7\,\text{A/m}^2$. Inserting appropriate values into Eq. 8.7, we obtain:

$$V_D = \left(\frac{A_m}{A_{cd}}\right)\frac{[J_{op}]_m^2 E_m}{I_{op} Z(T_i, T_f)} \qquad (8.7)$$

$$= \frac{\gamma_{c/s}[J_{op}]_m^2 E_m}{(\gamma_{c/s}+1)I_{op}Z(T_i,T_f)} = \frac{4(6.25 \times 10^7\,\text{A/m}^2)^2(10 \times 10^6\,\text{J})}{5(1500\,\text{A})(5 \times 10^{16}\,\text{A}^2\,\text{s/m}^4)} \qquad (S1.9)$$

$$= 417\,\text{V}$$

A discharge voltage of 417 V is safe and should not pose any undue difficulties.

d) With $E_m = 100\,\text{MJ}$ substituted into Eq. 8.7, we obtain a new value of V_D, which now must be 4170 V, a very dangerous level within the cryostat environment. One method widely used is to center-tap the dump resistor to ground, reducing the magnet-to-cryostat discharge voltage to $\pm 2085\,\text{V}$—a significant reduction. This of course does not reduce the magnet terminal voltage. Under certain conditions, voltage levels up to $\sim 5000\,\text{V}$ are considered manageable. As remarked above, Eq. 8.7 indicates one way to reduce V_D for a given E_m is to increase I_{op}. (Note that not much is gained from Z.) This is one of the reasons that for Class 1 magnets, e.g. for fusion reactors, operating currents as high as $50\sim100\,\text{kA}$ have been proposed. Increasing I_{op}, however, makes other aspects of design difficult, e.g. bulkier and generally more expensive conductors; higher heat inputs into the cryostat through gas-cooled leads; greater Lorentz forces acting on bus lines within the cryostat; higher dissipation rates at joints; and more expensive power supplies.

Comments on Z Functions for Magnet Protection

As evident from plots of Fig. 8.4, silver, included because it is the basic matrix material for most HTS, is the best material for this protection criterion. This is the result of good conductivity and "massiveness" as compared with aluminum which is ~ 4 times lighter than silver. Copper, almost universally used in LTS, is almost as good as silver.

Problem 8.2: Hot-spot temperatures in Hybrid III SCM

This problem deals with protection of the Hybrid III SCM. Table 8.2 lists appropriate conductor parameters. The magnet relies on the detect-and-dump protection method discussed in Problem 8.1.

The dump resistor R_D for Hybrid III is $0.3\,\Omega$. The inductance, L_m, of the SCM is $8.0\,H$. At 2230 A (the highest operating current used), the SCM stores a magnetic energy of 19.9 MJ. (The system's nominal operating current is 2100 A.)

a) Compute the final temperature, T_f, at each hot spot for each grade of Nb_3Sn and Nb-Ti composites when the SCM is dumped from 2230 A, assuming that each conductor remains superconducting during the dump except at the hot spot in each of the four conductors, and that adiabatic conditions prevail at each hot spot. Assume further that each hot spot contributes negligible electrical resistance to the circuit. Use the Z-function curve for copper (RRR=50) in Fig. 8.4.

What actually happens when the SCM dump is initiated at $t = 0$ is that the entire SCM is driven normal essentially at $t = 0$, primarily because of AC heating generated by a rapid field change within the winding. The winding is subsequently heated further by Joule dissipation. It is therefore more realistic to include $r(t)$ in the analysis of current decay. For the sake of simplicity, let us express $r(t)$ for this problem by:

$$r(t) = R_0 + \eta t \tag{8.8}$$

where R_0 and η are both constants.

b) Show that the SCM current $I(t)$ during the dump ($t \geq 0$) may be given by:

$$I(t) = I_{op} \exp\left[-\frac{(R_D + R_0)}{L_m}t - \frac{\eta}{2L_m}t^2\right] \tag{8.9}$$

Note that $I(t = 0) = I_{op}$.

c) Using the above model, compute the total energy dissipated in the SCM, E_{scm}, for the following set of values: $I_{op} = 2230\,A$; $L_m = 8\,H$; $R_D = 0.3\,\Omega$; $R_0 = 0.3\,\Omega$; $\eta = 0.04\,\Omega/s$.

Table 8.2: Hybrid III SCM Conductor Parameters

Superconductor		Nb_3Sn		Nb-Ti	
Conductor Grade		HF	LF	HF	LF
Overall width, a	[mm]	9.49	9.10	9.20	9.20
Overall thickness, b	[mm]	4.52	4.47	2.60	2.00
A_{cu}/A_{nc}, $\gamma_{c/s}$		4.1	5.3	3.0	10

Solution to Problem 8.2

a)   In general, when the discharge time constant is determined completely by the SCM inductance L_m and dump resistor R_D, we have, from Eq. $S1.5$:

$$Z(T_f) = \left(\frac{A_m}{A_{cd}}\right)[J_{op}]_m^2 \frac{L_m}{2R_D} = \left(\frac{\gamma_{c/s}}{\gamma_{c/s}+1}\right)[J_{op}]_m^2 \frac{L_m}{2R_D} \qquad (S2.1)$$

where $\gamma_{c/s}$ is the volumetric copper-to-superconductor ratio. $[J_{op}]_m$ is given by:

$$[J_{op}]_m = \frac{I_{op}}{A_m} = \frac{I_{op}(\gamma_{c/s}+1)}{ab\gamma_{c/s}} \qquad (S2.2)$$

Nb_3Sn Grade 1:   We have:

$$[J_{op}]_m = \frac{(2230\,\text{A})(5.1)}{[(9.49\times10^{-3}\,\text{m})(4.52\times10^{-3}\,\text{m})}(4.1)]$$

$$= 6.47\times10^7\,\text{A/m}^2$$

Using Eq. $S2.1$, we have:

$$Z(T_f) = \left(\frac{4.1}{5.1}\right)\frac{(8\,\text{H})(6.47\times10^7\,\text{A/m}^2)^2}{2(0.3\,\Omega)}$$

$$= 4.5\times10^{16}\,\text{A}^2\,\text{s/m}^4$$

From Fig. 8.4 (copper RRR=50), we find $T_f \sim 75\,\text{K}$.

Table 8.3 presents a summary for the four conductors. From Table 8.3 we note that because of excessive hot-spot temperatures both grades of the Nb-Ti conductors may be damaged severely.

b)   The circuit differential equation for $t \geq 0$ is given by:

$$L_m \frac{dI(t)}{dt} + (R_D + R_0 + \eta t)I(t) = 0 \qquad (S2.3)$$

Equation $S2.3$ may be solved as:

$$\frac{dI(t)}{I(t)} = -\frac{(R_D + R_0 + \eta t)}{L_m}dt \qquad (S2.4)$$

$$\ln\left[\frac{I(t)}{I_{op}}\right] = -\frac{(R_D + R_0)}{L_m}t - \frac{\eta}{2L_m}t^2 \qquad (S2.5)$$

Table 8.3: Z and T_f Values for Hybrid III Conductors

Conductor	A_m $[10^{-6}\,\text{m}^2]$	$[J_{op}]_m$ $[\text{MA/m}^2]$	$Z(T_f)$ $[10^{16}\,\text{A}^2\,\text{s/m}^4]$	T_f $[\text{K}]$
Nb_3Sn HF	34.5	64.7	4.5	~75
Nb_3Sn LF	34.2	65.2	4.7	~75
Nb-Ti HF	17.4	124.3	15.4	≫ 300
Nb-Ti LF	16.7	133.3	21.4	≫ 300

Solution to Problem 8.2

Solving Eq. $S2.5$ for $I(t)$, we obtain:

$$I(t) = I_{op} \exp\left[-\frac{(R_D + R_0)}{L_m}t - \frac{\eta}{2L_m}t^2\right] \qquad (8.9)$$

c)   There are two methods to solve this problem.

Method 1:   The easiest and quickest way to compute E_{scm} is to estimate an average value of $r(t)$ during the current decay, $\tilde{r}$, and use a simple "voltage divider" method to determine the energy dissipated in the SCM: $E_{scm} = E_m\tilde{r}/(\tilde{r} + R_D)$, where in this case $E_m = 19.9\,\mathrm{MJ}$.

Without $r(t)$, the circuit time constant, τ_D, is given by L_m/R_D, which is ~27 s. From Eq. 8.8 we have: $r(0) = 0.3\,\Omega$; $r(5\,\mathrm{s}) = 0.5\,\Omega$; $r(10\,\mathrm{s}) = 0.7\,\Omega$; $r(15\,\mathrm{s}) = 0.9\,\Omega$; $r(20\,\mathrm{s}) = 1.1\,\Omega$.

The average of $r(t)$ over this time period is $0.7\,\Omega$, or a new dump time constant of ~8 s $[= L_m/(R_D + 0.7\,\Omega)]$. This means the time average should be taken between 0 and ~10 s, or a new average value of $\tilde{r}$ of $0.5\,\Omega$. That is, $\sim63\%$ $[= 0.5/(0.3+0.5)]$ of $19.9\,\mathrm{MJ}$ is dissipated in the SCM: $E_{scm}\sim12\,\mathrm{MJ}$.

Method 2:   A more rigorous way to determine E_{scm} is to integrate $r(t)I^2(t)$. That is:

$$E_{scm} = \int_0^\infty r(t)I_0^2 \exp\left[-\frac{2(R_D + R_0)}{L_m}t - \frac{\eta}{L_m}t^2\right] dt \qquad (S2.6)$$

Because Eq. $S2.6$ cannot be solved in a closed form, it must be integrated graphically. Results are presented in Table 8.4.

Integrating $r(t)I^2(t)$ over the period from 0 to 20 s, we obtain E_{scm}: $\sim11.6\,\mathrm{MJ}$, which is $\sim60\%$ of the energy initially stored in the SCM.

Incidentally, $Z(T_f)$ for Grade 1 Nb-Ti now becomes $\sim8\times10^{16}\,\mathrm{A^2\,s/m^4}$ or $T_f\sim135\,\mathrm{K}$ and $Z(T_f)$ for Grade 2 Nb-Ti becomes $\sim9\times10^{16}\,\mathrm{A^2\,s/m^4}$ or $T_f\sim160\,\mathrm{K}$.

Table 8.4: Energy Dissipated in the Nb-Ti Coil

t [s]	$r(t)$ [Ω]	$I(t)$ [A]	$r(t)I^2(t)$ [MW]
0	0.3	2200	1.45
5	0.5	1420	1.01
10	0.7	809	0.46
15	0.9	407	0.15
20	1.1	181	0.04

Problem 8.3:   Quench-voltage detection (QVD)

1. Basic technique using a bridge circuit

This problem and the ones to follow discuss bridge circuit based quench-voltage detection (QVD) techniques particularly useful in "coupled" magnet systems, such as hybrid magnets consisting of two independently chargeable magnets.

Figure 8.5 shows the basic bridge circuit containing two coils, Coil 1 and Coil 2, connected in series. The two coils can really be one coil divided into two parts. L_1 is Coil 1's self inductance and L_2 is Coil 2's self inductance. (In this model when the two coils are connected in series, the mutual inductance between the two coils can be included in the self inductances.) r represents the resistance of a small normal zone created in Coil 1. R_1 and R_2 are the bridge circuit resistors and $V_{out}(t)$ represents the bridge output. In the following analysis, assume all circuit elements, including r, are constant; also assume that R_1 and R_2 are sufficiently large so that they do not "load" the bridge circuit.

a) Show that an expression for output voltage $V_{out}(t)$ is given by:

$$V_{out}(t) = \left(\frac{R_2}{R_1 + R_2}\right) L_1 \frac{dI(t)}{dt} - \left(\frac{R_1}{R_1 + R_2}\right) L_2 \frac{dI(t)}{dt} + \left(\frac{R_2}{R_1 + R_2}\right) rI(t)$$

$$(8.10)$$

b) Show that when the condition given by Eq. 8.11a is satisfied, $V_{out}(t)$ becomes proportional only to $rI(t)$. Namely:

$$R_2 L_1 = R_1 L_2 \qquad (8.11a)$$

$$V_{out}(t) = \left(\frac{R_2}{R_1 + R_2}\right) rI(t) \qquad (8.11b)$$

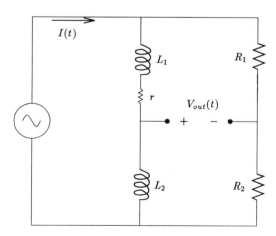

Fig. 8.5   Bridge circuit voltage detection technique.

Solution to Problem 8.3

a) For the case when $R_1 + R_2$ are "large," the total voltage across the two coils, $V_{cl}(t)$, is given by:

$$V_{cl}(t) = L_1 \frac{dI(t)}{dt} + rI(t) + L_2 \frac{dI(t)}{dt} \qquad (S3.1)$$

For the same condition, the current through the resistors R_1 and R_2, $i_R(t)$, is given by:

$$i_R(t) = \frac{V_{cl}(t)}{R_1 + R_2} \qquad (S3.2)$$

From the circuit shown in Fig. 8.5, we have:

$$V_{out}(t) = L_1 \frac{dI(t)}{dt} + rI(t) - R_1 i_R(t) \qquad (S3.3)$$

Combining Eqs. $S3.1$, $S3.2$, and $S3.3$, we obtain:

$$V_{out}(t) = L_1 \frac{dI(t)}{dt} + rI(t)$$
$$- \frac{R_1}{R_1 + R_2}\left[L_1 \frac{dI(t)}{dt} + rI(t) + L_2 \frac{dI(t)}{dt} \right] \qquad (S3.4)$$

$$V_{out}(t) = \left(\frac{R_2}{R_1 + R_2} \right) L_1 \frac{dI(t)}{dt} - \left(\frac{R_1}{R_1 + R_2} \right) L_2 \frac{dI(t)}{dt} + \left(\frac{R_2}{R_1 + R_2} \right) rI(t) \qquad (8.10)$$

b) To make $V_{out}(t)$ proportional only to $rI(t)$, the first two terms in the right-hand side of Eq. 8.10 must be equal to zero:

$$\left(\frac{R_2}{R_1 + R_2} \right) L_1 \frac{dI(t)}{dt} - \left(\frac{R_1}{R_1 + R_2} \right) L_2 \frac{dI(t)}{dt} = 0 \qquad (S3.5)$$

Equation $S3.5$ is simplified to give the required condition:

$$R_2 L_1 = R_1 L_2 \qquad (8.11a)$$

With the first two terms in the right-hand side of Eq. 8.10 eliminated, Eq. 8.10 becomes:

$$V_{out}(t) = \left(\frac{R_2}{R_1 + R_2} \right) rI(t) \qquad (8.11b)$$

As we shall see in the next problem, the condition in real hybrid magnets is far from ideal: it is generally very difficult to achieve the condition $R_2 L_1 = R_1 L_2$ independent of $I(t)$ and dI/dt.

Problem 8.4: Quench-voltage detection (QVD)
2. An improved technique

Figure 8.6 represents a schematic model for Hybrid II, another hybrid magnet operating at FBNML [3.5]. The superconducting magnet is a Nb-Ti coil comprised of 22 double pancakes (DP). In addition to the Nb-Ti coil, the water-cooled insert and copper radiation plates are included in the figure to emphasize that in a "real" system, magnetic coupling is not confined just to double pancakes; all three components are coupled. The magnetic coupling between these components makes "balancing" the bridge circuit studied in Problem 8.3 not straightforward.

Two QVD techniques are used for the Hybrid II, whose SCM is divided into four sections: section B' (DP 1 through 7); section A' (DP 8 through 11); section A (DP 12 through 15); and B (DP 16 through 22).

Technique I

In this technique, the magnet is divided into A'+A and B'+B. Although inductive voltage canceling achieved by this technique is slightly better than that achieved by a more conventional technique that divides the magnet into B'+A' and A+B, it is still not entirely satisfactory. The technique cannot completely eliminate all inductive voltages that are developed whenever either the water-cooled insert or the Nb-Ti coil is energized.

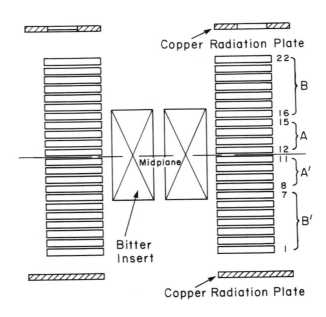

Fig. 8.6   Schematic arrangement for Hybrid II containing 22 double pancakes [3.5].

Problem 8.4: 2. An improved technique

Technique II

The second technique, developed by Ishigohka [8.23], divides the magnet into 22 double pancakes, and then combines them into two major components—one component, $V_{2n-1}(t)$, containing odd-numbered double pancakes and the other component, $V_{2n}(t)$, containing even-numbered double pancakes. By adjusting the gain of each of 22 separate amplifiers, we can adjust the voltage from each double pancake to give minimum $V_{out}(t)$ in the absence of a resistive voltage. Thus:

$$V_{out}(t) = \sum_{n=1}^{11} [\alpha_{2n-1} V_{2n-1}(t) - \alpha_{2n} V_{2n}(t)] \tag{8.12}$$

where α_{2n-1} is the amplifier gain for the $(2n-1)$th double pancake and α_{2n} is the amplifier gain for the $2n$th double pancake.

Discuss why Technique II is more effective than Technique I in eliminating unwanted inductive voltages.

Voltage Attenuation in Magnet Protection Circuit

In Hybrid III SCM, fifty voltage signals are monitored for protection and control of the magnet operation. Generally, each voltage signal, coming from one of either 32 Nb-Ti double pancakes or 18 Nb$_3$Sn layers, is brought out with a pair of twisted voltage leads from inside the cryostat to a data acquisition system located outside the cryostat. The voltage leads pass through a hermetic connector placed at the cryostat. The connector sometimes becomes the source of a lead burnout, caused by the arcing of its pins. Arcing can be induced by a high voltage that appears during a fast discharge.

A technique often used to limit the current through shorted leads is to insert a resistor, R_{lm}, close to each voltage tap on the coil, as schematically shown in Fig. 8.7. The induced current (i_j) through the *shorted* circuit, consisting of the coil (voltage V_j), two resistors, two lengths of voltage leads, and a short occurring across pins, is given by $V_j/2R_{lm}$; clearly we can make i_j arbitrarily small by choosing a large value for R_{ml}. In Hybrid III SCM, a 5-kΩ resistor is used for each R_{ml}. Note also that the presence of R_{ml} in each voltage tap lead makes it easier to attenuate voltage signals reaching amplifiers.

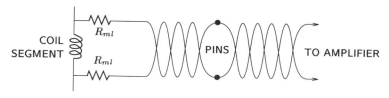

Fig. 8.7   Schematic drawing of the arrangement of voltage leads for Hybrid III SCM.

Solution to Problem 8.4

How successfully we can eliminate inductive voltages in Technique I $(A + A'$ vs $B + B')$ depends on the closeness of inductive voltages of the two parts irrespective of current level or current sweep rate. It turns out that in Hybrid II, because the cryostat housing the Nb-Ti coil is not perfectly symmetric with respect to the coil's midplane, voltage balancing cannot be maintained irrespective of current level or current sweep rate.

Also, more seriously, a bridge circuit setting optimized for charging the Nb-Ti coil alone is not optimized when the insert is charged, and the optimized setting shifts with self-field sweep rate as well as with insert sweep rate.

Technique II (odd-numbered pancakes vs even-numbered pancakes) minimizes spatial nonuniformity of the entire system, which includes the coil itself, insert, radiation shields, and other parts of the cryostat. This makes the total unbalanced inductive voltages significantly smaller than those of the bottom-top case.

Figure 8.8 shows three "balanced" voltages, $V_{AA'}$, $V_{BB'}$, and the odd-even difference voltage, V_δ for an insert trip from 25 kA [8.23]. Note that the peak value of $V_{BB'} - V_{AA'}$ is about ~100 times greater than the peak value of V_δ.

V_δ is thus a much more sensitive method for monitoring *only* the operating condition of the superconducting coil. The cost of improved sensitivity is the large number of differential amplifiers required.

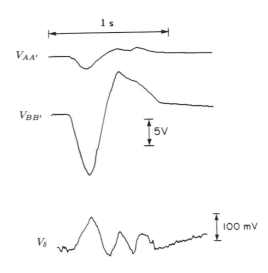

Fig. 8.8   Unbalanced voltages recorded at an insert trip in Hybrid II [8.23].

Problem 8.5: Quench-induced pressure in CIC conductors
1. Analytical approach

One of the most serious design issues for CIC conductors is the large rise in internal pressure that results when a CIC-wound magnet is driven normal.

This problem follows the analytical approach developed by Dresner [8.24]. We consider the case in which a CIC conductor of total length 2ℓ is driven normal at one instant over its entire length. The following equations are valid for fluid (helium) inside the CIC conductor:

$$\frac{d\rho}{dt} + \rho\frac{\partial u}{\partial z} = 0 \tag{8.13a}$$

$$\rho\frac{du}{dt} + \frac{\partial p}{\partial z} + \rho\frac{2f_\mu u^2}{D} = 0 \tag{8.13b}$$

$$\rho\frac{d}{dt}\left(e + \frac{u^2}{2}\right) + \frac{\partial(pu)}{\partial z} = g_{jf} \tag{8.13c}$$

Equations 8.13a, 8.13b, and 8.13c, respectively, state the mass continuity, force balance, and energy balance of fluid. ρ, u, and p are, respectively, helium's density, velocity, and pressure. f_μ is the frictional coefficient for helium flow through the CIC conductor; D is hydraulic diameter. g_{jf} is Joule power density generated in the conductor computed in terms of a unit *fluid* volume.

To derive an expression for the maximum pressure rise at the center ($z = 0$) of the normal zone, $(\Delta p_0)_{mx}$, it is necessary to proceed in several steps.

a) First, assuming further that $du/dt = 0$ (a condition nearly satisfied in this case because the friction forces greatly dominate inertial forces), show that an expression for the pressure rise at $z = 0$, Δp_0, is given by:

$$\Delta p_0 = \frac{2f_\mu}{D}\int_0^\ell \rho u^2\, dz \tag{8.14}$$

b) Next, assuming further that u increases linearly with z and $u = 0$ at $z = 0$ and $u = u_m$ at $z = \ell$, and that ρ is independent of z, show that an expression for Δp_0 is given by:

$$\Delta p_0 = \frac{2f_\mu \rho u_m^2 \ell}{3D} \tag{8.15}$$

c) Now, derive the following equation:

$$d\rho = \frac{1}{c^2}\,dp - \frac{\beta\rho}{C_p}T\,ds \tag{8.16}$$

from the following thermodynamics equation:

$$T\,ds = \frac{\kappa C_v}{\beta}\,dp - \frac{C_p}{\beta\rho}\,d\rho \tag{8.17}$$

Problem 8.5: 1. Analytical approach

where c is the velocity of sound, β is the thermal expansion coefficient, and κ is the isothermal compressibility, which is the reciprocal of the isothermal bulk modulus. c^2, β, and κ are given by:

$$c^2 = \left(\frac{\partial p}{\partial \rho}\right)_s \tag{8.18a}$$

$$\beta = \frac{1}{v}\left(\frac{\partial v}{\partial T}\right)_p = -\frac{1}{\rho}\left(\frac{\partial \rho}{\partial T}\right)_p \tag{8.18b}$$

$$\kappa = -\frac{1}{v}\left(\frac{\partial v}{\partial p}\right)_T = \frac{1}{\rho}\left(\frac{\partial \rho}{\partial p}\right)_T \tag{8.18c}$$

d) Next, combine Eqs. 8.13b, 8.13c, 8.15, and the 1st law of thermodynamics to show that an expression for $d\rho/dt$ when $du/dt = 0$ may be given by:

$$\frac{d\rho}{dt} = \frac{1}{c^2}\frac{dp}{dt} - \frac{\beta\rho}{C_p}\left(\frac{g_{jf}}{\rho} + \frac{2f_\mu u^3}{D}\right) \tag{8.19}$$

e) Show that for the present case where u depends linearly on z, the time rate of change of pressure at the center $(z = 0)$, $dp/dt|_0$, is given by:

$$\left.\frac{dp}{dt}\right|_0 = \frac{\beta c^2 g_{jf}}{C_p} - \frac{\rho c^2}{\ell}\sqrt{\frac{3D\Delta p_0}{2f_\mu \rho \ell}} \tag{8.20}$$

f) Finally, when the pressure rise reaches a maximum, we have $dp/dt|_0 = 0$. Show that an expression for $(\Delta p_0)_{mx}$ is given by:

$$(\Delta p_0)_{mx} = \frac{2f_\mu \ell^3 g_{jf}^2}{3D}\left(\frac{\beta^2}{\rho C_p^2}\right) \tag{8.21}$$

Using an appropriate dependance of $(\beta^2/\rho C_p^2)$ on p, Dresner reduces Eq. 8.21 to the following expression, valid when all the parameters are given in SI units.

$$(\Delta p_0)_{mx} \simeq 0.17\left(\frac{\ell^3 g_{jf}^2}{D}\right)^{0.36} \tag{8.22}$$

For limited experimental conditions, Eq. 8.22 has been shown to be quite accurate.

Solution to Problem 8.5

a)   From Eq. 8.13b, with $du/dt = 0$, we have:

$$\frac{\partial p}{\partial z} = -\rho \frac{2f_\mu u^2}{D} \qquad (S5.1)$$

Integrating Eq. $S5.1$ between 0 and ℓ, we have:

$$p(\ell) - p(0) = -\Delta p_0 = -\frac{2f_\mu}{D} \int_0^\ell \rho u^2 \, dz$$

Thus:

$$\Delta p_0 = \frac{2f_\mu}{D} \int_0^\ell \rho u^2 \, dz \qquad (8.14)$$

b)   With $u = u_m(z/\ell)$ and $\rho =$ constant, we obtain from Eq. 8.14:

$$\Delta p_0 = \frac{2f_\mu \rho u_m^2}{D\ell^2} \int_0^\ell z^2 \, dz$$

$$= \frac{2f_\mu \rho u_m^2 \ell}{3D} \qquad (8.15)$$

c)   We start with an equation for $d\rho$ in terms of variables p and s:

$$d\rho = \left(\frac{\partial \rho}{\partial p}\right)_s dp + \left(\frac{\partial \rho}{\partial s}\right)_p ds \implies d\rho = \frac{1}{c^2} dp + \left(\frac{\partial \rho}{\partial s}\right)_p ds \qquad (S5.2)$$

From Eq. 8.17 we obtain, with $dp = 0$:

$$\left(\frac{\partial \rho}{\partial s}\right)_p = -\frac{\beta\rho}{C_p}T \qquad (S5.3)$$

Substituting Eq. $S5.3$ into Eq. $S5.2$, we have:

$$d\rho = \frac{1}{c^2} dp - \frac{\beta\rho}{C_p}T \, ds \qquad (8.16)$$

d)   From Eq. 8.16, we have:

$$\frac{d\rho}{dt} = \frac{1}{c^2} \frac{dp}{dt} - \frac{\beta\rho}{C_p}T \frac{ds}{dt} \qquad (S5.4)$$

The 1st law of thermodynamics states:

$$de = T \, ds - p \, dv = T \, ds + \frac{p}{\rho^2} \, d\rho$$

$$T\frac{ds}{dt} = \frac{de}{dt} - \frac{p}{\rho^2} \frac{d\rho}{dt} \qquad (S5.5)$$

Substituting Eq. $S5.5$ into Eq. $S5.4$, we obtain:

$$\frac{d\rho}{dt} = \frac{1}{c^2} \frac{dp}{dt} - \frac{\beta\rho}{C_p}\left(\frac{de}{dt} - \frac{p}{\rho^2} \frac{d\rho}{dt}\right) \qquad (S5.6)$$

Solution to Problem 8.5

From Eq. 8.13c, we have:

$$
\frac{de}{dt} = \frac{g_{jf}}{\rho} - \frac{u}{\rho}\frac{du}{dt} - \frac{p}{\rho}\frac{\partial u}{\partial z} - \frac{u}{\rho}\frac{\partial p}{\partial z}
$$

$$
= \frac{g_{jf}}{\rho} - \frac{p}{\rho}\frac{\partial u}{\partial z} - \frac{u}{\rho}\frac{\partial p}{\partial z} \quad \left(\text{when } \frac{du}{dt} = 0\right) \tag{S5.7}
$$

Combining Eqs. S5.7 and S5.6, we obtain:

$$
\frac{d\rho}{dt} = \frac{1}{c^2}\frac{dp}{dt} - \frac{\beta\rho}{C_p}\left(\frac{g_{jf}}{\rho} - \frac{p}{\rho}\frac{\partial u}{\partial z} - \frac{u}{\rho}\frac{\partial p}{\partial z} - \frac{p}{\rho^2}\frac{d\rho}{dt}\right) \tag{S5.8}
$$

From Eq. 8.13a, we have:

$$
-\frac{p}{\rho}\frac{\partial u}{\partial z} = \frac{p}{\rho^2}\frac{\partial \rho}{\partial t} \tag{S5.9}
$$

and when $du/dt = 0$, we have, from Eq. 8.13b:

$$
-\frac{u}{\rho}\frac{\partial p}{\partial z} = \frac{2f_\mu u^3}{D} \tag{S5.10}
$$

Combining Eqs. S5.8, S5.9, and S5.10, we obtain:

$$
\frac{d\rho}{dt} = \frac{1}{c^2}\frac{dp}{dt} - \frac{\beta\rho}{C_p}\left(\frac{g_{jf}}{\rho} + \frac{2f_\mu u^3}{D}\right) \tag{8.19}
$$

e) With $u = (u_m/\ell)z$ we have, from Eq. 8.13a, $d\rho/dt = -\rho u_m/\ell$. Combining this and Eq. 8.19, we obtain:

$$
-\rho\frac{u_m}{\ell} = \frac{1}{c^2}\frac{dp}{dt} - \frac{\beta\rho}{C_p}\left(\frac{g_{jf}}{\rho} + \frac{2f_\mu u^3}{D}\right) \tag{S5.11}
$$

From Eq. 8.15, we have:

$$
u_m = \sqrt{\frac{3D\Delta p_0}{2f_\mu\rho\ell}} \tag{S5.12}
$$

At $z = 0$ we have, from symmetry, $u = 0$ and combining Eqs. S5.11 and S5.12, we obtain:

$$
\left.\frac{dp}{dt}\right|_0 = \frac{\beta c^2 g_{jf}}{C_p} - \frac{\rho c^2}{\ell}\sqrt{\frac{3D\Delta p_0}{2f_\mu\rho\ell}} \tag{8.20}
$$

f) With $dp/dt|_0 = 0$, Eq. 8.20 may be written as:

$$
0 = \frac{\beta c^2 g_{jf}}{C_p} - \frac{\rho c^2}{\ell}\sqrt{\frac{3D(\Delta p_0)_{mx}}{2f_\mu\rho\ell}} \tag{S5.13}
$$

Solving Eq. S5.13 for $(\Delta p_0)_{mx}$, we have:

$$
(\Delta p_0)_{mx} = \frac{2f_\mu\ell^3 g_{jf}^2}{3D}\left(\frac{\beta^2}{\rho C_p^2}\right) \tag{8.21}
$$

Problem 8.6: Quench-induced pressure in CIC conductors
2. CIC coil for the NHMFL's 45-T hybrid

As mentioned in Chapter 3, a 45-T hybrid magnet is presently under construction for the National High Magnetic Field Laboratory [3.13]. The system consists of a 24-MW 31-T resistive insert and a 610-mm room-temperature bore superconducting magnet. The SCM in turn consists of two Nb_3Sn coils and the outermost Nb-Ti coil, all wound with CIC conductors. FBNML is responsible for the design and construction of the 31-T resistive insert and the 8-T outermost Nb-Ti superconducting coil, while NHMFL is responsible for design and construction of the 6-T inner Nb_3Sn coils. The CIC coils are cooled by *static* (no forced flow) superfluid helium at 1.8 K.

Table 8.5 presents appropriate parameters for the Nb-Ti superconducting coil, which is composed of 29 double pancakes. Both terminals of each double pancake are connected to a manifold that supplies superfluid helium.

a) Using Eq. 8.22, make a rough estimate of the maximum internal pressure expected when one single double pancake in the Nb-Ti coil is driven normal at the SCM's normal operating current of 10 kA.

b) Williams estimates the maximum internal pressure using a completely different approach [8.25]. He assumes an isochoric (constant volume) process for the static helium inside the CIC conductor during which both the strands and helium are heated to a final temperature of 150 K. What will be the final pressure inside the CIC conductor under this assumption?

c) Williams assumes that a particular double pancake "soaks" up energies from other parts of the coils. If indeed a given double pancake is heated isochorically to 150 K, what will be the total energy deposited in the double pancake? Note that the energy stored in each double pancake at 10 kA is $\sim$3 MJ. Is there any "soaking" of energy from other parts of the coil by this double pancake? For thermal energy computation, consider only the conductor and the helium and neglect the conduit.

Table 8.5: The Nb-Ti Coil and Conductor Parameters
For NHMFL's 45-T Hybrid SCM

Nb-Ti Coil			Conductor		
Winding i.d., $2a_1$	[mm]	1150	# strands		135
Winding o.d., $2a_2$	[mm]	1680	Strand dia. d_{st}	[mm]	0.81
Winding length, $2b$	[mm]	923	Total A_{nc}	[mm^2]	11
# of double pancakes (DP)		29	Total A_{cu}	[mm^2]	59
Conductor length/DP	[m]	156	Total A_{he}	[mm^2]	44
B_{mx} @ $I_{op} = 10$ kA	[T]	9	ρ_{cu} @ T_{op}, B_{mx}	[nΩ m]	0.45

Solution to Problem 8.6

a) To determine the maximum pressure in the double pancake, let us first calculate g_{jf} (heat generation per unit *helium* volume). The total heat generation in the copper matrix per unit conductor length is given by: $G_{cu} = \rho_{cu} I_{op}^2 / A_{cu}$ [W/m]. In terms of per unit helium volume, we have: $g_{jf} = G_{cu}/A_{he}$ [W/m^3]. Thus:

$$g_{jf} = \frac{G_{cu}}{A_{he}} = \frac{\rho_{cu} I_{op}^2}{A_{cu} A_{he}} = \frac{(4.5 \times 10^{-10}\,\Omega\,\mathrm{m})(10^4\,\mathrm{A})^2}{(59 \times 10^{-6}\,\mathrm{m}^2)(44 \times 10^{-6}\,\mathrm{m}^2)} \qquad (S6.1)$$

$$= 1.73 \times 10^7\,\mathrm{W/m}^3$$

Because both ends of each layer are open to manifolds, ℓ to be inserted into Eq. 8.22 is given by one half of the conductor length for one double pancake: $\ell = 78\,\mathrm{m}$. The hydraulic diameter D is given by: $D = 4A_{he}/(135\pi d_{st}) = 5.36 \times 10^{-4}\,\mathrm{m}$. With $\ell = 78\,\mathrm{m}$, $g_{jf} = 1.66 \times 10^7\,\mathrm{W/m}^3$, and $D = 5.36 \times 10^{-4}\,\mathrm{m}$, we have, from Eq. 8.22:

$$(\Delta p_0)_{mx} = 0.17 \left(\frac{\ell^3 g_{jf}^2}{D} \right)^{0.36} = 0.17 \left[\frac{(78\,\mathrm{m})^3 (1.73 \times 10^7\,\mathrm{W/m}^3)^2}{5.36 \times 10^{-4}\,\mathrm{m}} \right]^{0.36} \qquad (S6.2)$$

$$= 4.6 \times 10^7\,\mathrm{Pa} \simeq 455\,\mathrm{atm}$$

From Eq. 8.20, we have $dp/dt|_0 < (\beta c^2 g_{jf}/C_p)$. β varies quite a bit, but it never exceeds a value of $0.1\,/\mathrm{K}$ above $p \sim 10\,\mathrm{atm}$. For supercritical helium we have: $c \sim 200\,\mathrm{m/s}$ and $C_p \sim 5 \times 10^3\,\mathrm{J/kg\,K}$. Now with $g_{jf} \sim 2 \times 10^7\,\mathrm{W/m}^3$, we find $dp/dt|_0 < (\beta c^2 g_{jf}/C_p) \sim 1 \times 10^7\,\mathrm{Pa/s}$ (= 100 atm/s), or it takes $\sim 4\,\mathrm{s}$ to reach $\sim 400\,\mathrm{atm}$. Because the effective discharge time for this SCM is $\sim 4\,\mathrm{s}$, if there is no energy transfer from other parts of the system, $(\Delta p_0)_{max}$ should be less than $\sim 400\,\mathrm{atm}$ [(4 s)×(200 atm/s)]. Note that a rate of pressure rise of $\sim 100\,\mathrm{atm/s}$ does not include the negative contribution in Eq. 8.20; a pressure of $\sim 400\,\mathrm{atm}$ should therefore be the upper limit.

b) From isochoric P vs T curves for helium given in Fig. A2.1 (Appendix II), we find that at $150\,\mathrm{K}$ a density of $147\,\mathrm{kg/m}^3$ ($1.8\,\mathrm{K}$, 1 atm) corresponds to a pressure of $\sim 800\,\mathrm{atm}$.

c) Copper has an enthalpy of $25.3\,\mathrm{kJ/kg}$ at $150\,\mathrm{K}$; with a total conductor mass (assumed to be all copper) in one double pancake of $100\,\mathrm{kg}$, the total energy absorbed by the conductor in one double pancake is $2.5\,\mathrm{MJ}$. Helium has an internal energy density of $\sim 500\,\mathrm{kJ/kg}$ at $150\,\mathrm{K}$ and $800\,\mathrm{atm}$ (Fig. A2.2, Appendix II) and a negligible amount at $1.8\,\mathrm{K}$. With a total helium mass of $1\,\mathrm{kg}$ in one double pancake, the total energy absorbed by the helium is $\sim 0.5\,\mathrm{MJ}$, bringing the total energy absorbed by one double pancake to be $\sim 3\,\mathrm{MJ}$. Because each double pancake stores $3\,\mathrm{MJ}$ to start with, there is no significant soaking of energy by any particular double pancake when it is assumed to be heated isochorically to a final temperature of $150\,\mathrm{K}$.

Problem 8.7: Normal-zone propagation (NZP)
1. Velocity in the longitudinal direction

In this problem we derive an expression for normal-zone propagation (NZP) velocity in the longitudinal (along the conductor axis) direction, U_ℓ, under adiabatic conditions. U_ℓ is an important parameter for protection of high-performance adiabatic magnets, *e.g.* MRI and NMR, because in these magnets, protection must rely to a great extent on NZP within each winding; it determines the speed with which this normal zone spreads out in the winding. In these adiabatic windings, NZP is not confined only along the conductor axis but spreads out three-dimensionally. It has been demonstrated experimentally that the "transverse" propagation velocity, U_t, is proportional to U_ℓ [8.17].

a) Consider a normal-superconducting boundary moving at a constant velocity, U_ℓ, along a current-carrying (I) superconducting wire in the $+x$-direction under adiabatic conditions ($g_q = 0$ in Eq. 6.1, p. 203), as illustrated schematically in Fig. 8.9. We shall first treat the case for a wire having *no* matrix metal. With no cooling term in Eq. 6.1, we can write two one-dimensional power density equations, one in the normal region ($x < 0$) and the other in the superconducting region ($x > 0$). Here ρ_n and J are, respectively, the normal-state resistivity and current density of the superconductor. Using the coordinate transformation, $z = x - U_\ell t$, show that an expression for the power density equation in the normal region may be given by:

$$\frac{d}{dz}\left(k_n \frac{dT_n}{dz}\right) + C_n U_\ell \frac{dT_n}{dz} + \rho_n J^2 = 0 \qquad (8.23a)$$

and that an expression for the power density equation in the superconducting region is given by:

$$\frac{d}{dz}\left(k_s \frac{dT_s}{dz}\right) + C_s U_\ell \frac{dT_s}{dz} = 0 \qquad (8.23b)$$

where the subscripts n and s refer to the normal and superconducting states.

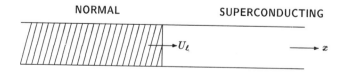

Fig. 8.9 One-dimensional normal-to-superconducting boundary moving at a constant velocity U_ℓ.

Problem 8.7: 1. Velocity in the longitudinal direction

b) Assuming k_n, k_s, C_n, and C_s are all constant and also $d^2T_n/dz^2 \simeq 0$ near $z = 0$ [8.3], show that an expression for U_ℓ is given by:

$$U_\ell = J\sqrt{\frac{\rho_n k_n}{C_n C_s (T_t - T_{op})}} \qquad (8.24)$$

where T_t and T_{op} are, respectively, the superconductor's transition and operating temperatures. (Note that $T_t = T_c$ when $I = 0$.) Although it is rarely necessary to use an exact expression of U_ℓ for which material properties are temperature dependent, it is given below for the sake of completeness [8.19]:

$$U_\ell = J\sqrt{\frac{\rho_n(T_t)k_n(T_t)}{\left[C_n(T_t) - \frac{1}{k_n(T_t)}\frac{dk_n}{dT}\Big|_{T_t}\int_{T_{op}}^{T_t}C_s(T)dT\right]\int_{T_{op}}^{T_t}C_s(T)\,dT}} \qquad (8.25)$$

For constant material properties, we may note that Eq. 8.25 reduces to Eq. 8.24. Also for the case $C_n = C_s = C_o$, Eq. 8.24 may be written as:

$$U_\ell = \frac{J}{C_o}\sqrt{\frac{\rho_n k_n}{(T_t - T_{op})}} \qquad (8.26)$$

Equations 8.24 through 8.26 are valid for superconductors having no matrix metal; in reality most superconductors are composites and we may replace material properties with those of the matrix metal. According to Joshi [8.17], T_t is replaced by $T_t' = (T_{cs} + T_c)/2$, where T_{cs} is the current sharing temperature. [In the absence of cooling, of course, there will be no current sharing. Nevertheless, it may still be defined as: $T_{cs} = T_c(I)$.] Thus, Eq. 8.26 becomes:

$$U_\ell = \frac{J_m}{C_{cd}}\sqrt{\frac{\rho_m k_m}{(T_t' - T_{op})}} \qquad (8.27)$$

where C_{cd} is the volumetric average of the superconductor's and the matrix's heat capacities and J_m is the current density over the matrix metal's cross sectional area. Because ρ_m (the matrix metal's electrical resistivity) is much smaller than ρ_n, and k_m (the matrix metal's thermal conductivity) is much greater than k_n, they are used in Eq. 8.27.

Note that U_ℓ may be generalized as:

$$U_\ell \propto \frac{\sqrt{g_j}}{C_{cd}} \qquad (8.28)$$

Further discussion of Eq. 8.28 will be given in connection with the next problem, which deals with the propagation velocity in the transverse direction, U_t. Also as will be discussed in Problem 8.12 in connection with the discussion of normal-zone propagation velocity in HTS, $U_\ell \propto \sqrt{g_j}$ is another reason why HTS's U_ℓ is much smaller than LTS's.

Solution to Problem 8.7

a) A power density equation in the x-direction for the adiabatic case of NZP in the normal region is given by:

$$C_n \frac{\partial T_n}{\partial t} = \frac{\partial}{\partial x}\left(k_n \frac{\partial T_n}{\partial x}\right) + \rho_n J^2 \qquad (S7.1a)$$

Similarly, a power density equation in the x-direction for the adiabatic case in the superconducting region is given by:

$$C_s \frac{\partial T_s}{\partial t} = \frac{\partial}{\partial x}\left(k_s \frac{\partial T_s}{\partial x}\right) \qquad (S7.1b)$$

When the normal-superconducting boundary moves at a constant velocity U_ℓ in the $+x$-direction, we may transform the x coordinate to z: $z = x - U_\ell t$. Thus:

$$C_n \frac{\partial T_n}{\partial t} = \frac{\partial}{\partial z}\left(k_n \frac{\partial T}{\partial z}\right) + \rho_n J^2 \qquad (S7.2a)$$

$$C_s \frac{\partial T_s}{\partial t} = \frac{\partial}{\partial x}\left(k_s \frac{\partial T_s}{\partial x}\right) \qquad (S7.2b)$$

We also have:

$$\frac{\partial T_n}{\partial t} = \frac{\partial T}{\partial z}\frac{\partial z}{\partial t} = -U_\ell \frac{\partial T}{\partial z} \qquad (S7.3)$$

Substituting Eqs. $S7.3$ into $S7.2a$ and $S7.2b$, we obtain for the normal region:

$$-C_n U_\ell \frac{dT_n}{dz} = \frac{d}{dz}\left(k_n \frac{dT_n}{dz}\right) + \rho_n J^2 \qquad (S7.4a)$$

$$-C_s U_\ell \frac{dT_s}{dz} = \frac{d}{dz}\left(k_s \frac{dT_s}{dz}\right) \qquad (S7.4b)$$

Rearranging Eqs. $S7.4a$ and $S7.4b$, we have an expression of the power density equation in the normal zone given by:

$$\frac{d}{dz}\left(k_n \frac{dT_n}{dz}\right) + C_n U_\ell \frac{dT_n}{dz} + \rho_n J^2 = 0 \qquad (8.23a)$$

Similarly, we have an expression of the power density equation in the superconducting zone given by:

$$\frac{d}{dz}\left(k_s \frac{dT_s}{dz}\right) + C_s U_\ell \frac{dT_s}{dz} = 0 \qquad (8.23b)$$

Solution to Problem 8.7

b) With k_n and k_s constant and noting $d^2T_n/dz^2 = 0$ near $z = 0$, we can rewrite Eqs. 8.23a and 8.23b:

$$\text{Normal } (x < 0) \qquad C_n U_\ell \frac{dT_n}{dz} + \rho_n J^2 = 0 \qquad (S7.5a)$$

$$\text{Superconducting } (x > 0) \quad k_s \frac{d^2T_s}{dz^2} + C_s U_\ell \frac{dT_s}{dz} = 0 \qquad (S7.5b)$$

An expression for $T_s(z)$ is given directly from Eq. $S7.5b$:

$$T_s(z) = Ae^{-cz} + T_{op} \qquad (S7.6)$$

where T_{op} is the operating temperature far away from the normal-superconducting boundary and $c = C_s U_\ell / k_s$.

We also know that $T_s(0) = T_t$, the critical temperature. Thus:

$$T_s(z) = (T_t - T_{op}) \exp\left(-\frac{C_s U_\ell}{k_s} z\right) + T_{op} \qquad (S7.7)$$

Another boundary condition is that the $k(dT/dz)$ of each region should be equal at $z = 0$—heat flow must be continuous across the boundary:

$$k_n \frac{dT_n}{dz}\bigg|_0 = k_s \frac{dT_s}{dz}\bigg|_0 \qquad (S7.8)$$

Combining Eqs. $S7.5a$, $S7.7$, and $S7.8$, we have:

$$-\frac{k_n \rho_n J^2}{C_n U_\ell} = -C_s U_\ell (T_t - T_{op}) \qquad (S7.9)$$

Solving Eq. $S7.9$ for U_ℓ, we obtain:

$$U_\ell = J\sqrt{\frac{\rho_n k_n}{C_n C_s (T_t - T_{op})}} \qquad (8.24)$$

Important points to be noted from Eq. 8.24 (and Eqs. 8.26 and 8.27) are that U_ℓ is directly proportional to current density J and inversely proportional to heat capacity. As stated above, Eq. 8.24 is valid for a bare superconductor under adiabatic conditions.

Problem 8.8: Normal zone propagation (NZP)
2. Transverse (turn-to-turn) velocity

This problem discusses turn-to-turn normal zone propagation velocity, U_t, for tape-wound pancakes under adiabatic conditions. Taking an appropriate limit, we can show that U_t becomes proportional to $\sqrt{k_i/k_m}U_\ell$, where k_i is the thermal conductivity of the insulating layer between adjacent tapes and k_m is the thermal conductivity of the matrix metal.

Figure 8.10 presents a section of a tape-wound pancake coil consisting of layers of composite superconducting tape with a turn-to-turn insulating layer between adjacent tapes. Also shown in the figure is an electrical analog of the lumped parameter thermal model between tape 1 at temperature T_1 and adjacent tape 2 at temperature T_2 with the insulating layer at temperature T_i [8.26]. The first $R_i/2$, with units of $\mathrm{m^2\,K/W}$, represents thermal resistance between tape 1 and insulator; the second $R_i/2$ represents thermal resistance between the insulator and tape 2. C_i, with units of $\mathrm{J/m^3\,K}$, represents the insulator's thermal capacitance.

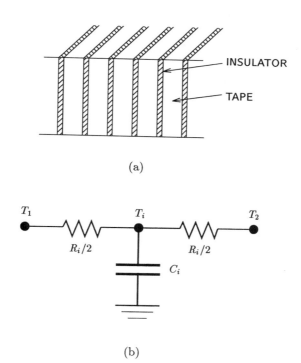

(a)

(b)

Fig. 8.10   (a) Section of a tape-wound pancake winding, showing tape-insulator-tape arrangement. (b) Lumped electrical analog for heat flow from tape 1 to tape 2.

Problem 8.8: 2. Transverse (turn-to-turn) velocity

The normal zone "transient time" τ_t may be defined as the time needed to raise tape 2's (thickness δ_{cd} and heat capacity C_{cd}) temperature from T_{op} to T_t', the "effective" transient temperature for composite conductors under adiabatic conditions. Thus, the following energy density equation may be written for tape 2:

$$\delta_{cd}\int_{T_{op}}^{T_t'} C_{cd}\,dT = \int_0^{\tau_t} q_2(t)\,dt \tag{8.29}$$

In Eq. 8.29, time $t = 0$ is taken to be the instant that tape 1 is driven normal and $q_2(t)$ is the heat flux to tape 2 from tape 1 through the insulator.

Using the electrical analog for heat flow between tape 1 and tape 2 shown in Fig. 8.10b, we can write an equation relating $q_2(t)$, T_1, and T_i:

$$\frac{T_1 - T_i}{(R_i/2)} = \delta_i C_i \frac{dT_i}{dt} + q_2(t) \tag{8.30}$$

Equation 8.30 states that heat flowing from tape 1 is divided into two parts, one part to "charge" the capacitor (insulator of thickness δ_i and heat capacity C_i) and the other $[q_2(t)]$ to heat tape 2.

a) Assuming that the rise in T_2 is small compared with the source temperature, T_i, we can make another approximation: $q_2(t) \simeq T_i/(R_i/2)$. For the case where T_1 increases linearly with time, i.e. $T_1 = \theta_1 t$, and $R_i = \delta_i/k_i$, where k_i is the insulator thermal conductivity, show that $q_2(t)$ is given by:

$$q_2(t) = \left(\frac{\theta_1 k_i}{4\delta_i}\right)\left\{4t - \tau_i\left[1 - \exp\left(-\frac{4t}{\tau_i}\right)\right]\right\} \tag{8.31}$$

where $\tau_i = \delta_i^2 C_i/k_i$ is the thermal time constant of the insulator.

b) Combining Eqs. 8.29 and 8.31, show that an expression containing transit time τ_t is given by:

$$\left(\frac{4\delta_{cd}\delta_i}{\theta_1 k_i}\right)\int_{T_{op}}^{T_t'} C_{cd}\,dT = 2\tau_t^2 - \tau_t\tau_i + \frac{\tau_i^2}{4}\left[1 - \exp\left(-\frac{4\tau_t}{\tau_i}\right)\right] \tag{8.32}$$

Note that T_t', which depends on field and transport current, does influence τ_t, though in a complicated manner.

c) Noting that $\tilde{C}_{cd}(dT_1/dt) = \rho_m J_m^2$, where $\tilde{C}_{cd}$ is the average heat capacity of the tape over the temperature range of interest, show that Eq. 8.32 can be expressed by:

$$\left(\frac{4\delta_{cd}\delta_i\tilde{C}_{cd}}{k_i\rho_m J_m^2}\right)\int_{T_{op}}^{T_t'} C_{cd}\,dT = 2\tau_t^2 - \tau_t\tau_i + \frac{\tau_i^2}{4}\left[1 - \exp\left(-\frac{4\tau_t}{\tau_i}\right)\right] \tag{8.33}$$

Problem 8.8: 2. Transverse (turn-to-turn) velocity

d) Note for $\tau_t \gg \tau_i$, which is the case in many practical applications, especially with HTS, show that an expression for τ_t may be given by:

$$\tau_t = \frac{1}{J_m} \sqrt{\frac{2\delta_{cd}\delta_i \tilde{C}_{cd}}{\rho_m k_i}} \int_{T_{op}}^{T_t'} C_{cd}\, dT \tag{8.34}$$

e) Show that an expression for U_t when $\delta_i \ll \delta_{cd}$ may be given by:

$$U_t = \frac{J_m \sqrt{\rho_m k_i \delta_{cd}}}{\sqrt{2\delta_i \tilde{C}_{cd} \int_{T_{op}}^{T_t'} C_{cd}\, dT}} \tag{8.35}$$

By further approximating Eq. 8.35 with $\int C_{cd}\, dT \simeq \tilde{C}_{cd}(T_t' - T_{op})$, we have an expression for U_t given by:

$$U_t = \frac{J_m}{\tilde{C}_{cd}} \sqrt{\frac{\rho_m k_i \delta_{cd}}{2\delta_i(T_t' - T_{op})}} \tag{8.36}$$

f) Combining the expression of U_ℓ given by Eq. 8.27 and recognizing that C_{cd} appearing in it should really be replaced by $\tilde{C}_{cd}$ as in Eq. 8.36, show that under the conditions assumed ($\tau_t \gg \tau_i$ and $\delta_i \ll \delta_{cd}$) the velocity ratio in the two directions are proportional to the square root of the ratio of "weighted" thermal conductivities in the respective directions:

$$\frac{U_t}{U_\ell} \propto \sqrt{\left(\frac{\delta_{cd}}{\delta_i}\right)\left(\frac{k_i}{2k_m}\right)} \tag{8.37}$$

In both directions the driving force is the generation term, g_j, and the process is controlled by thermal diffusion.

"Truth is not always in a well." —C. Auguste Dupin

Solution to Problem 8.8

a) Using the approximation $q_2(t) \simeq T_i/(R_i/2)$ and defining $\tau_i = \delta_i R_i C_i = \delta_i^2 C_i/k_i$ (because $R_i = \delta_i/k_i$), we can derive from Eq. 8.30 a differential equation for $q_2(t)$:

$$\frac{dq_2(t)}{dt} + \frac{4q_2(t)}{\tau_i} = \frac{4\delta_i C_i T_1}{\tau_i^2} \tag{S8.1}$$

With $T_1 = \theta_1 t$ and $q_2(0) = 0$, we obtain:

$$q_2(t) = \left(\frac{\theta_1 k_i}{4\delta_i}\right) \left\{ 4t - \tau_i \left[1 - \exp\left(-\frac{4t}{\tau_i}\right)\right] \right\} \tag{8.31}$$

b) By combining Eqs. 8.29 and 8.31, we have:

$$\delta_{cd} \int_{T_{op}}^{T_t'} C_{tp}\, dT = \int_0^{\tau_t} q_2(t)\, dt = \int_0^{\tau_t} \left(\frac{\theta_1 k_i}{4\delta_i}\right) \left\{ 4t - \tau_i \left[1 - \exp\left(-\frac{4t}{\tau_i}\right)\right] \right\} dt$$

$$= \left(\frac{\theta_1 k_i}{4\delta_i}\right) \left\{ 2\tau_t^2 - \tau_t \tau_i + \frac{\tau_i^2}{4} \left[1 - \exp\left(-\frac{4t}{\tau_i}\right)\right] \right\} \tag{S8.2}$$

From Eq. S8.2, we have:

$$\left(\frac{4\delta_{cd}\delta_i}{\theta_1 k_i}\right) \int_{T_{op}}^{T_t'} C_{cd}\, dT = 2\tau_t^2 - \tau_t \tau_i + \frac{\tau_i^2}{4} \left[1 - \exp\left(-\frac{4\tau_t}{\tau_i}\right)\right] \tag{8.32}$$

c) Substituting $\theta_1 = \rho_m J_m^2/\tilde{C}_{cd}$ into Eq. 8.32, we immediately obtain:

$$\left(\frac{4\delta_{cd}\delta_i \tilde{C}_{cd}}{k_i \rho_m J_m^2}\right) \int_{T_{op}}^{T_t'} C_{cd}\, dT = 2\tau_t^2 - \tau_t \tau_i + \frac{\tau_i^2}{4} \left[1 - \exp\left(-\frac{4\tau_t}{\tau_i}\right)\right] \tag{8.33}$$

d) When $\tau_t \gg \tau_i$, the right-hand side of Eq. 8.33 becomes $2\tau_t^2$, and we have:

$$\tau_t = \frac{1}{J_m} \sqrt{\frac{2\delta_{cd}\delta_i \tilde{C}_{cd}}{\rho_m k_i} \int_{T_{op}}^{T_t'} C_{cd}\, dT} \tag{8.34}$$

e) By noting that $U_t = (\delta_i + \delta_{cd})/\tau_t$, we have an expression for U_t when $\delta_i \ll \delta_{cd}$:

$$U_t = \frac{\delta_i + \delta_{cd}}{\tau_t} \simeq \frac{\delta_{cd}}{\tau_t} = \frac{J_m \sqrt{\rho_m k_i \delta_{cd}}}{\sqrt{2\delta_i \tilde{C}_{cd} \int_{T_{op}}^{T_t'} C_{cd}\, dT}} \tag{8.35}$$

Equation 8.36 (p. 353) was applied by Lim in his investigation of transverse normal zone propagation through Nb_3Sn tape-wound pancakes [8.27]. His values of U_t based on Eq. 8.36 agreed quite well with experimentally measured values.

f) By taking the ratio of U_t given by Eq. 8.36 and U_ℓ given by Eq. 8.27, we obtain:

$$\frac{U_t}{U_\ell} \propto \sqrt{\left(\frac{\delta_{cd}}{\delta_i}\right)\left(\frac{k_i}{2k_m}\right)} \tag{8.37}$$

Problem 8.9: Passive Protection of "isolated" magnets
1. Basic concepts

This problem deals with the basic concepts of a passive protection technique used for persistent-mode superconducting magnets such as for MRI and NMR systems. Unlike the active protection technique discussed in Problem 8.1, passive protection techniques generally do not or cannot rely on devices located outside the cryostat. The basic operational procedure for persistent-mode magnets is first described.

Figure 8.11 presents a circuit showing essential elements for operation of persistent-mode magnets. Here the magnet is represented by two inductors in series to model, in the simplest way, a real MRI or NMR magnet that is generally comprised of as many as a dozen coils. We need at least two coils because the technique relies on coupling of coils within the system. The entire magnet is shunted by a superconducting switch, SW, in which a heater can make it either in the "normal" (heater current on) or superconducting (heater current off) state. During the charging sequence, the switch is "open" (normal state) and current from the power supply flows chiefly into the magnet. At a designated operating current, the switch heater current is turned off and the switch is "closed" (superconducting), placing the magnet in persistent mode. Now the supply current can be reduced to zero and, in order to minimize heat input through the current leads, they are removed from the magnet, leaving the magnet "isolated." As indicated with the dots (•) in the figure, this particular magnet system requires at least 3 "superconducting joints" to keep the current passage essentially lossless and thus to ensure persistent operation. Each coil is shunted with a resistor (R_1 or R_2) and this shunting of each coil, as we shall study here, protects these isolated magnets.

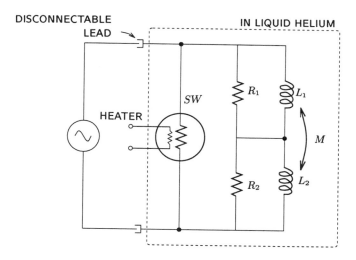

Fig. 8.11 Circuit for an "isolated," persistent-mode 2-coil magnet.

Problem 8.9: 1. Basic concepts

It is noted here that the circuit of Fig. 8.11 is representative of a variety of circuits, some of which include diodes, used for isolated, persistent-mode magnets.

Regardless of whether the magnet is in persistent mode or being charged, its terminals are always shorted either by switch SW or by the power supply. When it is in persistent mode, SW is closed and the terminals are shorted; during charge up the terminals are "shorted" by the power supply.

For this particular example, for the sake of simplicity, we shall assume that the self inductance of each coil is identical: $L_1 = L_2 = L$. The mutual inductance between the two coils is M, where $M = k\sqrt{L_1 L_2} = kL$. k is the coupling coefficient. Also each coil is shunted with the same resistor, namely, $R_1 = R_2 = R$. Each coil is carrying a constant transport current I_0. At time $t = 0$, a small normal zone is created in Coil 1; it is represented by a resistor r, which for this analysis is constant. The total magnetic energy stored by the system, E_m, is $LI_0^2 + MI_0^2$.

Through a laborious process of computation—don't try it—we have the ratio of the total energy dissipated in resistor r, E_r, to E_m, with $\zeta = r/R$, given by:

$$\frac{E_r}{E_m} = \frac{0.5\zeta(1-k) + (1+k)}{\zeta + (1+k)} \tag{8.38}$$

For $r \gg R(1 + k)$—a condition that can be met in most high-performance coils—the analysis also gives expressions for $I_1(t)$ and $I_2(t)$:

$$\frac{I_1(t)}{I_0} = \frac{R(1+k)^2}{2r} \exp\left(-\frac{R}{2L}t\right)$$

$$+ \left[1 - \frac{R(1+k)^2}{2r}\right] \exp\left[-\frac{r}{(1-k^2)L}t\right] \tag{8.39a}$$

$$\frac{I_2(t)}{I_0} = (1+k)\exp\left(-\frac{R}{2L}t\right) - k\exp\left[-\frac{r}{(1-k^2)L}t\right] \tag{8.39b}$$

a) Equation 8.38 gives $E_r/E_m \to 1$ as $\zeta \to 0$. Explain this result and comment on an important protection implication based on this technique.

b) Equation 8.38 also gives $E_r/E_m \to 0.5(1 - k)$ as $\zeta \to \infty$, which for $k = 1$ means $E_r/E_m = 0$. Again explain this result. Also, how is the rest of the energy divided between the two shunt resistors?

c) Equation 8.39b indicates that when $2L/R \gg (1-k^2)L/r$, $I_2(t)$, the current in Coil 2 that is still superconducting, initially increases. Explain this increase in terms of the conditions imposed by the circuit.

d) The increase in $I_2(t)$ is another key phenomenon on which this protection technique relies. Explain how this increase is used for protection.

e) The increase in $I_2(t)$ may also spell trouble for Coil 2. Discuss what this trouble might be and how this trouble might be quantified.

Solution to Problem 8.9

a) In order to transfer energy into the shunt resistors—the only other elements capable of absorbing dissipative energy—it is necessary to create a voltage across each shunt resistor. If the normal zone has a very small value of r, a very small voltage appears across Coil 1's shunt resistor, R_1. Note that this voltage is the algebraic *difference* between the resistive voltage drop, $\sim rI_0$, and the inductive voltage, LdI/dt, which is negative. Thus, for an infinitesimally small r, no voltage appears across each shunt resistor, and the entire energy, even that stored in Coil 2, is dissipated by the normal zone, giving rise to $E_r/E_m \to 1$.

The implication of the above result on coil protection is ominous; however, fortunately, once a normal zone is created in the adiabatic environment of these windings, r never remains small; it rapidly increases, providing a sufficient voltage across each shunt resistor.

b) Under this condition, a large voltage appears across each shunt resistor and most of the total energy is dissipated in the shunt resistors. If the two coils are well coupled ($k \to 1$), the energy stored in Coil 2 is transferred to Coil 1 and then dissipated through the shunt resistors.

Note that because the total voltage across the two shunt resistors must remain 0 throughout, *equal* (but opposite-polarity) current always flows through each shunt resistor, making dissipation through it also identical. That is, each shunt resistor dissipates an identical amount of energy—this is so because $R_1 = R_2$.

c) The insertion of r in Coil 1 clearly forces a current to flow through shunt resistor R_1. Because the change in voltage across the entire magnet terminal must be zero—the power supply is equivalent to a short—an equal but opposite current must flow in shunt resistor R_2. The current flowing in each shunt resistor ends up in Coil 2, increasing $I_2(t)$ and decreasing $I_1(t)$ by the same amount. This also means flux is preserved.

d) The increase in $I_2(t)$ can continue until it reaches the critical current of the conductor at the innermost winding radius of the coil midplane, *inducing* a quench in Coil 2 thus contributing to a rapid expansion of the normal zone. This process is further explored in the next problem in which a real coil situation is studied in more detail.

e) A large increase in $I_2(t)$, beneficial in triggering a quench, may spell trouble because, as mentioned above, the flux remains essentially constant and thus there will be a large increase in stresses in the winding. This means that in designing coils that are to be protected by shunt resistors, they must be designed to withstand stresses that appear during quenching. An important parameter here is $I_2(t) \times B_2(t)$ during quenching.

Another important parameter—the internal voltage appearing in each coil—will be discussed in Problem 8.11.

Problem 8.10: Passive Protection of "isolated" magnets

2. Two-section test coil

This problem studies normal-zone propagation in an epoxy-impregnated, two-section test solenoidal coil [8.17]. Each coil is wound with a round insulated Nb-Ti composite wire in a close-packed hexagonal pattern.

In the analysis, it is assumed that the normal-zone propagation is dominated by transverse heat conduction. The normal-zone growth is three dimensional—axial and radial over the entire circumference. Although $U_\ell \gg U_t$, because $2\pi a_{1_1} U_\ell \ll d_1 U_t$, where a_{1_1} and d_1 are, respectively, the inner winding radius and conductor diameter of Section 1, longitudinal propagation along the conductor axis may be neglected compared with transverse propagation for both sections.

Table 8.6 gives the coil's appropriate parameters. The total inductance is 1 H. As Section 2 is wound directly over Section 1, the boundaries of the two sections are in good thermal contact. When considering normal-zone propagation, therefore, the entire coil may be considered as one homogeneous thermal unit. (Note that, as indicated in Table 8.6, the conductor diameter is different for the two sections and thus normal-zone propagation velocities are different in the two sections.)

A Nichrome heater, placed at the midplane of the innermost radius of Section 1, is used to initiate a quench. We can therefore assume that the normal zone starts as a ring at the midplane of the innermost radius of Section 1 and spreads as depicted in Fig. 8.1.

The two sections are connected in series, each shunted with a 0.5-Ω resistor. The power supply may be modeled as a constant current source for voltages up to 10 V. Figure 8.12 gives a circuit model for the system.

Figure 8.13 shows current and voltage vs time traces for a heater-driven quench event in which the solenoid is initially energized at 100 A. Note that the terminal voltage limit of the power supply is 10 V. Both current and voltage plots consist of four traces. The solid traces are experimental and dashed traces are analytical. For both current and voltage traces, the curves labeled 1 are for Section 1 and the curves labeled 2 are for Section 2. In answering the following questions, you may ignore the analytical curves.

Table 8.6: Coil Parameters
(Coil Inductance: 1 H)

Parameter	Sec. 1	Sec. 2
winding i.d. [mm]	76	112
winding o.d. [mm]	112	135
winding length [mm]	71	71
wire diameter [mm]	0.90	0.70
Cu/Nb-Ti ratio	3	2

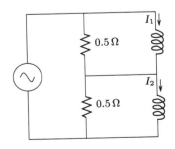

Fig. 8.12 Circuit for the coil.

Problem 8.10: 2. Two-section test coil

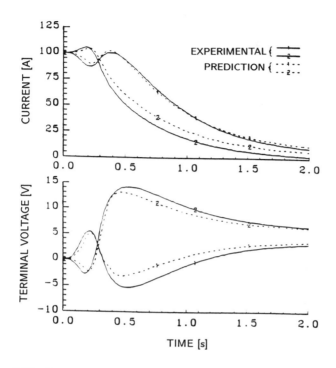

Fig. 8.13 Current and voltage *vs* time traces for a heater-driven quench event with the coil initially at 100 A. Note that both current and voltage traces labeled 1 and 2 correspond, respectively, to those in Section 1 and Section 2. Solid curves: experimental; dashed curves: computational.

a) Describe qualitatively the behaviors of the four experimental traces shown in Fig. 8.13. Include a discussion of the peaks and valleys of the traces.

b) Make a rough estimate of the total energy dissipated in the magnet. A value computed for the time period between $t = 0$ and $t = 2\,\mathrm{s}$ may be used for this estimate.

c) How would the current traces given in Fig. 8.13 be modified if the initial coil current is 50 A?

d) The analytical program also computes the average winding temperature at the end of the event when current has decayed to zero. It predicts an average temperature of 45 K for the entire coil. Is this value reasonable?

e) Suppose aluminum (99.99% purity) instead of copper is used for both conductors. Estimate a final average temperature for the entire coil.

Solution to Problem 8.10

a) The following observations may be made from the traces shown in Fig. 8.13:

- I_1 decreases initially because Section 1 is where a quench is initiated.

- I_2 increases initially 1) to satisfy the circuit requirement of zero voltage across the coil and 2) to keep the flux constant.

- The behaviors of I_1 and I_2 are also reflected in V_1 and V_2. V_1 increases because ΔI_1, not flowing through Section 1, is now flowing through R_1.

- To keep the terminal voltage zero (at least initially), V_2 swings negative. These initial responses are consistent with results discussed in Problem 8.9. Eventually $V_1 + V_2$ climbs up to $10\,\text{V}$, the power supply limit.

- At $t \sim 0.2\,\text{s}$, V_2 starts climbing up, a definite indication that the normal zone is induced in or has reached Section 2.

- I_2 thus begins to drop, and I_1 increases, trying to keep the flux constant.

- At $t \sim 0.4\,\text{s}$, $V_1 + V_2$ reaches $10\,\text{V}$ and I_1 must start decreasing.

- $V_1 + V_2 = 10\,\text{V}$ for $t > 0.4\,\text{s}$.

b) The total energy dissipated in the magnet, E_d, may be given by:

$$E_d = E_m + E_s - E_{R1} - E_{R2} \qquad (S10.1)$$

where E_m is the total energy stored in the magnet initially, E_s is the energy supplied by the power supply between $t = 0$ and $t = 2\,\text{s}$, and E_{R1} and E_{R2} are, respectively, energies dissipated in resistor R_1 and R_2. E_m is $5000\,\text{J}$ [$= (0.5)(1\,\text{H})(100\,\text{A})^2$]. E_s is given by $V_s(t)I_s(t)$ integrated for $0 \leq t \leq 2\,\text{s}$. $V_s(t)$ and $I_s(t)$ are, respectively, the power supply voltage and current. The power supply may be modeled as a constant current supply ($100\,\text{A}$) for $0 \leq t \leq 0.4\,\text{s}$ and a constant voltage supply ($10\,\text{V}$) for $t \geq 0.4\,\text{s}$. We have, for $0 \leq t \leq 0.4\,\text{s}$, $V_s(t) = V_1(t) + V_2(t)$ and, for $t \geq 0.4\,\text{s}$, $I_s(t) = I_1(t) - V_1(t)/R_1$ (a proof of a relationship similar to this involving more coils is a question in Problem 8.11). Using traces shown in Fig. 8.13, we can compute E_s, E_{R1}, and E_{R2}:

$$E_s = (100\,\text{A})\int_0^{0.4\,\text{s}} [V_1(t) + V_2(t)]\,dt$$

$$+ (10\,\text{V})\int_{0.4\,\text{s}}^{2\,\text{s}} \left[I_1(t) + \frac{V_1(t)}{R_1} \right] dt \qquad (S10.2a)$$

$$\simeq 200\,\text{J} + 650\,\text{J} \simeq 850\,\text{J}$$

$$E_{R1} = \frac{1}{R_1}\int_0^{2\,\text{s}} V_1(t)^2\,dt \simeq 50\,\text{J} \qquad (S10.2b)$$

$$E_{R2} = \frac{1}{R_2}\int_0^{2\,\text{s}} V_2(t)^2\,dt \simeq 300\,\text{J} \qquad (S10.2c)$$

The total energy dissipated in the magnet is thus about $5500\,\text{J}$.

Solution to Problem 8.10

c) Both currents start at 50 A. The normal zone should reach Section 2 at ~0.4 s or later because $U_t \propto U_\ell \propto I_t$. Also, because I_t and B are one half the previous values, T_c is higher, making the arrival time even later than 0.4 s. The terminal voltage should reach 10 V later than 0.4 s, perhaps as late as ~0.8 s, because U_t is slower by at least a factor of 2 and also because shunt voltages will be lower by a factor of 2 and it takes longer for the terminal voltage to reach 10 V.

d) The total coil's winding (conductor and epoxy filler) volume is 694 cm³. Assuming the entire winding heat capacity, C_{wd}, can be given by that of copper—volumetrically copper's and epoxy's heat capacities are roughly equal—we have:

$$\mathcal{V}_{cd}[h_{cu}(T_f) - h_{cu}(T_{op})] \simeq (643\,\text{cm}^3)[h_{cu}(T_f)] = 5500\,\text{J} \qquad (S10.3)$$

where $\mathcal{V}_{cd}$ is winding volume and h_{cu} is copper's volumetric enthalpy. For $T_f > T_{op} = 4.2\,\text{K}$, $h_{cu}(T_f) \gg h_{cu}(T_{op})$. From Fig. A3.3, we find $T_f = 45\,\text{K}$, which agrees with the value computed from simulation.

e) When aluminum is substituted for copper, Fig. A3.3 gives $T_f = 57\,\text{K}$. A simulation gives a temperature of 58 K [8.28].

Figure 8.14 shows spatially averaged temperature plots for Section 1 and Section 2 of this magnet. The solid curves correspond to Nb-Ti/copper wires, while the dotted curves correspond to Nb-Ti/aluminum wires [8.28].

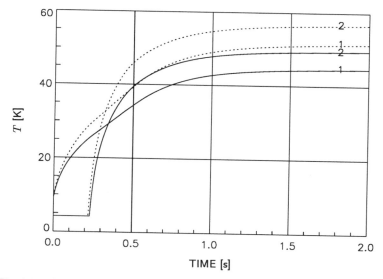

Fig. 8.14   Spatially averaged temperature *vs* time plots for each section of the coil. Solid curves: Nb-Ti/copper wires; dotted curves: Nb-Ti/aluminum wires [8.28].

Problem 8.11: Passive Protection of "isolated" magnets
3. Multi-coil NMR magnet

This problem studies important protection issues for multi-coil magnets such as those used in MRI and NMR systems. We will use experimental results recorded for the Nb-Ti coils of a 750-MHz (17.6 T) superconducting magnet developed at FBNML [8.29] as examples.

The full magnet system is comprised of 12 nested solenoidal coils, of which, counting from the innermost coil, the first 7 coils are wound with Nb$_3$Sn conductors and the remaining 5 outer coils are wound with Nb-Ti conductors. Each coil is impregnated with epoxy resin and incorporates the "floating-winding" technique. Figure 8.15 indicates the locations of the 12 coils.

As may be inferred from Fig. 8.15, Coils 10, 11, and 12 are so-called "correction" coils, whose primary function is to improve field homogeneity at the magnet center; they form a variant of notched solenoid studied in Problem 3.6 (p. 67). Note that Coil 9, wound on one coil form, has two sections, each shunted, and Coils 11 and 12 share one shunt resistor, as shown in Fig. 8.16.

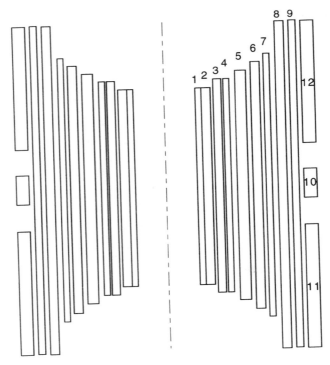

Fig. 8.15   Drawing showing the locations of 12 coils in a 750-MHz (17.6 T) magnet [8.29]. Note that the horizontal scale in this sketch is 4.5 times the axial scale.

Problem 8.11: 3. Multi-coil NMR magnet

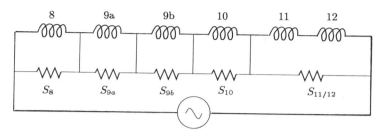

Fig. 8.16   Circuit for the Nb-Ti coils.

Figure 8.16 does not include a persistent switch that shunts the entire system because discussion here chiefly concerns premature quenching that occurs during system charge up when the switch is open. Because of the presence of the power supply, the entire system is "shorted," as is the case when the system is in persistent mode.

The values of the inductance matrix and shunt resistors are given, respectively, in Tables 8.7 and 8.8. As is evident from Table 8.7, although ideally identical, Coils 11 and 12 actually have slightly different values of inductance.

Our rule of thumb in determining values of shunt resistors in these multi-coil system is to first choose a total value of shunt resistance from voltage considerations. In this particular example, a value of ~1.5 Ω was selected because at an operating current of 310 A, it would translate to a voltage level of ~500 V—a safe level. Each shunt resistor is then determined to be roughly proportional to each coil's total stored energy.

Table 8.7: Inductance Matrices [H] for the Nb-Ti Coils

Coil	8	9a	9b	10	11	12
8	4.413	2.268	2.243	0.715	2.747	2.755
9a	2.268	1.344	1.343	0.427	1.645	1.649
9b	2.243	1.343	1.404	0.450	1.737	1.742
10	0.715	0.427	0.450	0.606	0.378	0.379
11	2.747	1.645	1.737	0.378	5.382	0.368
12	2.755	1.649	1.742	0.379	0.368	5.410

Table 8.8: Shunt Resistors [mΩ] for the Nb-Ti Coils

S_8	S_{9a}	S_{9b}	S_{10}	$S_{11/12}$
288	156	165	58	868

Problem 8.11: 3. Multi-coil NMR magnet

Figure 8.17 shows a set of voltage traces recorded when the magnet quenched prematurely at 227 A. As evident from the traces, the quench started in Coil 9a; signals from AE sensors (not shown here) indicated the premature quench was caused by a mechanical event occurring in the magnet system. Because a resistive voltage first appeared in Coil 9a, it is most likely that the mechanical event took place in Coil 9a. Note that between $t = 1.6\,\mathrm{s}$ and $t = 2.25\,\mathrm{s}$, $V_{11/12}$, the sum of recorded voltages from Coils 11 and 12, is saturated.

Figure 8.18 shows a set of *computed* current traces through the coils based on the voltage traces shown in Fig. 8.17; Table 8.9 shows a set of dI/dt values at selected times for the current traces shown in Fig. 8.18.

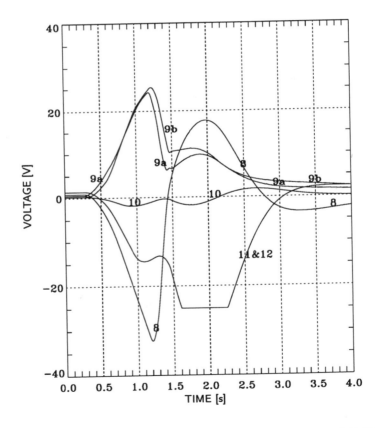

Fig. 8.17   Voltage traces recorded across Coils 8, 9a, 9b, 10, and 11/12 following a quench at 227 A [8.30].

Problem 8.11: 3. Multi-coil NMR magnet

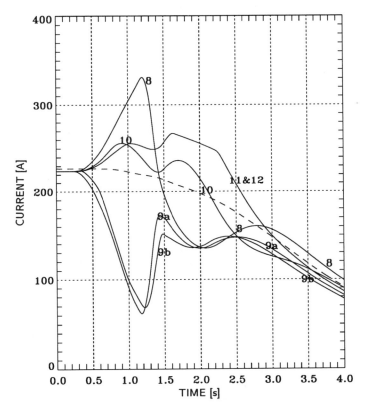

Fig. 8.18   Current traces through Coils 8, 9a, 9b, 10, and 11/12 corresponding to the voltage traces of Fig. 8.17. The dotted curve is the supply current, I_0.

Table 8.9: Values of dI/dt at Selected Times

	dI/dt [A/s]			
	$t = 0.5\,\text{s}$	$t = 1.0\,\text{s}$	$t = 1.5\,\text{s}$	$t = 4.0\,\text{s}$
Coil 8	84.5	147.2	−252.8	−56.1
Coil $9a$	−154.1	−234.7	−62.1	−48.9
Coil $9b$	−107.1	−198.1	−57.8	−42.8
Coil 10	41.3	−44.8	81.5	−47.4
Coils 11/12	33.6	19.3	113.1	−48.9

Problem 8.11: 3. Multi-coil NMR magnet

a) Show that the current in each coil may be determined from the coil's V trace according to the following equations:

$$I_8 = I_0 - \frac{V_8}{S_8} \qquad (8.40a)$$

$$I_{9a} = I_0 - \frac{V_{9a}}{S_{9a}} \qquad (8.40b)$$

$$I_{9b} = I_0 - \frac{V_{9b}}{S_{9b}} \qquad (8.40c)$$

$$I_{10} = I_0 - \frac{V_{10}}{S_{10}} \qquad (8.40d)$$

$$I_{11/12} = I_0 - \frac{V_{11/12}}{S_{11/12}} \qquad (8.40e)$$

where I_0 is the power supply current.

b) Almost immediately after Coil 9a is driven normal, which is evident from its rising voltage trace (Fig. 8.17), Coil 9b follows, inducing excess currents in the rest of the coils (Fig. 8.18). Coil 10 is the next to go normal, followed by Coil 8, whose voltage starts decreasing at $t = 0.3\,$s. Compute the sum of *inductive* voltages appearing across Coil 8 at $t = 0.5\,$s and show that Coil 8 is still completely superconducting at that time.

c) Show that, although its voltage is still decreasing, Coil 8 already has a normal zone at $t = 1.0\,$s and estimate its resistance. Also discuss how we might determine the precise moment at which a normal zone appears in Coil 8.

d) Compute an *approximate* value of the net Joule heating [W] generated by the entire magnet (Coils 8~12) at $t = 4.0\,$s. Pay attention to the word approximate.

e) At $t{\sim}1\,$s, when Coil 8 is already driven normal, its peak field is ~6 T and the conductor's critical current (at 4.2 K) is ~900 A, well above the observed quench current of ~270 A (an average of Coil 8 currents at $t = 0.5\,$s and $t = 1.0\,$s, from Fig. 8.18). Offer plausible sources of the seemingly improbable quench initiated in Coil 8 at this low current.

f) Explain why Coils 11 and 12 are not shunted separately.

g) Make a general comment about the risk of not shunting the two coils (11 and 12) separately.

"Half a man's life is devoted to what he calls improvements, yet the original had some quality which is lost in the process." —E.B. White

Solution to Problem 8.11

a) From Fig. 8.16, it is clear that $I_8 = I_0 - I_{r8}$. I_{r8} is the current flowing in shunt resistor 8: $I_{r8} = V_8/S_8$. Thus:

$$I_8 = I_0 - I_{r8}$$

$$= I_0 - \frac{V_8}{S_8} \tag{8.40a}$$

Similarly, we have:

$$I_{9a} = I_8 + I_{r8} - I_{r9a} = I_0 - \frac{V_8}{S_8} + \frac{V_8}{S_8} - \frac{V_{9a}}{S_{9a}}$$

$$= I_0 - \frac{V_{9a}}{S_{9a}} \tag{8.40b}$$

$$I_{9b} = I_{9a} + \frac{V_{9a}}{S_{9a}} - \frac{V_{9b}}{S_{9b}} = I_0 - \frac{V_{9a}}{S_{9a}} + \frac{V_{9a}}{S_{9a}} - \frac{V_{9b}}{S_{9b}}$$

$$= I_0 - \frac{V_{9b}}{S_{9b}} \tag{8.40c}$$

$$I_{10} = I_{9b} + \frac{V_{9b}}{S_{9b}} - \frac{V_{10}}{S_{10}} = I_0 - \frac{V_{9b}}{S_{9b}} + \frac{V_{9b}}{S_{9b}} - \frac{V_{10}}{S_{10}}$$

$$= I_0 - \frac{V_{10}}{S_{10}} \tag{8.40d}$$

$$I_{11/12} = I_{10} + \frac{V_{10}}{S_{10}} - \frac{V_{11/12}}{S_{11/12}} = I_0 - \frac{V_{10}}{S_{10}} + \frac{V_{10}}{S_{10}} - \frac{V_{11/12}}{S_{11/12}}$$

$$= I_0 - \frac{V_{11/12}}{S_{11/12}} \tag{8.40e}$$

b) Voltage across Coil 8, V_8, is given by:

$$V_8 = V_r|_8 + L_8 \frac{dI_8}{dt} + M_{8,9a} \frac{dI_{9a}}{dt}$$

$$+ M_{8,9b} \frac{dI_{9b}}{dt} + M_{8,10} \frac{dI_{10}}{dt}$$

$$+ M_{8,11} \frac{dI_{11}}{dt} + M_{8,12} \frac{dI_{12}}{dt} \tag{S11.1}$$

where $V_r|_8$ is the resistive voltage across Coil 8 due to the presence of a normal zone. Inserting appropriate values at $t = 0.5$ s, taken from Tables 8.7 and 8.9, into the right-hand side of Eq. S11.1, we obtain:

$$V_8 \simeq V_r|_8 + (4.41\,\text{H})(84.5\,\text{A/s}) + (2.27\,\text{H})(-154.1\,\text{A/s})$$

$$+ (2.24\,\text{H})(-107.1\,\text{A/s}) + (0.72\,\text{H})(41.3\,\text{A/s})$$

$$+ (2.75\,\text{H})(33.6\,\text{A/s}) + (2.76\,\text{H})(33.6\,\text{A/s}) \tag{S11.2a}$$

Solution to Problem 8.11

$$V_8 = V_r|_8 + 372.6 - 349.8 - 239.9 + 29.7 + 92.4 + 92.7$$

$$= V_r|_8 - 2.3\,\text{V} \qquad (S11.2b)$$

From Fig. 8.17, we find $V_8 \simeq -2.3\,\text{V}$ at $t = 0.5\,\text{s}$, which is equal to the net inductive voltage given by Eq. $S11.2b$, making $V_r|_8 = 0\,\text{V}$ at $t = 0.5\,\text{s}$. Coil 8 thus is still superconducting at $t = 0.5\,\text{s}$.

c)   Again, inserting appropriate values into Eq. $S11.1$, we obtain V_8 at $t = 1.0\,\text{s}$:

$$V_8 = V_r|_8 + (4.41\,\text{H})(147.2\,\text{A/s}) + (2.27\,\text{H})(-234.7\,\text{A/s})$$

$$+ (2.24\,\text{H})(-198.1\,\text{A/s}) + (0.72\,\text{H})(-44.8\,\text{A/s})$$

$$+ (2.75\,\text{H})(19.3\,\text{A/s}) + (2.74\,\text{H})(19.3\,\text{A/s}) \qquad (S11.3a)$$

$$V_8 = V_r|_8 + 649.2 - 532.8 - 443.7 - 32.3 + 53.1 + 52.9$$

$$= V_r|_8 - 253.6\,\text{V} \qquad (S11.3b)$$

According to the voltage trace of Fig. 8.17, $V_8 = -23\,\text{V}$ at $t = 1.0\,\text{s}$, thus from Eq. $S11.3b$, we have $V_r|_8 \simeq 231\,\text{V}$. From Fig. 8.18, we find $I_8 \simeq 306\,\text{A}$, and thus $R_8 = 231\,\text{V}/306\,\text{A} = 0.75\,\Omega$.

We can determine the precise moment when a normal zone appears in Coil 8 by finding the time at which $V_r|_8$ just begins to become nonzero.

d)   The Joule heating generated by the entire magnet, P_{mg}, is given by:

$$P_{mg} = \sum_{n=8}^{12} V_r|_n \times I_n \qquad (S11.4)$$

It is thus necessary to compute V_r for each coil as in b) or c). However, at $t = 4.0\,\text{s}$, we note that each coil has nearly the same values of: 1) voltage, $\tilde{V} \sim 0\,\text{V}$ (Fig. 8.17); 2) current, $\tilde{I} \sim 90\,\text{A}$ (Fig. 8.18); and 3) time rate of change of current, $d\tilde{I}/dt \sim -50\,\text{A/s}$ (Table 8.9). Thus, for this particular time Eq. $S11.4$ may be *approximated* by:

$$P_{mg} \simeq \left(\tilde{V} - \sum_{m,n=8}^{12} L_{m,n} \frac{d\tilde{I}}{dt} \right) \times \tilde{I} \qquad (S11.5)$$

Note that the term within the parentheses in Eq. $S11.5$ is equal to the approximate resistive voltage across the entire magnet. From Table 8.7, we obtain the sum of the inductances to be 60.25 H. With $d\tilde{I}/dt \simeq -50$ A/s and $\tilde{I} \simeq 90$ A, we have:

$$P_{mg} \simeq [0 - (60.25\,\text{H})(-50\,\text{A/s})](90\,\text{A}) \simeq 270,000\,\text{W} \qquad (S11.6)$$

Note that because $\tilde{V} \sim 0\,\text{V}$, the inductive and resistive voltages are nearly balanced and at this time, the total resistive voltage has an amplitude of $\sim 3000\,\text{V}$ ($\sim 270,000\,\text{W}/90\,\text{A}$). This also means the total magnet resistance has grown to $\sim 33\,\Omega$ ($\sim 3000\,\text{V}/90\,\text{A}$).

Solution to Problem 8.11

e) According to our criterion, and incorporated in quench codes developed over the past ~10 years at FBNML and successfully applied [8.17, 8.19~8.21], an induced quench in a target coil occurs when the target coil's transport current [in this case $I_8(t)$] reaches the critical current corresponding to temperature T_{op} (4.2 K) and maximum field within the coil. Based on this criterion, the observed premature quench should never have taken place at 270 A; it is much lower than 900 A, an estimated critical current of Coil 8's conductor at the time of the quench. Apparently, the criterion works well for premature quenches in the source coil occurring at relatively high currents, near designed operating currents, so that they are closer to the critical currents (at T_{op}) of the target coils.

A recent study by Yunus [8.30] indicates that the condition of constant conductor temperature at T_{op} is not valid, particularly in adiabatic windings subjected to time-varying magnetic field and current. A new criterion proposed by Yunus includes AC losses in the computation of real-time local conductor temperature, which in turn gives rise to a reduced critical current in the target coil. The new criterion indeed makes it possible to have a target coil quenching at a current close to the observed value of ~270 A.

That coupling loss is an additional heating source in a quenching adiabatic winding is quite significant. Because filament twist pitch length (ℓ_p) is a key parameter in controlling coupling loss, it implies that ℓ_p is another critical design parameter relating to protection of these high-performance magnets; within a reasonable extent, ℓ_p should be specified to be purposefully long for protection purposes.

f) Both Coils 11 and 12 are situated off the magnet midplane. There is thus a net axial force acting on Coil 11 (located below the midplane) towards the midplane ($+z$-directed) and a net axial force acting on Coil 12 (located above the midplane) towards the midplane ($-z$-directed). As long as the currents through Coil 11 and Coil 12 are identical, there will be no net unbalancing axial force acting on the system. This force-balance condition can be achieved only when the two coils are connected in series with a common shunt resistor across them.

If each coil is shunted independently, the current induced in each coil will be different, potentially creating a massive net unbalanced force on the system. At $t = 1$ s, suppose, instead of both coils carrying a current of 250 A as is the case according to Fig. 8.18, Coil 11 carried 275 A and Coil 12 carried 225 A. Under this condition, the force pushing Coil 11 upward would be 581 kN and the force pushing Coil 12 down would be 525 kN, with a resulting net unbalanced upward axial force of 56 kN or almost 6 tons!

g) Although not evident from these sets of voltage and current traces, particularly as the voltage trace for Coil 11/12 is saturated between $t = 1.3$ s and $t = 2$ s, the most critical danger in connecting Coils 11 and 12 in series and having a common shunt resistor is that in case of a quench in either Coil 11 or Coil 12, very high inductive voltages can be generated within the combined coil.

Problem 8.12: NZP velocity in HTS magnets

This problem studies normal-zone propagation velocity in HTS magnets, demonstrating, as remarked in Chapter 1, that protection is one issue that increases in difficulty with rising operating temperature. Indeed the slowness of NZP velocity in HTS magnets, verified recently by measurements and simulation study [8.21], suggests that HTS magnets are unlikely to be self protecting and need to be protected by some other innovative technique.

Figure 8.19 presents typical schematic I_c vs T plots, one for a low-T_c superconductor and the other for a high-T_c superconductor, with the zero abscissa point being T_{op} for both conductors. For the same transport current I_t flowing in each conductor, as indicated in the figure, the temperature difference, $T_c - T_{cs}$, is significantly greater for the HTS than that for the LTS. This is valid when the external field each conductor is exposed to is relatively small, perhaps ~5 T or less, and T_c for the LTS is typically near 10 K and that for the HTS is ~80 K. Note that as is the case with low-T_c superconductors, the $I_c(T)$ plot for the HTS may be quite accurately represented by a linearly decreasing function of temperature [5.12].

a) Show that $G_j(T)$ for the HTS, unlike that for the LTS, which increases linearly with T between T_{cs} and T_c (Eq. 6.5b, p. 212), generally has a component that increases with T^2. Namely:

$$G_j(T) = R_m I_t^2 \left(\frac{T - T_{cs}}{T_c - T_{cs}} \right) \quad \text{(for LTS)} \quad (6.5b)$$

$$G_j(T) \propto T^2 \quad \text{(for HTS)} \quad (8.41)$$

b) Make plots of $G_j(T)$ for both LTS and HTS whose $I_c(T)$ plots are given in Fig. 8.19.

c) Using the $G_j(T)$ plots, make general comments on NZP velocity (U_ℓ) for the HTS. In discussing this, also consider that an HTS generally has a temperature span $T_c - T_{cs}$ that is much greater than that of an LTS.

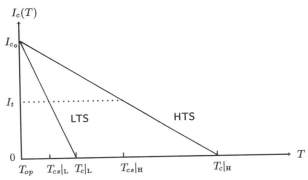

Fig. 8.19 I_c vs T plots for an LTS and an HTS with the same I_t, I_{c_o}, and T_{op}.

Solution to Problem 8.12

a) The quadratic temperature dependence of $G_j(T)|_{HTS}$ occurs because over the likely operating temperature range of HTS magnets, from ~20 K to ~80 K, the matrix metal's resistivity no longer remains temperature independent, as is the case for LTS which operate well below 20 K. Electrical resistivities of pure metals such as silver and copper increase roughly linearly with temperature above ~20 K. Combined with this linear dependance and the usual linear dependance of $G_j(T)$ on temperature between T_{cs} and T_c, $G_j(T)$ has a T^2 component.

b) Figure 8.20 presents $G_J(T)$ plots for the conductors whose $I_c(T)$ plots are shown in Fig. 8.19. As remarked above, $G_j(T)$ for the HTS increases as T^2. Note that because matrix resistivity is constant up to ~30 K and increases linearly with temperature, LTS's $G_j(T)$ in the normal state is constant and smaller than HTS's $G_j(T)$, which, unlike LTS's, does not remain constant.

c) Because of the complex nature of the partial differential power density equations that govern the propagating normal zone (Eqs. $S7.1a$ and $S7.1b$, Problem 8.7, p. 349), it becomes increasingly difficult (and inaccurate) to use an expression for U_ℓ such as that given by Eq. 8.24 when covering a large temperature span as is the case with HTS. Equation 8.28, which states that U_ℓ is proportional to the square root of generation and inversely proportional to C_{cd}, suggests, because HTS's temperature span between T_{cs} and T_c is much wider than LTS's, HTS's U_ℓ will be slower than LTS's by a factor even greater than the ratio of heat capacities.

Table 8.10 presents values of U_ℓ, computed by Bellis [8.21], for two superconducting tapes of similar dimensions, one copper-sheathed Nb₃Sn and the other silver-sheathed BiPbSrCaCuO(2223), under adiabatic conditions. The matrix current density, J_m, is kept constant at 159 A/mm² for the four field-temperature combinations studied, results of which are presented in the table. The three field-temperature combinations—20 T and above at 4.2 K; 3~6 T in the range 25 K; and essentially zero-field at 65 K—are considered to be most appropriate combinations for HTS magnets at the present time [8.31]. Progress towards the use of HTS magnets in the first combination (20 T at 4 K) is moving well [8.32].

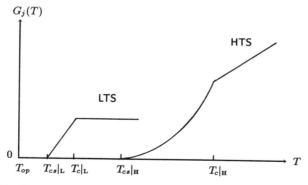

Fig. 8.20   Heat generation *vs* temperature for the LTS and HTS shown in Fig. 8.19.

Solution to Problem 8.12

Table 8.10: Computed U_ℓ Values for HTS and LTS [8.21]

Material	B	T_{op}	T_c	I_t	J_m	T_{cs}	$T_{cs} - T_{op}$	U_ℓ
	[T]	[K]	[K]	[A]	[A/mm^2]	[K]	[K]	[cm/s]
BSCCO	20	4.2	51	70	159	15.3	11.1	10.8
	5	25	63	70	159	33.6	8.6	2.8
	0	65	93	70	159	65.8	0.8	2.5
Nb$_3$Sn	12	4.2	7.5	111.4	159	~5.6	~1.4	400

Two conclusions may be made from Table 8.10:

- For the Nb$_3$Sn tape, U_ℓ, as expected, is much faster than values for the BiPb-SrCaCuO tape.

- For the BiPbSrCaCuO tape, U_ℓ decreases with T_{op} and as expected the values are much smaller than that of the Nb$_3$Sn.

The high-performance magnets such as the one studied in Problem 8.11 depend to a great extent on high values of U_ℓ (and hence correspondingly high values of U_t) for protection of their individual coils, and they may be called self-protecting. Based on values of U_ℓ listed in Table 8.10 for BSCCO tapes, we may conclude that it is unlikely that HTS coils will be self-protecting. Protection is thus an important and *difficult* magnet issue for HTS magnets as was remarked in Chapter 1.

"Dry" High-Field HTS Magnets Operating at 20 K

Design and operational issues of stability and disturbance, as discussed in Chapters 6 and 7, are not as critical in HTS magnets as they are in LTS magnets. It appears that the time has come to seriously consider building and operating high-field magnets of *substantial* parameters, wound entirely of HTS and operating at or close to 20 K. Also, promising progress has been made towards the goal of producing long (at least 100 m) and high-operating-current HTS magnet-grade conductors [8.33]. An HTS magnet, if solenoidal, would have a room-temperature bore of at least 100 mm, generate an induction of at least 10 T, and, coupled to a cryocooler and cooled by conduction, operate literally "dry." Such dry magnets, demonstrated as early as the mid 1960's with a Nb-Zr magnet by McEvoy and others [8.34], have recently become practical options for both Nb-Ti and Nb$_3$Sn magnets [8.35~8.37]. Protection considerations require that the 10-T HTS magnet operate at a current in excess of ~1000 A [8.38], thereby challenging the conductor manufacturers. The crucial design issue, besides protection, will be mechanical integrity [8.38], which as pointed out in Chapter 1, is virtually independent of operating temperature. Before the end of this century, "dry" 20-K HTS magnets will undoubtedly be incorporated in MRI, NMR, hybrid magnets, maglev, generators and motors, SMES, and perhaps, on an experimental basis, in high energy physics (dipoles and quadrupoles) and even fusion.

References

[8.1] D.B. Montgomery, "Review of fusion system magnet problems," *Proc. IEEE 13th Sympo. Fusion Engr.* (IEEE Catalogue No. 89CH2820-9, 1989), 27.

[8.2] R.J. Thome, J.B. Czirr, and J.H. Schultz, "Survey of selected magnet failures and accidents," *Am. Nucl. Soc. 7th Topical Conf. on Fusion Engr.* (1986).

[8.3] W.H. Cherry and J.I. Gittleman, "Thermal and electrodynamic aspects of the superconductive transition process," *Solid State Electronics* **1**, 287 (1960).

[8.4] Z.J.J. Stekly, "Theoretical and experimental study of an unprotected superconducting coil going normal," *Adv. Cryogenic Eng.* **8**, 585 (1963).

[8.5] P.F. Smith, "Protection of superconducting coils," *Rev. Sci. Instrum.* **34**, 368 (1963).

[8.6] C.N. Whetstone and C. Roos, "Thermal transitions in superconducting NbZr alloys," *J. Appl. Phys.* **36**, 783 (1965).

[8.7] V.V. Altov, M.G. Kremlev, V.V. Sytchev and V.B. Zenkevitch, "Calculation of propagation velocity of normal and superconducting regions in composite conductors," *Cryogenics* **13**, 420 (1973).

[8.8] D. Hagendorn and P. Dullenkopf, "The propagation of the resistive region in high current density coils," *Cryogenics* **14**, 264 (1974).

[8.9] L. Dresner, "Propagation of normal zones in composite superconductors," *Cryogenics* **16**, 675 (1976).

[8.10] K. Ishibashi, M. Wake, M. Kobayashi and A. Katase, "Propagation velocity of normal zones in a sc braid," *Cryogenics* **19**, 467 (1979).

[8.11] B. Turck, "About the propagation velocity in superconducting composites," *Cryogenics* **20**, 146 (1980).

[8.12] Z.J.J. Stekly, "Behavior of superconducting coil subjected to steady local heating within the windings," *J. Appl. Phys.* **37**, 324 (1966).

[8.13] M.N. Wilson, "Computer simulation of the quenching of a superconducting magnet," (Rutherford High Energy Physics Laboratory Memo RHEL/M151, 1968).

[8.14] J.E.C. Williams, "Quenching in coupled adiabatic coils," *IEEE Trans. Magn.* **MAG-21**, 396 (1985).

[8.15] K. Funaki, K. Ikeda, M. Takeo, K. Yamafuji, J. Chikaba and F. Irie, "Normal-zone propagation inside a layer and between layers in a superconducting coil," *IEEE Trans. Magn.* **MAG-23**, 1561 (1987).

[8.16] V. Kadambi and B. Dorri, "Current decay and temperatures during superconducting magnet coil quench," *Cryogenics* **26**, 157 (1986).

[8.17] C.H. Joshi and Y. Iwasa, "Prediction of current decay and terminal voltages in adiabatic superconducting magnets," *Cryogenics* **29**, 157 (1989).

[8.18] K. Kuroda, S. Uchikawa, N. Hara, R. Saito, R. Takeda, K. Murai, T. Kobayashi, S. Suzuki, and T. Nakayama, "Quench simulation analysis of a superconducting coil," *ibid.*, 814 (1989).

[8.19] Zong-Ping Zhao, "Thermo-electrodynamics of the resistive transition of superconductors in epoxy-resin impregnated superconducting magnets," (Dr. Eng. thesis, Dept. of Electrical Engineering, Tokyo Denki University, 1990; unpublished.)

[8.20] A. Ishiyama, H. Matsumura, W. Takita, and Y. Iwasa, "Quench propagation analysis in adiabatic superconducting windings," *IEEE Trans. Magn.* **27**, 2092 (1991).

[8.21] R.H. Bellis and Y. Iwasa, "Quench propagation in high T_c superconductors," *Cryogenics* **34**, 129 (1994).

[8.22] B.J. Maddock and G.B. James, "Protection and stabilisation of large superconducting coils," *Proc. Inst. Electr. Eng.* **115**, 543 (1968).

[8.23] T. Ishigohka and Y. Iwasa, "Protection of large superconducting magnets: a normal-zone voltage detection method," *Proc. 10th Sympo. Fusion Eng.* (IEEE CH1916-6/83/0000-2050, 1983), 2050.

[8.24] L. Dresner, "Superconductor stability '90: a review," *Cryogenics* **31**, 489 (1991).

[8.25] J.E.C. Williams (an internal report, FBNML, unpublished 1992).

[8.26] Chandrashekhar Haihar Joshi, "Thermal and electrical characteristics of adiabatic superconducting solenoids during a spontaneous transition to the resistive state," (Sc. D. Thesis, Dept. of Mech. Engineering, MIT, 1987; unpublished).

[8.27] H. Lim, Y. Iwasa, J.L. Smith, Jr., "Normal zone propagation in a cryocooler-cooled Nb_3Sn tape-wound magnet," *Cryogenics* **35**, (1995).

[8.28] J.B. Kim (an internal report, FBNML, unpublished 1994).

[8.29] A. Zhukovsky, Y. Iwasa, E.S. Bobrov, J. Ludlam, J.E.C. Williams, R. Hirose, Z. Ping Zhao, "750 MHz NMR Magnet Development," *IEEE Trans. Magn.* **28**, 644 (1992).

[8.30] Mamoon I. Yunus, Yukikazu Iwasa, and John E.C. Williams, "AC-loss-induced quenching in multicoil adiabatic superconducting magnets," *Cryogenics* **35**, (1995).

[8.31] See, for example, Y. Iwasa, "HTS magnets," *Advances in Superconductivity – V (ISS92)*, Eds. Y. Bando and H. Yamauchi (Springer-Verlag Tokyo, 1993), 1205.

[8.32] See, for example, P. Haldar, J.G. Hoehn, Jr., L.R. Motowildo, U. Balachandran, Y. Iwasa, "Fabrication and characteristics of a test magnet from HTS Bi-2223 silver-clad tapes," *Adv. Cryogenic Eng.* **40**, 313 (1994).

[8.33] T. Hikata, K. Muranaka, S. Kobayashi, J. Fujikami, M. Ueyama, T. Kato, T. Kaneko, H. Mukai, K. Ohkura, N. Shibuta, and K. Sato, "1 km-class Ag-sheathed Bi-based superconducting wires and applications," *1994 Int'l Workshop on Superconductivity* (Kyoto, Japan, June 1994), 69.

[8.34] J.P. McEvoy, Jr., L.C. Morris, and J.F. Panas, "Conduction cooling of a traveling wave maser superconducting magnet in a closed-cycle refrigerator," *Adv. Cryogenic Eng.* **10**, 486 (1965).

[8.35] M.T.G. van der Laan, R.B. Tax, H.H.J. ten Kate, L.J.M. van de Klundert, "The cryogenic system of a conduction cooled 12 K superconducting magnet," *Cryogenics* **30 September Supplement**, 163 (1990).

[8.36] S. Masuyama, H. Yamamoto, and Y. Matsubara, "A NbTi split magnet directly cooled by a cryocooler," *IEEE Trans. Appl. Superconduc.* **3**, 262 (1993).

[8.37] Mark E. Vermilyea and Constantinos Minas, "A cryogen-free superconducting magnet design for maglev vehicle applications," *ibid.*, 444 (1993).

[8.38] Y. Iwasa, "High-field HTS magnets operating at 20 K," *7th Conf. on Superconductivity and Applications* (Buffalo, NY, September 1994).

CHAPTER 9
CONCLUDING REMARKS

This book has presented important design and operational issues for superconducting magnets. It is hoped that the book will be useful for the designer of both LTS and HTS magnets. The reader may now understand that a higher operating temperature doesn't necessarily make the magnet designer's task easier. The cost of running the cryogenic system surely decreases with operating temperature, but cryogenics should not be an overriding issue in HTS magnets. Equally important are mechanical integrity, stability, protection, and conductor specification; on these, as we have studied, the impact of increased operating temperature is mixed.

9.1 Enabling Technology vs Replacing Technology

A new technology is either enabling or replacing. An enabling technology expands the technical limit of an existing technology; here competition from the existing technology is virtually nonexistent and economics is often a secondary issue. If a product based on this enabling technology is commercially viable, the technology will succeed. A replacing technology offers an approach different from an existing technology; here competition from the existing technology is fierce and economics often dictates the fate of the new technology. Table 9.1 presents principal applications of superconducting magnet technology, classifying SCM technology for each application to be either enabling or replacing. Although SCM technology is enabling for fusion, because fusion itself is replacing technology for energy conversion, it is still too early to tell how they will fare in the commercial world; by comparison, MHD, another enabling technology, has not to date fared well.

Table 9.1: Principal Applications of
Superconducting Magnet Technology

Enabling Technology
Fusion
High-field research magnets
MHD
MRI
NMR
Replacing Technology
Generators & motors
Maglev
Separation (ore, recycling)
SMES
Transformers

9.2 Outlook for the HTS

Since its discovery in 1911, superconductivity has fascinated physicists, engineers, and entrepreneurs alike. But to date, with the exception of MRI, NMR, and to a limited degree in ore separation, the commercial impact of superconducting magnet technology during the LTS era has been disappointing. Note that the commercial success of MRI and NMR owes a great measure to SCM technology, because SCM technology far exceeds the technical limits possible by conventional techniques.

The same real-world economic scrutiny awaits SCM technology based on HTS. For applications such as generators and motors, transformers, and many others that HTS promises to benefit [9.1], it is necessary to upgrade SCM technology for each application from replacing to enabling. HTS must offer something else that was missing in LTS.

It is worth noting that what a superconductor promises to deliver is no more free or rewarding than oil, rain water, solar power, or anything else that nature provides. Oil is free as it lies deep beneath the surface of the earth. To utilize it to power machinery, however, has consumed tremendous resources—technical and financial; it still does. Rain water can be harnessed to generate electric power, but it too does so at a price. Likewise, to reap benefits from superconductors requires much research and development effort. Today (1994), this appreciation of the need for the *continued* commitment of resources by all participants is particularly appropriate as we are about to complete the initial, material-oriented phase of the HTS era and are poised to surge forward into the 21st century in which principal large-scale applications will undoubtedly involve magnets [9.2, 9.3].

References

[9.1] Thomas P. Sheahen, *Introduction to High-Temperature Superconductivity*, (Plenum Press, New York, 1994).

[9.2] Martin Wood, "Superconductivity – the evolution of a new technology," *Advances in Superconductivity – V (ISS92)*, Eds. Y. Bando and H. Yamauchi (Springer-Verlag, Tokyo, 1993), 9.

[9.3] Peter Komarek, "HTSC application in power engineering and nuclear fusion," *ibid.*, 1199.

"Il resto nol dico, ..." —Figaro

APPENDIX I

PHYSICAL CONSTANTS AND CONVERSION FACTORS

Table A1.1: Selected Physical Constants*

Speed of light	c	3.00×10^8 m/s
Permeability of free space	μ_o	$4\pi \times 10^{-7}$ H/m
Permittivity of free space	ϵ_o	8.85×10^{-12} F/m
Avogadro's number	N_A	6.02×10^{26} particle/kg-mole
Electronic charge	e	1.60×10^{-19} C
Electron rest mass	m_o	9.11×10^{-31} kg
Proton rest mass	M_{p_o}	1.67×10^{-27} kg
Planck's constant	h	6.63×10^{-34} J s
Boltzmann's constant	k_B	1.38×10^{-23} J/K
Gas constant	R	8.32×10^3 J/kg-mole K
Molal gas volume	V_R	22.4 m^3/kg-mole
Stefan-Boltzmann constant	σ_{SB}	5.67×10^{-8} W/m^2 K^4
Acceleration of gravity	g	9.81 m/s^2
Wiedemann-Franz number	Λ	2.45×10^{-8} W Ω/K^2

* Except for the permeability of free space, values are approximate.

Table A1.2: Selected Conversion Factors

"Common" Non-SI Units*	SI Units*
Electromagnetic	
1 oersted	$250/\pi$ A/m
1 gauss	10^{-4} T
1 emu/cm^3	1000 A/m
Pressure	
1 mmHg (1 torr)	*133* Pa
1 atm (760 torr)	*101* kPa
1 bar (*750* torr)	0.1 MPa
1 psi (*52* torr)	*6.9* kPa
Viscosity	
1 poise	0.1 Pa s (0.1 kg/m s)
Energy & Power	
1 eV	1.6×10^{-19} J
1 cal	*4.18* J
1 BTU	*1055* J
1 hp	*746* W
Temperature	
0°C	*273* K
1 eV	*11600* K
Mass	
1 lb	*0.452* kg
1 metric ton	1000 kg
Dimension	
1 in	25.4 mm
1 French league	*4* km
1 liter	0.001 m^3
1 ft^3 (*28.3* liter)	*0.0283* m^3

* Values in italics are approximate.

APPENDIX II

THERMODYNAMIC PROPERTIES OF CRYOGENS

Table A2.1: Helium at 1 Atm

T [K]	ρ^* [kg/m^3]	C_p^* [kJ/kg K]	h^* [kJ/kg]
1.6	146.9	1.56	1.09
1.7	147.0	2.15	1.28
1.8	147.0	2.95	1.53
1.9	147.1	4.10	2.20
2.0	147.3	5.24	2.32
2.18	148.1	123.91	3.49
2.2	148.2	2.94	3.87
3.0	143.3	2.28	5.64
3.5	138.0	2.94	6.97
4.0	130.1	4.08	8.70
4.22	125.0	4.98	9.71
	16.9	*9.78*	*30.13*
10	*5.02*	*5.43*	*64.9*
20	*2.44*	*5.25*	*118*
30	*1.62*	*5.22*	*170*
40	*1.22*	*5.21*	*222*
50	*0.97*	*5.20*	*274*
75	*0.65*	*5.20*	*404*
100	*0.49*	*5.20*	*534*
150	*0.33*	*5.19*	*794*
200	*0.24*	*5.19*	*1054*
250	*0.20*	*5.19*	*1313*
273	*0.18*	*5.19*	*1433*
293	*0.17*	*5.19*	*1537*
300	*0.16*	*5.19*	*1573*

* Italics are for the vapor phase.

Table A2.2: Helium at Saturation

T [K]	p^* [torr]	ρ† [kg/m³]		h† [kJ/kg]	
1.60	5.59	0.23;	145	23.1;	0.39
1.65	6.90	0.27;	145	23.3;	0.48
1.70	8.45	0.33;	145	23.6;	0.58
1.75	10.2	0.38;	145	23.8;	0.70
1.80	12.3	0.45;	145	24.0;	0.84
1.85	14.6	0.52;	145	24.2;	1.00
1.90	17.2	0.60;	145	24.5;	1.18
1.95	20.2	0.69;	145	24.7;	1.38
2.00	23.4	0.78;	146	24.9;	1.63
2.05	27.0	0.89;	146	25.1;	1.92
2.10	31.0	0.99;	146	25.3;	2.23
2.18	38.0	1.18;	146	25.4;	—
2.40	63.6	1.81;	145	26.3;	3.82
2.60	94.0	2.52;	144	27.0;	4.27
2.80	133	3.40;	143	27.7;	4.73
3.00	182	4.46;	141	28.3;	5.23
3.25	257	6.08;	139	20.0;	5.93
3.50	352	8.09;	136	29.5;	6.72
3.75	470	10.5;	133	29.9;	7.62
4.00	615	13.6;	129	30.1;	8.65
4.22	760	16.9;	125	30.1;	9.78
4.30	815	18.2;	124	30.1;	10.1
4.40	892	20.0;	121	30.0;	10.7
4.50	974	22.1;	119	29.8;	11.3
4.75	1202	28.7;	112	29.0;	13.0
5.00	1466	39.3;	101	27.3;	15.4
5.20‡	1706	69.6		21.36	

* Below 4.2 K, there are slight discrepancies in saturation pressure among data sources listed on p. 384, *e.g.* 12.26~12.56 torr at 1.80 K.

† Italics are for the vapor phase.

‡ Critical point.

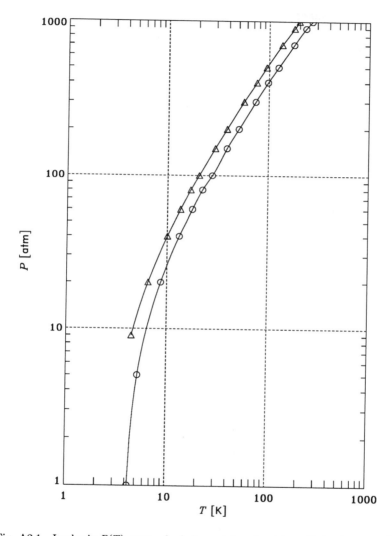

Fig. A2.1   Isochoric $P(T)$ curves for helium at two densities. Circles [125 kg/m^3 (4.22 K, 1 atm)]; triangles [147 kg/m^3 (1.8 K, 1 atm; 4.5 K, 9 atm)].

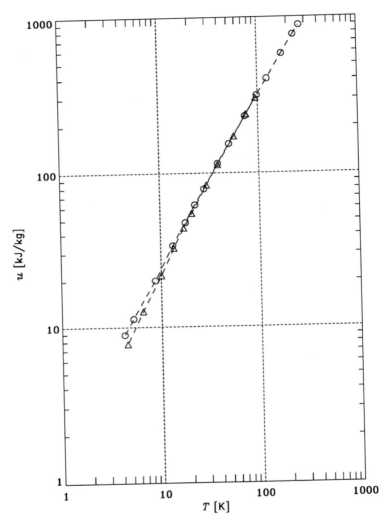

Fig. A2.2   Isochoric $u(T)$ curves for helium at two densities. Circles [125 kg/m^3 (4.22 K, 1 atm)]; triangles [147 kg/m^3 (1.8 K, 1 atm; 4.5 K, 9 atm)].

Table A2.3: Selected Properties of Cryogens at 1 Atm

Property		He	H_2*	Ne	N_2	O_2
Boiling temperature, T_s	[K]	4.22	20.39	27.09	77.39	90.18
Triple point†	[K]	—	13.96	24.56	63.16	54.36
Heat of vaporization	[kJ/kg]	20.9	443	85.9	199.3	213
	[J/cm³]‡	2.6	31.3	104	161	243
Density(T_s, liq.)	[kg/m³]	125	70.8	1206	807	1141
Density(T_s, vap.)	[kg/m³]	16.9	1.33	9.37	4.60	4.47
Density (293 K)	[kg/m³]	0.167	0.084	0.840	1.169	1.333
Density(T_s, liq.)/density(293 K)		749	843	1436	690	856

* Normal hydrogen.
† At the triple point pressure.
‡ Unit liquid volume.

Table A2.4: Heat Transfer Properties of Cryogen Gases at 1 Atm

T [K]	He	H_2*	Ne	N_2	O_2
Viscosity [μPa s]					
10	2.26	—	—	—	—
25	4.13	1.38	—	—	—
30	4.63	1.61	5.12	—	—
50	6.36	2.49	8.18	—	—
100	9.78	4.21	14.4	6.98	7.68
200	15.1	6.81	23.8	13.0	14.8
300	19.9	8.96	31.7	17.9	20.7
Thermal Conductivity [mW/m K]					
10	17.5	—	—	—	—
25	30.6	19.4	—	—	—
30	34.1	22.9	—	—	—
50	46.8	36.2	14.1	—	—
100	73.6	68.0	22.2	9.3	9.0
200	118	128	36.9	18.0	18.3
300	155	177	49.2	25.8	26.6
Prandtl Number Pr					
10	0.699	—	—	—	—
25	0.706	0.757	—	—	—
30	0.708	0.751	0.565	—	—
50	0.707	0.722	0.617	—	—
100	0.690	0.696	0.666	0.797	0.793
200	0.665	0.719	0.663	0.750	0.737
300	0.667	0.725	0.663	0.722	0.717

* Normal hydrogen.

Property Data Sources

Because most sets of property data presented in this Appendix are compiled from many sources, the sources are grouped together and presented below.

Randall F. Barron, *Cryogenic Systems* 2nd Ed., (Clarendon Press, Oxford, 1985).

R.D. McCarty, *Thermophysical Properties of Helium-4 from 2 to 1500 K with Pressures to 1000 Atmospheres* (NBS Technical Note 631, 1972).

Robert D. McCarty, *The Thermodynamic Properties of Helium II from 0 K to the Lambda Transitions* (NBS Technical Note 1029, 1980).

Russell B. Scott, *Cryogenic Engineering* (1963 Edition reprinted in 1988 by Met-Chem Research, Boulder, CO).

Steven W. Van Sciver, *Helium Cryogenics* (Plenum Press, New York, 1986).

APPENDIX III

PHYSICAL PROPERTIES OF MATERIALS

In using property data presented in this Appendix, we should be well aware that although most data are given to at least three significant figures, implying these particular data are quite accurate, they do not necessarily accurately represent the property value of the specific material for which we seek information. Among property data presented here, those that are subject to considerable degrees of variation from one material batch to another include: thermal conductivity data (Fig. A3.1); mechanical property data (Table A3.1); thermal expansion data (Table A3.2), particularly of non-metals.

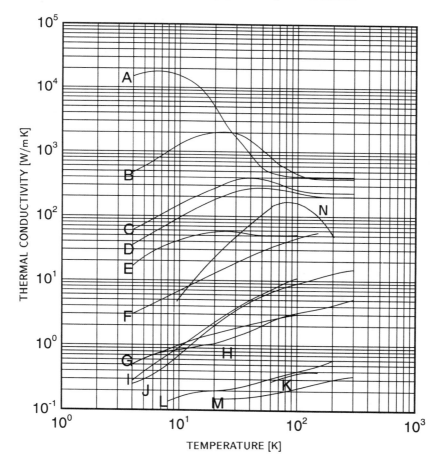

Fig. A3.1 Thermal conductivity *vs* temperature plots. A. Silver; B. Copper; C. Aluminum (1100F); D. Aluminum (6063-T5); E. Solder (Sn50-Pb50); F. Brass; G. Teflon; H. GE varnish; I. Inconel; J. Stainless steel; K. Phenolic; L. G-10; M. Epoxy; N. Alumina.

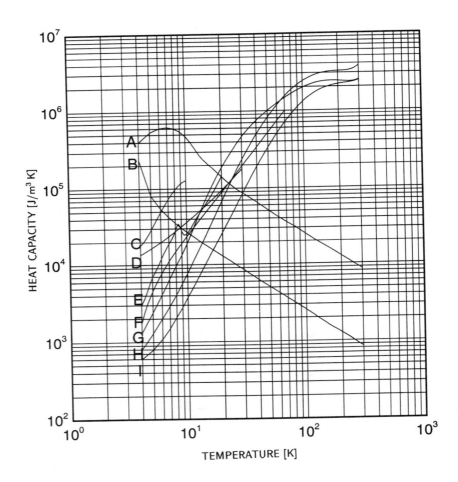

Fig. A3.2   Heat capacity *vs* temperature plots. A. Helium (10 atm);
B. Helium (1 atm); C. GE varnish; D. Stainless steel; E. Nb-Ti; F.
Epoxy; G. Silver; H. Copper; I. Aluminum.

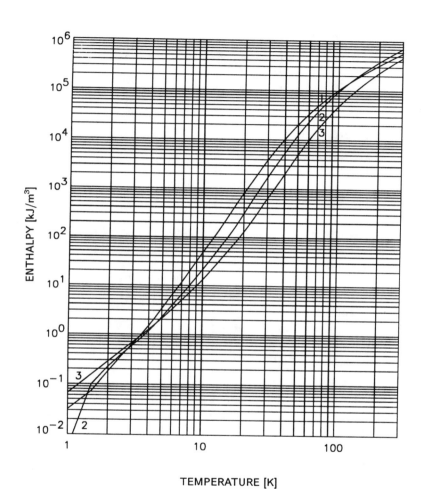

Fig. A3.3 Volumetric enthalpy *vs* temperature plots.
1. Silver; 2. Copper; 3. Aluminum.

Table A3.1: Mechanical Properties of Materials*

Material	T [K]	σ_U [MPa]	σ_Y [MPa]	E [GPa]
Aluminum 6061 (T6)	295	315	280	70
	77	415	380	77
Copper (annealed)	295	160	70	70
	77	310	90	100
Copper (1/4 hard)	295	250	240	130
	77	350	275	150
Copper (1/2 hard)	295	300	280	130
	77	—	13	77
Silver (annealed)	295	—	—	71
	77	310	90	100
Incoloy 908	295	1500	1280	179
	4	1900	1490	182
Inconel	295	900	850	220
	77	1160	1040	230
Stainless steel 304 (low limit)	295	550	200	190
	77	1450	260	200
Stainless steel 304 (high limit)	295	1030	620	190
	77	1860	1050	200
Stainless steel 316LN	295	1290	1100	185
	77	1790	1400	195
Epoxy	295	540	—	27
	77	890	—	28
G-10 lengthwise	295	280	—	18
crosswise	295	245	—	14
Mylar	295	145	—	7
	77	215	—	13
Teflon	295	14	—	0.4
	77	105	—	5
Nb$_3$Sn	4	420	—	162
Nb-Ti	4	2200	—	82

* For Sn-Pb solders, see Table 7.8 (p. 300).

Table A3.2: Mean Linear Thermal Expansion of Materials
$[L(T) - L(293\,\mathrm{K})]/L(293\,\mathrm{K})$ in 10^{-3}

Material	T [K]				
	20	80	140	200	973†
Aluminum	−4.15*	−3.91	−3.12	−2.01	—
Brass (70Cu-30Zn)	−3.69	−3.37	−2.60	−1.63	+13.0
Bronze	−3.30	−3.04	−2.37	−1.50	+13.3
Copper	−3.24	−3.00	−2.34	−1.48	+13
Silver	−4.09	−3.60	−2.70	−1.71	+15.0
Incoloy 908	−1.73	−1.54	−1.22	−0.81	+9.23
Inconel 718	−2.38	−2.23	−1.77	−1.44	+15
Stainless steel 304	−3.06	−2.81	−2.22	−1.40	+13.2
Epoxy	−11.5	−10.2	−8.99	−5.50	—
G-10‡	−7.06	−6.38	−5.17	−3.46	—
Phenolic‡	−7.30	−6.43	−5.13	−3.41	—
Teflon (TFE)	−21.1	−19.3	−16.6	−12.4	—
Nb₃Sn	−1.71	−1.41	−1.02	−0.67	+5.5
Nb-45Ti	−1.87	−1.67	−1.24	−0.78	—
Solder (Sn50-Pb50)	—	−4.98	−3.65	−2.29	—

* An aluminum bar 1-m long at 293 K *shrinks* by 4.15 mm when cooled to 20 K.
† A common Nb₃Sn reaction temperature.
‡ crosswise.

Property Data Sources

A.F. Clark, "Low temperature thermal expansion of some metallic alloys," *Cryogenics* **8**, 282 (1968).

Robert J. Corruccini and John J. Gniewek, *Specific Heats and Enthalpies of Technical Solids at Low Temperatures* (NBS Monograph 21, 1960).

Cryogenic Materials Data Handbook Volumes 1 and 2 (Martin Marietta Corp. Air Force Materials Laboratory, 1970).

C.C. Koch and D.S. Easton, "A review of mechanical behavior and stress effects in hard superconductors," *Cryogenics* **17**, 391 (1977).

J. H. McTaggart and G.A. Slack, "Thermal conductivity data of General Electric No. 7031 varnish," *Cryogenics* **9**, 384 (1969).

T. Nishio, Y. Itoh, F. Ogasawara, M. Suganuma, Y. Yamada, U. Mizutani, "Superconducting and mechanical properties of YBCO-Ag composite superconductors," *J. Materials Science* **24**, 3228 (1989).

Russell B. Scott, *Cryogenic Engineering* (1963 Edition reprinted in 1988 by Met-Chem Research, Boulder CO).

N.J. Simon, E.S. Drexler, and R.P. Reed, *Properties of Copper and Copper Alloys at Cryogenic Temperatures* (NIST Monograph 177, 1992).

N.J. Simon and R.P. Reed, *Structural Materials for Superconducting Magnets* (Preliminary Draft, 1982).

L.S. Toma, M.M. Steeves, R.P. Reed, *Incoloy Alloy 908 Data Handbook* (Plasma Fusion Center Report PFC/RR-94-2, MIT, Cambridge MA, 1994).

Y.S. Touloukian, R.W. Powell, C.Y. Ho, and P.G. Klemens, *Thermophysical Properties of Matter, Volume 2*, (IFI/Plenum, New York-Washington, 1970).

Y.S. Touloukian and R.K. Kirby, *Thermophysical Properties of Matter, Volume 12* (IFI/Plenum, New York-Washington, 1975).

APPENDIX *IV*

ELECTRICAL PROPERTIES OF NORMAL METALS

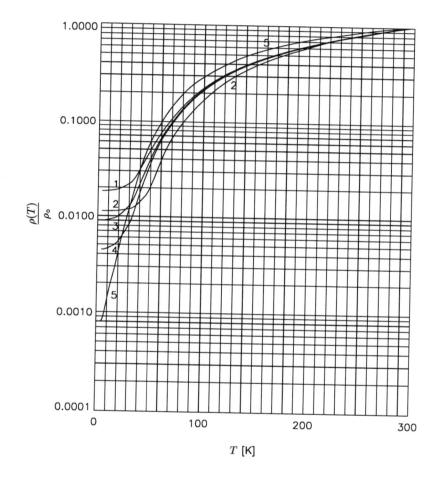

Fig. A4.1 Normalized zero-field electrical resistivity *vs* temperature plots. 1. Copper (RRR 50, ρ_0: 17.14 nΩ m); 2. Aluminum (99.99%, ρ_0: 26.44 nΩ m); 3. Copper (RRR 100, ρ_0: 17.03 nΩ m); 4. Copper (RRR 200, ρ_0: 16.93 nΩ m); 5. Silver (99.99%, ρ_0: 17.00 nΩ m). For each metal, $\rho(T)$ is normalized to ρ_0, its zero-field resistivity at 293 K.

391

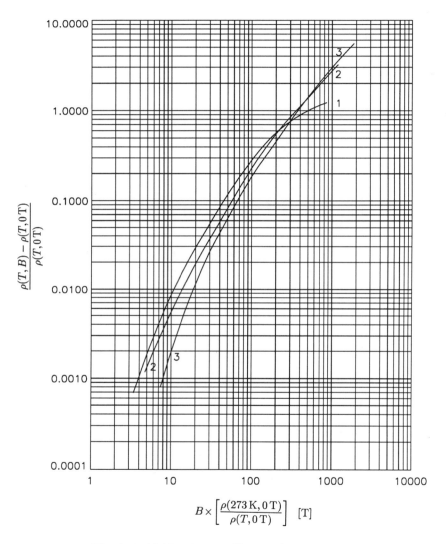

Fig. A4.2   Kohler plots. 1. Silver; 2. Copper; 3. Aluminum.

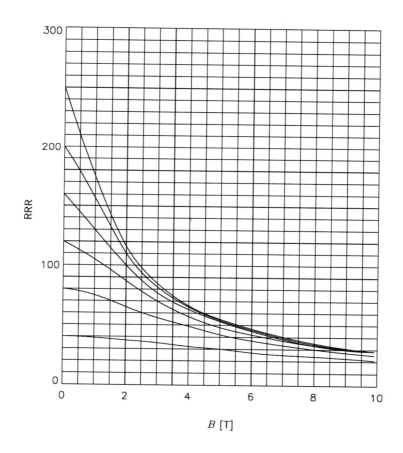

Fig. A4.3 Copper Residual Resistivity Ratio (RRR) *vs* magnetic induction plots. [Based on data of Taylor, Woolcock, and Barber.]

Table A4.1: Electrical Resistivity of Heater Metals

Metal	ρ [nΩ m]		
	300 K	77 K	≤ 20 K
Berylco 25	105	85	82
Brass (70Cu-30Zn)	72	50*	43
Naval brass	47	40	27
Constantan	525	450*	440
Cu-10Ni	184	169	167
Manganin	400	390*	370
Stainless steel 304L	704	514	496
310	873	724	685
316	765	585	553
Nb-Ti	—	—	600†

* At 90 K.
† In the normal state.

Property Data Sources

A.F. Clark, G.E. Childs, and G.H. Wallace, "Electrical resistivity of some engineering alloys at low temperatures," *Cryogenics* **10**, 295 (1970).

F.R. Fickett, "Oxygen-free copper at 4 K: resistance and magnetoresistance," *IEEE Trans. Magn.* **MAG-19**, 228 (1983).

Y. Iwasa, E.J. McNiff, R.H. Bellis, and K. Sato, "Magnetoresistivity of silver over the range 4.2~159 K," *Cryogenics* **92** (1992).

G.T. Meaden, *Electrical Resistance of Metals* (Plenum Press, New York 1965).

Materials at low temperatures, Eds. Richard P. Reed and Alan F. Clark (American Society For Metals, 1983).

M.T. Taylor, A. Woolcock, and A.C. Barber, "Strengthening superconducting composite conductors for large magnet construction," *Cryogenics* **8**, 317 (1968).

Guy Kendall White, *Experimental Techniques in Low-Temperature Physics* (Clarendon Press, Oxford, 1959).

APPENDIX V

PROPERTIES OF SUPERCONDUCTORS

Critical properties presented in this Appendix, like properties of materials given in Appendix III are *representative* and are to be used for only zeroth or first order estimates. Critical current data of Nb-Ti, Nb$_3$Sn, and BiPbSrCaCuO(2223) are given, respectively, in Figs. A5.1, A5.2, and A5.3. Selected physical, thermal, and electrical properties of YBCO and BSCCO are given in Table A5.4.

Critical Field *vs* Critical Temperature Data

Nb-Ti Conductors B_{c2} *vs* T_c may best be given by an expression of given by Lubell who derived it from various data:

$$B_{c2} = B_0 \left[1 - \left(\frac{T}{T_c} \right)^n \right] \tag{A5.1}$$

Lubell finds $B_0 = 14.5\,\text{T}$, $T_c = 9.2\,\text{K}$, and $n = 1.7$ give the best fit to the data. Table A5.1 gives values of T_c for selected values of B_{c2}; the temperature ranges are based on the original data.

Table A5.1: B_{c2} *vs* T_c Data for Nb-Ti

B_{c2} [T]	T_c [K]	B_{c2} [T]	T_c [K]
1	8.6~9.1	7	5.9~6.4
2	8.0~8.6	8	5.4~5.7
3	7.9~8.2	9	4.9~5.4
4	7.4~7.8	10	4.5~4.8
5	6.9~7.3	11	3.9~4.3
6	6.5~6.9	12	3.1~3.6

Nb$_3$Sn Conductors B_{c2} *vs* T_c may also best be given by Eq. A5.1. For Nb$_3$Sn conductors, appropriate values are: $B_0 = 27.9\,\text{T}$, $T_c = 18.3\,\text{K}$, and $n = 0.62 \sim 0.72$. Table A5.2 gives values of T_c for selected values of B_{c2}; the temperature ranges are bounded by values corresponding to $n = 0.62$ and $n = 0.72$.

Table A5.2: B_{c2} *vs* T_c Data for Nb$_3$Sn

B_{c2} [T]	T_c [K]	B_{c2} [T]	T_c [K]
8	10.6~11.4	14	5.9~7.0
10	8.9~9.9	16	4.6~5.6
12	7.4~8.4	18	3.4~4.3

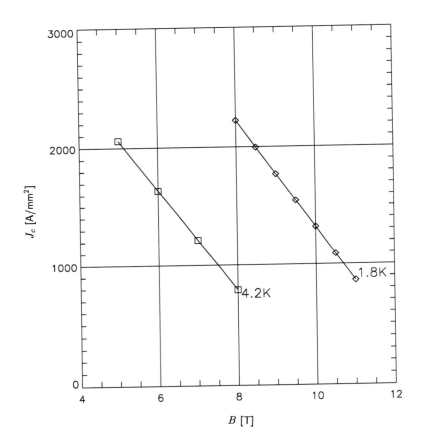

Fig. A5.1 J_c *vs* B plots for Nb-Ti at 1.8 and 4.2 K.
Only the Nb-Ti area is used to compute J_c.

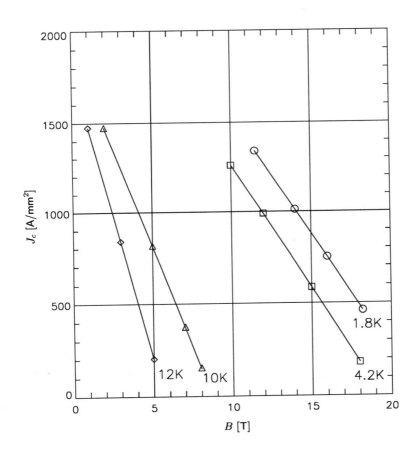

Fig. A5.2 J_c *vs* B plots for Nb$_3$Sn at 1.8, 4.2, 10, and 12 K. Only the non-copper area is used to compute J_c.

BiPbSrCaCuO (2223) Data

$J_c(B \parallel)$ data presented below are "representative," measured for a silver-sheathed BiPb-SrCaCuO tape, comprised of 61 "filamentary" strips, at 4.2 and 27 K (liquid neon) with field orientation parallel to the tape's broad surface. $J_c(B \perp)/J_c(B \parallel)$ data at the same temperatures for a tape *nearly* identical to Sample E1, are shown in Fig. 7.33 (p. 318). $I_c(T, B)$ data of Fig. 5.26 (p. 200) belong to mono-filament tapes with a J_c value (based on the superconductor area only) of 299 A/mm² at 77 K and 0 T.

Table A5.3: Parameters of BiPbSrCaCuO (2223)
at 77 K and 0 T

	I_c [A]	J_c [A/mm²]*
Sample E1	33.1	300

* Based on the BiPbSrCaCuO cross sectional area only.

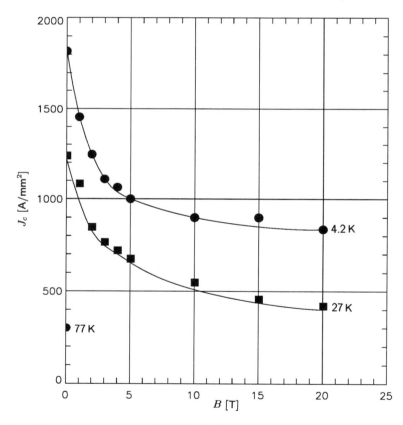

Fig. A5.3 J_c *vs* B plots for BiPbSrCaCuO (2223) at 4.2 and 27 K, with field directed parallel to tape's broad surface. See Fig. 7.33 for $J_c(B \perp)/J_c(B \parallel)$ data.

Table A5.4 presents selected properties of YBCO and BSCCO; corresponding values for silver are also included for comparison. The principal purpose of presenting these HTS properties is to give the magnet engineer, who, at least at the present moment, is generally much less familiar with these property values than the material scientist engaged in HTS research, ball-park figures to orient his "feel" for the values. For a specific design purpose, it is recommended that he consult more up-to-date references.

Table A5.4: Selected Properties of YBCO and BSCCO
– Comparison with Silver –

Property	T [K]	YBCO	BSCCO (2212)	BSCCO (2223)	BPSCCO (2223)	Silver (99.99%)
E [GPa]	300	97	39	96	54	76
	77	97			54	
σ_U [MPa]	300	50			70	370
σ_Y [MPa]	300		20			55
ϱ [kg/m^3]	300	6380	5350		4350	10490
c_p [J/kg K]	120	245	220	240	192	200
k [W/m K]	200	2.9	6.0	1.1	1.4	400
	T_c	3.0	5.5	1.0	1.1	
	77	3.0	5.6	1.0	1.9	420
ρ [$\mu\Omega$ m]	300	10	60		8	0.017
	$T_c{}^*$	10	60		8	0.004†

* Just above the critical temperature.
† At 77 K.

Property Data Sources

Nb-Ti B_{c2} vs T Data

M.S. Lubell, "Empirical scaling formulas for critical current and critical field for commercial NbTi," *IEEE Trans. Magn.* **MAG-19**, 754 (1983).

Nb-Ti and Nb₃Sn J_c vs B Data

J.E.C. Williams (personal communication, 1993).

BiPbSrCaCuO (2223) J_c vs B Data

J. Fujikami, K. Sato, Y. Iwasa, M. Yunus, H. Lim, and J.B. Kim (preliminary data, FBNML, 1994).

HTS Property Data (listed chronologically)

M.P. Boiko, V.G. Kantser, L.A. Konopko, A.S. Sidorenko, "Sign-reversal thermopower in $Bi_2Sr_2Ca_2Cu_3O_y$," *Physica C* **162–164**, 506 (1989).

Y.S. He, J. Xiang, F.G. Chang, J.C. Zhang, A.S. He, H. Wang, B.L. Gu, "Anomalous structural changes and elastic properties of bismuth oxide superconductors," *Physica C* **162–164**, 450 (1989).

T. Nishio, Y. Itoh, F. Ogasawara, M. Suganuma, Y. Yamada, and U. Mizutani, "Superconducting and mechanical properties of YBCO-Ag composite superconductors," *J. Mat. Science* **24**, 3228 (1989).

W. Schnelle, E. Braun, H. Broicher, H. Weiss, H. Geus, S. Ruppel, M. Galffy, W. Braunisch, A. Waldorf, F. Seidler, and D. Wohlleben, "Superconducting fluctuations in $Bi_2Sr_2Ca_2Cu_3O_x$," *Physica C* **161**, 123 (1989).

Ctirad Uher, "Review: Thermal conductivity of high-T_c superconductors," *J. Superconduc.* **3**, 337 (1990).

S. Jin, "Progress in bulk high-temperature superconductors," *Superconductivity and its Application*, (AIP Conference 251, 1991), 241.

S. Ochiai, K. Hayashi, and K. Osamura, "Influence of thermal cycling on critical current of superconducting silver-sheathed high T_c oxide wires," *Cryogenics* **31**, 954 (1991).

N.X. Tan, A.J. Bourdillon, and W.H. Tai, "A precursor method for reacting and aligning $Bi_2Sr_2Ca_2Cu_3O_{10}$," *Superconductivity and its Application* (AIP Conference, 1991), 251.

H. Ledbetter, "Elastic constants of polycrystalline $Y_1Ba_2Cu_3O_x$," *J. Mat. Sci. Res.* **7**, 11 (1992).

S. Ochiai, K. Hayashi, and K. Osamura, "Strength and critical density of Bi(Pb)-Sr-Ca-Cu-O and Y-Ba-Cu-O in silver-sheathed superconducting tapes," *Cryogenics* **32**, 799 (1992).

W. Schnelle, O. Hoffels, E. Braun, H. Broicher, and D. Wohlleben, "Specific heat and thermal expansion anomalies of high temperature superconductors," *Physics and Materials Science of High Temperature Superconductors – II* (Kluwer Publishers, Dordrecht, The Netherlands, 1992), 151.

A. Marino, T. Yasuda, E. Hoguin, and L. Rinderer, "Preparation and electrical properties of Bi(Pb)-Sr-Ca-Cu-O single phase high T_c films," *Physica C* **210**, 16 (1993).

J. Tenbrink, M. Wilhelm, K. Heine, H. Krauth, "Development of technical high-T_c superconductor wires and tapes," *IEEE Trans. Applied Superconduc.* **3**, 1123 (1993).

APPENDIX VI

GLOSSARY*

A-15: The cubic crystalline structure of most low-T_c intermetallic compound superconductors, *e.g.* Nb$_3$Al, Nb$_3$Sn, V$_3$Ga. Also known as the beta-tungsten (β-W) structure.

AC loss: Energy dissipation in a conductor caused by a time-varying magnetic field, transport current, or both. HYSTERESIS, COUPLING, and EDDY-CURRENT are AC losses.

Active protection: A magnet protection technique that uses devices located outside the CRYOSTAT; it generally involves two steps: detection of a QUENCH followed by activation of a DUMP. Drawback: the protection system itself is subject to malfunction.

Active shielding: A technique to minimize a FRINGING FIELD by means of a set of magnets that generates a field in the direction opposite to that of the main field. It is used in some MRI systems having a main field 1.5 T and above.

Adiabatic magnet: A magnet whose entire winding is not in direct contact with CRYOGEN. CRYOCOOLER-cooled magnets behave as adiabatic magnets.

AE (Acoustic Emission): Acoustic signals emitted by sudden mechanical events in a body being loaded or unloaded, *e.g.* a magnet being charged or discharged; useful for detection and location of a PREMATURE QUENCH caused by conductor motion or epoxy fracture event. The KAISER EFFECT is often observed in AE signals.

Anisotropy: Property exhibiting different values in different directions, *e.g.* magnetic field orientation effect on superconductor's CRITICAL CURRENT DENSITIY.

"Baseball" (tennis ball) magnet: A "mirror" magnet for linear fusion machines; its winding resembles the seam on a baseball (tennis ball). The largest versions, superconducting, were tested in the early 1980s at the Lawrence Livermore National Laboratory.

Bath-cooled magnet: A superconducting magnet immersed in a bath of CRYOGEN, *e.g.* boiling liquid helium, as opposed to a forced-cooled magnet. See CIC CONDUCTOR.

BCS theory: A microscopic theory of superconductivity by J. Bardeen, L. Cooper, and R. Schrieffer of the University of Illinois published in 1957. The theory explains electromagnetic and thermodynamic properties of LTS. The theory incorporates a hypothesis that special pairs of electrons called COOPER PAIRS carry the supercurrent.

Bean slab: A one-dimensional model for a bulk TYPE II SUPERCONDUCTOR introduced by C. Bean of the General Electric Research Laboratory to study magnetic behavior.

Bitter magnet: A high-field, water-cooled magnet built of BITTER PLATES. Named for Francis Bitter, who developed the first 10-T electromagnets in the 1930s.

Bitter plate: Annular disc used in BITTER MAGNETS, with either punched holes or etched radial channels for cooling water. Basically of copper, alloyed for strength with beryllium, chromium, aluminum oxide, or recently, silver inclusions.

Breakdown voltage: The voltage at which an insulator fails. An important parameter for magnet operation at DISCHARGE VOLTAGES above ~1 kV.

Bronze process: A process for making MULTIFILAMENTARY CONDUCTORS, primarily Nb$_3$Sn. Nb rods in bronze (a Cu-Sn alloy) are processed to form Nb filaments, each with a Nb3Sn surface layer. Developed by A.R. Kaufman of the Brookhaven National Laboratory in 1970. K. Tachikawa of the National Research Institute for Metals developed a similar process in 1970 for V$_3$Ga multifilamentary conductors.

---

* Includes terms discussed only briefly or not at all in the main text but which are of general interest in superconducting magnet technology and its areas of application. The list of laboratories and organizations, however, is limited to those whose acronyms are used in the main text.

BSCCO: A bismuth-based HTS; its basic chemical formula is $Bi_2Sr_2Ca_{n-1}Cu_nO_{2n+4}$, with $n = 2$ resulting in BSCCO (2212) having T_c of $\sim$85 K and $n = 3$ in BSCCO (2223) with T_c of $\sim$110 K. A fraction of lead is often substituted for bismuth, giving rise to BiPbSrCaCuO (2223), which is processed into silver-sheathed COMPOSITE SUPERCONDUCTOR tapes. BSCCO appears more promising than YBCO for magnets. H. Maeda of the National Research Institute for Metals is credited with its discovery.

Bubble chamber: A chamber filled with liquid hydrogen and surrounded by a magnet, usually superconducting; a tool for the study of high-energy particle interactions.

Carnot cycle: A reversible thermodynamic cycle, composed of two adiabatic and two isothermal processes, in which a working fluid operates between two thermal reservoirs to produce work or refrigeration at the most efficient level. N.S. Carnot published it in 1824, *"Reflexions sur la Puissance Motrice du Feu"* (*"On the Motive Power of Fire"*).

Chevrel phase: A molybdenum-based LTS phase. $PbGd_{0.2}Mo_6S_6$ has the highest B_{c_2} among LTS: 54 T at 4.2 K. Chevrel phase materials appear unsuitable for MAGNET-GRADE SUPERCONDUCTORS. Discovered in 1981 by R. Chevrel of Université de Rennes.

CIC (*Cable-in-conduit*) conductor: A cable of transposed MULTIFILAMENTARY CONDUCTORS encased in a conduit that provides strength and rigidity, and through which cooling fluid is forced. Generally the forced fluid is SUPERCRITICAL HELIUM.

Class 1 magnets: Superconducting magnets for "large-scale" applications, *e.g.* fusion, SMES; generally they are CRYOSTABLE or stable against transient disturbances.

Class 2 magnets: Another designation for HIGH-PERFORMANCE MAGNETS.

Coercive force: The magnetic field required to bring the magnetization to zero in a magnetized material. Materials for PERMANENT MAGNETS have high coercive forces.

Coherence length (ξ): The distance over which the superconducting-normal transition takes place; introduced by B. Pippard (Cambridge University) in the early 1950s. In the BCS THEORY, ξ represents the size of the wave function for the COOPER PAIRS.

Composite superconductor: A conductor comprised of strands or tapes of superconductor in a matrix of normal metal. Some are MULTIFILAMENTARY CONDUCTORS.

Cooper pair: The paired superelectrons responsible for superconductivity.

Copper-to-superconductor ratio: The volumetric ratio of copper to superconductor in a COMPOSITE SUPERCONDUCTOR (like Nb-Ti), or the ratio of copper to *non-copper* in a composite superconductor (like Nb_3Sn) which contains other metals.

Coupling loss: AC LOSS generated by field-induced currents between filaments in MULTIFILAMENTARY SUPERCONDUCTORS or between strands in CIC CONDUCTORS.

Coupling time constant: The predominant decay time constant of field-induced currents in MULTIFILAMENTARY SUPERCONDUCTORS or CIC CONDUCTORS.

Critical current (I_c): The maximum current a conductor can carry and still remain superconducting at a given temperature and magnetic field.

Critical current density (J_c): One of the three material-specific parameters that defines the critical surface for superconductivity. In TYPE II SUPERCONDUCTORS, J_c is sensitive to metallurgical processing.

Critical field (H_c): One of the three material-specific parameters that defines superconductivity; insensitive to metallurgical processing. In TYPE II SUPERCONDUCTORS, there are two critical fields: LOWER (H_{c1}) and UPPER (H_{c2}).

Critical state model: A model for BEAN SLAB's magnetic behavior in the MIXED STATE; the slab is in the critical state in that every part of the slab is carrying a current equivalent to the material's CRITICAL CURRENT DENSITY. The model can be used to derive a criterion for FLUX JUMP.

Critical temperature (T_c): One of the three material-specific parameters that defines the critical surface for superconductivity. Like B_c, T_c is insensitive to metallurgy. The parameter that most clearly distinguishes HTS from LTS.

Cryocooler: A device, which generally uses the G-M CYCLE, to provide refrigeration typically at 80 K and 20 K; it is an indispensable component of an HTS-magnet system.

Cryogen: A liquid that boils at CRYOGENIC TEMPERATURES. Examples include liquid helium and liquid nitrogen; both are commonly used in superconducting magnet systems.

Cryogenic temperatures: Temperatures below 150~200 K.

Cryopump: A pump that uses a CRYOCOOLER to create and maintain a high vacuum.

Cryostable magnet: A magnet satisfying the STEKLY or "EQUAL-AREA" CRITERION.

Cryostat: An enclosed container that maintains a cryogenic environment; it uses the basic design principle developed by Dewar. Much more rugged than a DEWAR and made generally of stainless steel, or for AC applications, of insulating materials.

Cryotribology: The study of friction and wear of materials (tribology) at CRYOGENIC TEMPERATURES. Applied in the study of conductor-motion disturbances in superconducting magnets; it is also useful for evaluating components of cryogenic devices.

Current sharing temperature (T_{cs}): The temperature at which the transport current (I_t) in the conductor equals the conductor's CRITICAL CURRENT. $I_t = I_c(T_{cs})$.

Cyclotron magnet: A magnet, usually superconducting, in a cyclotron that produces energetic protons for medical treatment.

Dewar: A double-walled flask with a vacuum between silvered glass walls for storing CRYOGENS. Developed by J. Dewar in 1892, who was first to liquefy hydrogen in 1898.

Dilution refrigerator: A millikelvin-range refrigerator that makes use of the difference in entropies of He^3 and He^4; it typically provides a refrigeration of $100\,\mu W$ at 50 mK.

Dip stick: A narrow, long tube for measuring liquid helium levels.

Dipole magnet: A magnet that generates a uniform field transverse to its axis over most of its bore; it deflects charged particles in accelerators and MHD systems.

Discharge voltage: The maximum voltage appearing across the terminals of a superconducting magnet during a DUMP. Also called DUMP VOLTAGE.

Double pancake: See PANCAKE WINDING.

DPC (Demonstration Poloidal Coil): A set of experimental coils, including the United States Demonstration Poloidal Coil (US-DPC) and Japan's Experimental Demonstration Poloidal Coil (DPC-EX), developed to demonstrate the feasibility of producing CIC CONDUCTORS suitable for superconducting PULSE MAGNETS in TOKAMAK machines. The coils were tested at JAERI in the late 1980s and early 1990s.

Dry magnet: A magnet wound without epoxy and other filler material in the winding. The term may also be used for a magnet coupled to a cryocooler and cooled by conduction.

Dump: A forced discharge of the current in a superconducting magnet in an emergency. A dump often involves opening a switch connecting the power supply and the magnet and discharging the stored magnetic energy through a DUMP RESISTOR.

Dump resistor: A resistor through which the current in the superconducting magnet is forced to flow during a DUMP; it speeds up the current decay in the magnet.

Dump voltage: Another name for a DISCHARGE VOLTAGE.

Eddy-current loss: A loss generated by eddy currents induced in conductive metal by a time-varying magnetic field; a nuisance as a source of extraneous heat in high-field, low-temperature (millikelvin range) experiments.

Energy margin: The maximum input pulse energy density to a small region of the winding without it leading to a QUENCH; it is a useful parameter to quantify stability against transient disturbances, *e.g.* conductor motion, occurring in the winding.

Epoxy-impregnated magnet: An ADIABATIC MAGNET having its winding filled with an epoxy resin to minimize incidents of conductor motion responsible for PREMATURE QUENCHES. The FLOATING WINDING method applied to these magnets is effective in reducing epoxy fracture responsible for premature quenches.

"Equal-area" criterion: A design criterion for CRYOSTABLE MAGNETS that incorporates thermal conduction along the conductor axis to achieve OVERALL CURRENT DENSITIES greater than those achieved by magnets meeting STEKLY CRITERION.

Fault mode: A failure of the system, *e.g.* a PREMATURE QUENCH of a superconducting magnet at or near the designed operating current. Magnet designers use anticipated fault modes that result in the most extreme conditions as guides to the system requirements.

FBNML (Francis Bitter National Magnet Laboratory): Founded in 1960, the first national laboratory devoted to research and development in magnetism and magnet technology. Located on the MIT campus, Cambridge, MA.

FEA (Finite Element Analysis): A computer-based technique for analyzing complex stress distributions, for example, in a magnet system.

Floating winding: A type of winding used in EPOXY-IMPREGNATED MAGNETS to allow the windings to separate freely from the coil forms as the magnets are energized and thereby to minimize PREMATURE QUENCH incidents induced by epoxy fracture.

Flux jump: A thermal instability in a superconductor in which flux motion induces heat generation, which in turn causes additional magnetic flux motion.

Flux pinning: A mechanism that inhibits FLUXOID motion. The flux pinning force is thought to be supplied by the crystal structure; it can be increased by material impurities and inhomogeneities caused by cold working and heat treatment to produce large J_c in materials used in MAGNET-GRADE SUPERCONDUCTORS.

Fluxoid: A quantized flux line ($\simeq 2.0 \times 10^{-15}$ Wb) having a normal-state core whose radius is on the order of the COHERENCE LENGTH; it is the basic unit for field measuring sensors based on a SQUID.

Formvar: Trade name for polyvinylformal, a venerable (over 100 years old) insulator material for conductors; it is particularly effective for low temperature applications.

Fringing field: An unwanted magnetic field outside a magnet system.

G-10: A laminated, thermosetting composite with glass fibers as base materials, used in magnets and CRYOSTATS as a high-strength material where metals are unsuitable.

G-M (Gifford-McMahon) cycle: A variant of the STIRLING CYCLE suitable for miniature refrigerators. Developed by W. Gifford and H. McMahon in the early 1960s.

Gas-cooled lead: A current lead connecting the magnet inside the CRYOSTAT to the power source outside; cold vapor rising from the boiling helium is funneled through the lead, reducing the heat load (by conduction and Joule heating) on the cryostat.

GGG (Gadolinium-Gallium-Garnet): A PARAMAGNETIC SALT ($Gd_3Ga_5O_{12}$) for MAGNETIC REFRIGERATORS operating below $\sim$10 K.

GLAG theory: A phenomenological theory developed in the 1950s by V. Ginzburg, L. Landau, A. Abrikosov, and L. Gorkov that explains the magnetic behavior of TYPE II SUPERCONDUCTORS; it describes the relationship between COHERENCE LENGTH and PENETRATION DEPTH, as well as the MIXED STATE and UPPER CRITICAL FIELD.

Gradient coil: A solenoid that generates a linearly varying axial magnetic field. A pulse gradient coil used in MRI generates a field gradient typically of $\sim$10 mT/m.

Hall probe: A sensor for measuring magnetic field. Based on the principle of the Hall effect, its output voltage, for a given supply current, is proportional to field strength.

He I: Another name for the non-superfluid liquid of ordinary helium (He^4).

He II: Another name for SUPERFLUID HELIUM.

He^3: A helium isotope with atomic weight 3; it boils at 3.2 K at 1 atm. Naturally, $\sim 1/10^7$ as abundant as He^4. A key element in a DILUTION REFRIGERATOR.

He^4: A natural isotope of helium with atomic weight 4; it boils at 4.2 K at 1 atm.

HF (High Field): The winding section of a magnet exposed to a high-field. See LF.

High-performance magnets: Magnets having high OVERALL CURRENT DENSITIES, usually ADIABATIC; they are required for NMR and MRI magnets.

Hot spot: The small region in the winding that attains the highest temperature after a QUENCH; generally it is at a quench initiation point.

Hot-spot temperature: The highest temperature reached at the HOT SPOT in a quenching magnet; hot-spot temperatures below 150 K are considered safe.

HTS [High-T_c Superconductor(s)]: A new class of superconductors having T_c higher than ~ 25 K. All HTS discovered to date are PEROVSKITE oxides, the first of which was $La_{1.85}Ba_{0.15}CuO_4$ with T_c of 35 K, discovered in 1986 by K.A. Müller and J.G. Bednorz of the IBM Zürich Research Laboratory.

Hybrid (magnet): A magnet comprised of both WATER-COOLED and superconducting magnets to enhance the net magnetic field in the center.

Hybrid shielding: A MAGNETIC SHIELDING technique which combines both the ACTIVE SHIELDING and PASSIVE SHIELDING techniques; it is used in some MRI magnets.

Hysteresis loss: An energy loss due to the hysteretic effect of a material property, *e.g.* magnetization of TYPE II SUPERCONDUCTORS.

Incoloy 908: Trade name for a nickel-iron based superalloy with low thermal expansion coefficient; it has been specifically developed by the U.S. Department of Energy's Magnetic Fusion Program for the conduit of Nb_3Sn-based ITER CIC CONDUCTORS.

Inconel: Trade name for a family of nickel based superalloys.

Index number (n): An exponent appearing in the voltage *vs* current relationship for a superconductor; $n = \infty$ for ideal superconductors. It is an important conductor parameter for PERSISTENT-MODE magnets; n values as low as 20 are considered acceptable.

Induction heating: Heating generated in conductive metal by a time-varying magnetic field—called induction heating when such heating is beneficial and EDDY-CURRENT LOSS when it is not.

Insert: A coil or a set of coils placed in the bore of another coil or set of coils.

Internal diffusion process: A modified BRONZE PROCESS for Nb_3Sn developed in 1974 by Y. Hashimoto of the Mitsubishi Electric Corp.

Irreversible field: A magnetic field above which a superconductor carries no transport current of significance to magnet operation. It is smaller than UPPER CRITICAL FIELD; an important parameter for magnet design.

Isolated magnet: Generally refers to a PERSISTENT-MODE superconducting magnet whose connection (through current leads) to the environment outside the CRYOSTAT is removed to minimize refrigeration load. MRI and NMR systems use these magnets.

ITER (International Thermonuclear Experimental Reactor): An international project involving the European Union, Japan, Russia, and the United States to construct a break-even TOKAMAK, which will be equipped with CIC CONDUCTOR magnets.

JAERI (*Japan Atomic Energy Research Institute*): Japan's principal research and development center for superconducting magnet technology, at Naka-machi, Ibaraki.

"Jelly-roll" process: A process for making Nb-Ti, Nb$_3$Sn, and other LTS, in which "foiled" conductor ingredients are rolled to form the basic ingot. Developed in 1976 by W.K. McDonald of Teledyne Wah Chang; the process is also used to manufacture HTS.

Josephson effect: A quantum effect characterized by the tunnelling of superelectrons through the insulator of a JOSEPHSON JUNCTION, and observable as current flow without a driving potential. Based on the theoretical work (1964) of B. Josephson.

Josephson junction: A device with two superconducting plates separated by an oxide layer. Josephson junctions are used in SQUID and other micro-scale electronic devices.

J-T (*Joule-Thomson*) **valve:** A valve, usually a needle valve, across which the working fluid expands adiabatically and isenthalpically; the expansion process is irreversible.

Kaiser effect: A mechanical behavior under cyclic loading in which events, *e.g.* conductor motion and epoxy fracture in a magnet, appear only when the loading exceeds the level achieved in the previous cycle. Discovered by J. Kaiser in the early 1950s.

Kapitza resistance: The thermal boundary resistance which occurs at the interface when heat flows from a solid to liquid helium. Discovered by P. Kapitza in 1941.

Kapton: Trade name for a polyimide. An insulating material having a high BREAK-DOWN VOLTAGE in the range of 100 kV/mm.

Kohler plot: A plot that combines the effects of magnetic field, material purity, and temperature on electrical resistivity of conductive metal.

Lambda point (T_λ): The temperature below which ordinary liquid helium (He^4) becomes SUPERFLUID HELIUM; it is 2.18 K at a pressure of 38 torr (0.050 atm).

Layer winding: A winding technique to create a solenoid. A conductor is wound along the length of a mandrel, one layer at a time.

LCT (*Large Coil Test Project*): A multinational project involving Euro-Atom, Japan, Switzerland, and the U.S. in which a TOROIDAL MAGNET consisting of 6 superconducting coils was tested in the mid 1980s at the Oak Ridge National Laboratory.

Level indicator: A sensor for measuring the liquid helium level in a CRYOSTAT or STORAGE DEWAR; often the sensor element is a "heated" superconducting wire.

LF (*Low Field*): The winding section of a magnet exposed to a low field, usually the magnet's outer region. Conductors for the HF and LF regions are often graded.

LHD (*Large Helical Device*): A Stellarator-like experimental plasma machine near completion at the National Institute for Fusion Science, Toki-shi, Japan. The MAGNETIC CONFINEMENT is achieved by superconducting magnets.

Lorentz force: A force arising from the interaction of current and magnetic field; it is the most important force in magnets. Named for H. Lorentz.

Lorenz number: A conductive metal property equal to the ratio of thermal conductivity times electrical resistivity, to temperature. See WIEDEMANN-FRANZ-LORENZ LAW.

Lower critical field (H_{c_1}): The magnetic field at which a superconductor's magnetic behavior departs from perfect diamagnetism.

LTS [*Low-T_c Superconductor(s)*]: Mostly metallic superconductors, all with $T_c < 25$ K; the first LTS, mercury ($T_c = 4.152$ K), was discovered in 1911 by H. Kamerlingh Onnes (Leiden University). During the 1950s, B.T. Matthias and his group at the Bell Telephone Laboratories discovered many LTS important for magnet applications, *e.g.* Nb$_3$Sn, Nb$_3$Al, and V$_3$Ga; J.K. Hulm and R.D. Blaugher at the Westinghouse Research Laboratories are credited with basic studies (1961) of alloys of Nb-Ti and Nb-Zr.

Maglev (*Magnetic levitation*):   A levitation phenomenon created by opposing magnetic fluxes. Commonly it refers to levitated high-speed trains equipped with superconducting magnets, proposed by J. Powell and G. Danby of the Brookhaven National Laboratory in the late 1960s. Pursued since 1970 by the Japan Railway Technical Research Institute, which is presently building a second maglev test track 40-km long.

Magnet-grade superconductor: A superconductor meeting rigorous specifications required for use in a magnet and readily available commercially; presently (1994) only Nb-Ti and Nb3Sn qualify as magnet-grade superconductors.

Magnetic confinement: A fusion technique that uses magnetic fields to confine hot plasma, best exemplified by the TOKAMAK.

Magnetic cycle: A thermodynamic cycle that uses a material, *e.g.* a PARAMAGNETIC SALT, whose entropy depends on magnetic field and temperature; it is a viable refrigeration cycle particularly for temperatures below 20 K. There are magnetic equivalents to the conventional CARNOT and STIRLING cycles. See MAGNETIC REFRIGERATION.

Magnetic drag force: A drag force that arises from the EDDY-CURRENT LOSS, an important force for MAGLEV vehicles.

Magnetic pressure: A pressure equal to the magnetic energy density.

Magnetic refrigeration: Refrigeration based on a MAGNETIC CYCLE, generally uses helium as a working fluid for heat transport between the two reservoirs. The magnet system required to provide the field is often superconducting.

Magnetic shielding: Shielding of space occupied by persons or field-sensitive equipment from a FRINGING FIELD. See ACTIVE SHIELDING and PASSIVE SHIELDING.

Meissner effect: A phenomenon of complete expulsion of magnetic flux from the interior of a superconductor. This perfect diamagnetism is the most important property of a superconductor. Discovered in 1933 by W. Meissner and R. Ochsenfeld.

"Melt" process: Short for the melt-powder-melt-growth (MPMG) process that yields uniform high-quality YBCO. Developed in 1991 by M. Murakami of the International Superconductivity Technology Center (ISTEC), Tokyo.

MHD (*Magnetohydrodynamic*):   The study of the motion of an electrically conducting fluid in the presence of a magnetic field. For magnet applications, important quantities are an electric potential and a LORENTZ FORCE, both developed in the direction transverse to the fluid flow direction.

MHD generation: Conversion of the thermal energy of hot ionized gas into DC electric power by MHD. For large-scale systems, superconducting magnets are required.

MHD ship propulsion: Ship propulsion achieved by the acceleration of sea water through a magnetic field. In 1992 the experimental ship *Yamato*, equipped with a set of six superconducting DIPOLE MAGNETS, cruised in Kobe harbor, Japan.

Microslip: A STICK-SLIP of $\sim 10\,\mu$m; a microslip in the winding of a magnet may cause a PREMATURE QUENCH. The KAISER EFFECT is observed in microslips.

Mixed state: The state consisting of many hexagonally-arranged normal-state "islands" (vortices) in a superconducting sea. A TYPE II SUPERCONDUCTOR is in the mixed state over most of its magnetic field range. See FLUXOID and GLAG THEORY.

MPZ (*Minimum Propagating Zone*):   The largest stationary volume in a magnet that can generate Joule heating, while the rest of the winding remains superconducting. An important concept for analyzing the stability of ADIABATIC MAGNETS.

MRI (*Magnetic Resonance Imaging*):   A large-scale application in which superconducting magnets are used to help create visual images of the brain and other body parts for diagnostic purposes; it is based on the phenomenon of NMR.

Multifilamentary conductor: A COMPOSITE SUPERCONDUCTOR in which the superconductor is in the form of many filaments that are twisted to eliminate FLUX JUMPS. Strands in CIC CONDUCTORS are multifilamentary conductors.

Mylar: A polyester, used in the form of thin sheets ($25 \sim 150 \, \mu$m thickness) in magnets as an insulating material. Less suitable than KAPTON for high voltage applications.

Nb_3Al: An intermetallic compound of niobium and aluminum, a promising alternative to Nb_3Sn principally because of its tolerance to higher strains than Nb3Sn; still under development. The JELLY-ROLL process has proven effective in making the conductor.

Nb_3Sn: An intermetallic compound of niobium and tin. Aside from Nb-Ti, it is the only other MAGNET-GRADE SUPERCONDUCTOR.

Nb-Ti: Alloys of niobium and titanium, typically $\sim$50wt.%Ti: exact compositions vary among manufacturers; it is the MAGNET-GRADE SUPERCONDUCTOR most widely used.

Nb-Zr: Alloys of niobium and zirconium, the first MAGNET-GRADE SUPERCONDUCTORS. Replaced by Nb-Ti, which is easier to co-process with copper.

NET (*Next European Torus*): A large TOKAMAK machine being developed by the European Union, with headquarters in Garching, Germany; it has a team dedicated to research and development of superconducting magnet technology.

NHMFL (*National High Magnetic Field Laboratory*): A national magnet laboratory founded in 1990; most of its user facilities are at Florida State University, Tallahassee.

NMR (*Nuclear Magnetic Resonance*): A quantum effect characterized by the absorption of radio waves by nuclei in the presence of a magnetic field. NMR spectroscopy is used to study the molecular structures of chemical compounds including proteins. The higher the field, the higher will be the resolution and signal-to-noise ratio. Superconducting magnets are best suited for creating "NMR-quality" fields above 2 T.

Nusselt number: A dimensionless heat transfer coefficient equal to the ratio of the convective heat transfer coefficient to the thermal conductivity, times a characteristic length. An important number for convective heat transfer. Named for W. Nusselt.

NZP (*Normal Zone Propagation*) *velocity:* The velocity of the normal-superconducting boundary propagation; it decreases precipitously with increasing temperature. SELF-PROTECTING MAGNETS generally have high NZP velocities.

OFHC (*Oxygen-Free-High-Conductivity*) *copper:* A high-purity copper used widely in normal-metal conductors, particularly for low-temperature applications.

Ohmic heating magnet: A PULSE MAGNET situated in a TOKAMAK bore that heats as well as stabilizes the plasma.

Overall current density (J_{ov}): Total ampere-turns divided by the winding cross sectional area. An important design parameter for magnets. See SPACE FACTOR.

Pancake winding: A winding technique in which a conductor is wound from the center outward, each turn lying further away radially but in the same plane as the previous turn; the finished coil resembles a pancake. A DOUBLE-PANCAKE is wound with a conductor that spirals in from the outside of one pancake and then spirals back out in the other. The transition turn at the inside diameter is continuous.

Paper magnet: A magnet available only on paper (or in a computer); when elaborate or large-scale, it is often the only version ever completed. Useful for parametric study.

Paramagnetic salt: A paramagnetic material for MAGNETIC REFRIGERATION. Most are rare-earth based compounds, *e.g.* GGG, DAG ($Dy_3Al_5O_{12}$), $ErNi_2$, EuS, Gd_5Si_4.

Passive protection: A magnet protection technique that does not use devices located outside the CRYOSTAT; a key parameter, NZP VELOCITY, must be *high* to expand the energy absorbing normal zone fast, and keep HOT-SPOT TEMPERATURES below 150 K.

Passive shielding: MAGNETIC SHIELDING that generally uses ferromagnetic materials such as steel; HTS is considered a viable material. For shielding a time-varying field, conductive metal is effective (SKIN EFFECT).

Peak field: The maximum field for a given field distribution in a magnet.

Penetration depth: The depth within which a surface supercurrent flows to exclude a magnetic field from the interior of a TYPE I SUPERCONDUCTOR. The concept was introduced by F. London and H. London (brothers) in the 1930s.

Permanent magnet: A magnet made of magnetized ferromagnetic material with a high COERCIVE FORCE; it can provide inductions up to ~2 T over a small volume.

Perovskite: The cubic structure of all known HTS, also of ferroelectric, ferromagnetic, and antiferromagnetic materials. Named for Count L.A. von Perovski, a crystallographer.

Persistent-mode: Operation of a superconducting magnet where the generated magnetic field remains virtually constant with time. Usually achieved by means of a SUPERCONDUCTING SWITCH which shunts the magnet, allowing it to become ISOLATED.

Phenolic: A laminated composite consisting of woven linen (sometimes cotton or even paper) layers bonded in a thermosetting adhesive.

Piezoelectric effect: The coupling of mechanical and electric effects in which a strain in a certain class of crystals, *e.g.* quartz, induces an electric potential and vice versa. A piezoelectric crystal is the key element in an AE sensor. Discovered by P. Curie in 1880.

Poloidal magnet: A PULSE MAGNET that generates a field in the axial (vertical) direction for plasma stabilization in a MAGNETIC CONFINEMENT fusion machine.

Polyhelix: A high-field, WATER-COOLED MAGNET consisting of nested single-layer coils, each wound with a conductor matched to its stress requirement.

ppm: Parts per million.

Prandtl number: A dimensionless property coefficient equal to the ratio of kinematic viscosity to thermal diffusivity, a measure of the relative diffusion rates of momentum and heat, and an important number in convective heat transfer. Named for L. Prandtl.

Premature quench: A QUENCH of a superconducting magnet below the designed operating current. A premature quench usually disappoints magnet engineers.

Pulse magnet: A magnet that generates a magnetic field of a short duration, ranging from microseconds to a fraction of a second, presently the only type capable of generating fields above ~40 T. Also used as an OHMIC HEATING MAGNET and POLOIDAL MAGNET. Large superconducting versions are generally wound with CIC CONDUCTORS.

QA (*Quasi-Adiabatic*) **magnet:** A magnet that is nearly adiabatic; a small amount of coolant in the winding keeps it superconducting in the presence of *transient* disturbances. Its OVERALL CURRENT DENSITY ranges between those of CRYOSTABLE MAGNETS and those of ADIABATIC MAGNETS; the HYBRID III SCM is an example.

Quadrupole magnet: A magnet that generates a linear gradient field transverse to its axis over the central region of its bore; it focuses particles in particle accelerators.

Quench: The superconducting-to-normal transition, specifically, the rapid irreversible process in which a magnet is driven fully normal.

Racetrack magnet: A magnet resembling a racetrack, wound in a plane with each turn consisting of two parallel sides and two semi-circles at each end; used by MAGLEV vehicles. Two magnets may be assembled to approximate the field of a DIPOLE MAGNET.

React-and-wind: The term used for a coil winding technique for a "reacted" conductor such as A-15 LTS and HTS that is wound with the conductor already reacted, applicable only when strains induced in the conductor during the winding process can be kept low.

Regenerator: A key element in STIRLING and G-M cycles, storing and releasing heat during isochoric processes; its effectiveness diminishes as 0 K is approached. For operation below ~10 K, rare-earth based compounds, *e.g.* Er_3Ni, have proven quite effective.

Reynolds number: The dimensionless ratio of inertial effects to viscous effects; it characterizes the viscous flow of fluids. Named for O. Reynolds.

RRR (Residual Resistivity Ratio): The ratio of a metal's electrical resistivity at 273 K to that at 4.2 K, a parameter to express the metal's purity. OFHC COPPER has an RRR of ~200; it can reach 10,000 and higher in pure aluminum.

"Saddle" magnet: A magnet with a winding that resembles a saddle, generally with a long winding length. Two saddle magnets can be assembled to produce a DIPOLE MAGNET field, four saddles a QUADRUPOLE field, and six saddles a sextuple field.

Saturation magnetization: The maximum magnetization of a ferromagnetic material. An important material property for PASSIVE SHIELDING.

SCM: Superconducting magnet(s).

Search coil: A coil for measuring a magnetic field; it requires a very stable integrator to convert the search coil output voltage to field.

Self-field loss: A hysteresis AC LOSS in a superconductor due to the magnetic field generated by a transport current.

Self-protecting magnet: A magnet with the built-in PASSIVE PROTECTION features. Ideally all ISOLATED MAGNETS should be self-protecting.

Skin depth: The distance from the metal surface at which the amplitude of a sinusoidally time-varying magnetic field, due to the SKIN EFFECT, is 0.37 the applied field's.

Skin effect: A phenomenon in which the amplitude of a sinusoidally time-varying magnetic field decreases exponentially with distance from the surface of a conductive metal into its interior. See SKIN DEPTH.

SMES (Superconducting Magnetic Energy Storage): Generally refers to the magnetic storage of electrical energy for future use by an electric utility. Because energy must be stored until needed, SMES must use PERSISTENT-MODE magnets. Other non-utility applications of SMES have also been proposed, especially for HTS magnets.

SOR (Synchrotron Orbital Radiation): A device to produce x-rays by means of accelerating electrons in the presence of a magnetic field; SOR magnets are superconducting.

Space factor: The fraction of the winding cross sectional area occupied by the current-carrying conductor. An important design parameter for magnets.

SQUID (Superconducting Quantum Interference Device): A device using the JOSEPHSON EFFECT to measure the smallest possible change in magnetic flux. See FLUXOID.

SSC (Superconducting Supercollider): Terminated in 1993, it was to have been the largest (super) "atom smasher" (collider) for high-energy physics research; the prefix superconducting was used because its nearly 10,000 DIPOLE and QUADRUPOLE MAGNETS were to be superconducting. SSC would have been the largest single consumer of superconductor in history—nearly 1000 tons of Nb-Ti MULTIFILAMENTARY CONDUCTOR.

Stekly criterion: A design criterion for CRYOSTABLE MAGNETS; the cooling flux at the conductor surface must match the conductor's full normal-state Joule heating flux.

Stick-slip: A sudden motion induced by a frictional force that decreases with sliding velocity; it may cause a PREMATURE QUENCH. See KAISER EFFECT and MICROSLIP.

Stirling cycle: A cycle used in CRYOCOOLERS, consisting of two isothermal and two isochoric processes; it incorporates a REGENERATOR. Invented by a Scottish minister, R. Stirling, in the late 1810s, before Carnot published his paper.

Storage dewar: An insulated container, sometimes portable, of CRYOGENS.

Subcooled superfluid helium: SUPERFLUID HELIUM at atmospheric pressure. Most superconducting magnets operating in this mode, *e.g.* HYBRID III, use the ingenious cryostat design principle developed by G. Claudet, P. Roubeau, and J. Verdier.

Submicron conductor: A MULTIFILAMENTARY CONDUCTOR having filaments of diameter $0.1{\sim}0.5\,\mu$m to keep HYSTERESIS LOSSES manageable in 60-Hz applications.

Superconducting generator: A generator having a rotor which is superconducting. The basic engineering adopted in today's machines was developed and demonstrated in the late 1960s with a 45-kVA model by J.L. Smith and others at MIT.

Superconducting motor: A motor having a rotor which is superconducting. An important application of HTS magnets; prototype 100-hp motors incorporating HTS magnets (77 K) are being developed in the U.S.

Superconducting power transmission: A superconductive transmission of electrical power, AC or DC, for an electric utility. A 1000-MVA AC prototype based on Nb_3Sn tape was demonstrated in the early 1980s by E.B. Forsyth and others at the Brookhaven National Laboratory. Potentially an important application for HTS.

Superconducting switch: A switch that shunts a PERSISTENT-MODE MAGNET; it has an open (normal state) position activated by a heater in the switch. Switch superconductors are matrixed in an alloy (Cu-Ni) to make their normal-state resistance large so that the current through the switch can be limited during magnet charging.

Supercritical helium: Helium above the critical point (2.26 atm and 5.20 K); often helium is called supercritical even below 5.20 K if the pressure is above 2.26 atm.

Superfluid helium: A liquid helium state with extraordinary properties that exists below 2.18 K, the LAMBDA POINT (T_λ). See SUBCOOLED SUPERFLUID HELIUM.

Superinsulation: Aluminized MYLAR sheets; it is placed in the vacuum space of a CRYOSTAT to reduce radiation heat input.

TBCCO: Thallium-based HTS discovered in 1988 by Z.Z. Sheng and A.M. Hermann of the University of Arkansas. $Tl_2Ba_2Ca_{n-1}Cu_nO_{2n+4}$ form is structurally similar to BSCCO system, with a (2223) material having T_c of ${\sim}125$ K; it appears to be more promising than BSCCO for high-field applications at 77 K.

Teflon: Trade name for polytetraflouroethylene (TFE), consisting of $[C\text{-}F_2]_n$ chains with strong C-F bonds. A common material used, both as a bulk and a coating material, in cryogenic systems for its insulating properties and low coefficient of friction.

Tokamak (*Toroidal Coil Magnetic Confinement*): A toroidal-shaped thermonuclear fusion machine that incorporates MAGNETIC CONFINEMENT to contain hot plasma; it was conceived in the 1950s by L.A. Artsimovich and A.D. Sakharov of the Kurchatov Institute of Atomic Energy, Moscow. Power-generating fusion reactors will likely use the Tokamak configuration with magnets that will all be superconducting. See ITER.

TORE SUPRA: A TOKAMAK with superconducting magnets operated in a bath of SUBCOOLED SUPERFLUID HELIUM; it is located at Cadarache, France.

Toroidal magnet: A magnet that generates a field in the toroidal (azimuthal) direction to confine hot plasma; the superconducting version is DC.

Training: Behavior of a HIGH-PERFORMANCE MAGNET that quenches prematurely but improves its performance towards the design operating current after each quench. Because most of these PREMATURE QUENCHES are caused by mechanical disturbances, training is considered as a manifestation of the KAISER EFFECT.

Transfer line: A double-walled vacuum insulated line used to transfer liquid helium from a STORAGE DEWAR to a CRYOSTAT.

Transposition: A configuration of strands in a CIC CONDUCTOR. In a transposed cable, every strand spirals to occupy every radial location in the cable's overall diameter along its TWIST PITCH LENGTH; by contrast, in a simply twisted cable the center strand remains at the center. Transposition minimizes flux linkage and thus the COUPLING LOSS. In MULTIFILAMENTARY CONDUCTORS, because filaments are arranged in fixed radial locations in the matrix at the billet-making stage, transposition is not possible; the conductor is simply twisted during the production stage.

Triple point: The equilibrium state for solid, liquid, and vapor phases. Triple points of hydrogen (13.8033 K), neon (24.5561 K), oxygen (54.3584 K), argon (83.8058 K), mercury (234.3156 K), and water (273.16 K) are fixed points in the International Temperature Scale of 1990 (ITS-90). Helium does not have a triple point.

Triplet: A set of three transposed strands. A standard building block for CIC CONDUCTORS containing hundreds of strands.

Twist pitch length: The linear distance over which a strand (or filament) of transposed cables (or twisted multifilamentary conductor) makes one complete revolution. An important parameter for CIC CONDUCTORS and MULTIFILAMENTARY CONDUCTORS.

Type I superconductors: Superconductors exhibiting the MEISSNER EFFECT up to their critical fields. Also called "soft" superconductors because most are mechanically soft, e.g. lead, mercury, indium. Unsuitable for MAGNET-GRADE SUPERCONDUCTORS.

Type II superconductors: Superconductors possessing a MIXED STATE. Also called "hard superconductors," they include Nb-Ti and Nb_3Sn, and have critical properties suitable for magnet applications. An alloy of lead-bismuth was the first Type II superconductor, discovered in 1930 by W. de Haas and J. Voogd.

Upper critical field $[H_{c_2}$ (or $B_{c_2} = \mu_0 H_{c_2})]$**:** The field at which a TYPE II SUPERCONDUCTOR loses its superconductivity completely.

Void fraction: The fraction of the space bounded by the interior wall of a CIC CONDUCTOR that is occupied by coolant; it ranges from 33 to 40%.

Water-cooled magnet: A magnet, usually made of copper or copper alloys, that is cooled by water forced through the winding.

Wiedemann-Franz-Lorenz law: A relationship stating that for conductive metals, the product of thermal conductivity and electrical resistivity is proportional to temperature with the proportionality constant being the LORENZ NUMBER.

Wind-and-react: A coil preparation technique consisting of two stages: winding of a coil with an "unreacted" conductor, followed by heat treatment of the wound coil. It is used for a conductor such as A-15 LTS and HTS that must be reacted at a high temperature (700~900°C). Required when strains induced in the conductor during the winding process could exceed the superconductor's strain limit. It is a more difficult process than the REACT-AND-WIND process, primarily because of the high reaction temperature that precludes the use of many materials in the pre-reacted coil.

YBCO: An yttrium-based HTS ($YBa_2Cu_3O_{7-\delta}$) with a T_c of 93 K discovered in 1987 by groups led by P.W. Chu at the University of Alabama and the University of Houston.

Yin-Yang magnet: A magnet consisting of two (Yin and Yang) coils that produces a "mirror" field for fusion devices; a relative of the "BASEBALL" (TENNIS BALL) MAGNET.

Appendix VII
Quotation Sources and Character Identification

Jonathan Livingston Seagull (Preface, p. vi):
A seagull in Richard Bach's *Jonathan Livingston Seagull, A Story* (Avon Books, 1970).

Michael Faraday (p. 10):
P.E. Andrew, *Michael Faraday* (Wheaton & Co., 1937).

Victor F. Weisskopf (p. 18):
American J. Phys. **45**, 422 (1977).

Ty Ty Walden (p. 32):
A religious Georgia farmer who has set aside one acre of his farm for digging gold for his church in Erskine Caldwell's *God's Little Acre* (A Penguin Books Edition, 1946).

Mathias J. Leupold (p. 60):
Personal communication (1991).

Lady Bracknell (p. 66):
A guardian in Oscar Wilde's trivial comedy for serious people *The Importance of Being Earnest* (Avon Books, 1965).

Obi Wan Kenobi (p. 79):
A Jedi master, mentor to Luke Skywalker, and later an outlaw in the Tatooine mountains in George Lucas's *Star Wars* (A Del Rey Book, 1976).

Anonymous (p. 102):
Posted in a paint shop in Boston; a universally apt reminder to the sponsor of a major project, particularly a superconducting one.

Owen Warland (p. 118):
A youthful watchmaker in Nathaniel Hawthorne's allegorical tale *The Artist of the Beautiful* (A Fawcett Premier Book, 1966).

An exasperated cryogenic engineer (p. 144):
Look around, there are quite a few like him.

Henry D. Thoreau (p. 154):
Walden (A Signet Classic, 1960).

Heike Kamerlingh Onnes (p. 171):
K. Mendelssohn, *The Quest for Absolute Zero, The Meaning of Low Temperature Physics* (World University Library, 1966). In the original Dutch: *"Door meten tot weten."*

Imanuel Kant, Jean-Paul Sartre, Frank Sinatra (p. 176):
A 5-word epitome of each philosopher's work.

Bertrand Russell (p. 209):
Great Essays in Science, edited by Martin Gardner (A Washington Square Press Book, 1957).

Robert Jordan (p. 236):
A Montanan mercenary in the Spanish Civil War in Ernest Hemingway's *For Whom the Bell Tolls* (Penguin Books, 1966).

Captain Nemo (p. 244):
The captain of the *Nautilus* in Jules Verne's science fiction *20,000 Leagues Under the Sea* (Translated from the French by Anthony Bonner, A Bantam Pathfinder Edition, 1962).

Arthur C. Clarke (p. 244):
Profiles of the Future, An Inquiry into the Limits of the Possible (Popular Library, 1977).

Lewis Thomas (p. 253):
The Lives of a Cell, Notes of a Biology Watcher (A Bantam Book, 1975).

O'Hara (p. 273):
Played by Peter Lorre, a member of the sophisticated beachcombers headed by Billy Dannreuther (Humphrey Bogart) in John Huston's whimsical thriller *Beat the Devil* (Santana-Romulus Production; released by United Artists, 1954).

Ada (p. 273):
Adelaida Durmanov Veen, the cousin of the protagonist in Vladimir Nabokov's poetic, antiworld novel *Ada or Ardor: A Family Chronicle* (A Fawcett Crest Book, 1969).

Ned Land (p. 292):
A Canadian harpooner on board the *Abraham Lincoln* and later a crew member of the *Nautilus* in *20,000 Leagues Under the Sea, ibid.*

Holly Golightly (p. 313):
A rural Texan turned New York's amoral playgirl in Truman Capote's *Breakfast at Tiffany's* (A Signet Book, 1959).

Ambrose Bierce (p. 327):
The Devil's Dictionary (Dover Publications, 1958).

C. Auguste Dupin (p. 353):
An eccentric genius in Edgar Allan Poe's *The Murders in the Rue Morgue* (An Airmont Classic, 1962).

E.B. White (p. 366):
One Man's Meat (Perennial Library, 1966).

Figaro (p. 376):
An original and witty liar, a valet in W.A. Mozart's comic opera *Le Nozze di Figaro* (Libretto by Lorenzo da Ponte). A rough translation: "The rest I need not say, . . ."

INDEX

A page number is italicized when that page contains a definition, numerical examples, or data. For HTS and Hybrid III only, most subjects are cross-indexed. Note that not every entry in the Glossary is indexed.

415